Computer-Aided Engineering Drawing

Computer-Aided Engineering Drawing

SEVENTH EDITION

Dr. Subrata Pal
BME, MME, PhD. (Engg.),
(J.U.), P.E.,
FASME (US), FIE,
Republic Day Awardee, 1998
Professor, Bioengineering
Jadavpur University
Kolkata - 700 032

Dr. Madhusudan Bhattacharyya
PhD. (Engg.), BME (Hons.) (Jad. Univ.)
Dip. Prod. E (Manchester Univ.)
FAE, MIE (India), MISTE, MISNT, Chartered Engineer
Ex-Head & Professor, Mechanical Engg. Department
Jadavpur University
Kolkata - 700 032

Viva Books
New Delhi | Mumbai | Chennai | Kolkata
Bengaluru | Hyderabad | Kochi | Guwahati

Sixth Edition 2005
This Seventh Edition published in 2009
Reprinted 2013

VIVA BOOKS PRIVATE LIMITED

- 4737/23 Ansari Road, Daryaganj, New Delhi 110 002
  E-mail: vivadelhi@vivagroupindia.net, Tel. 42242200
- 76, Service Industries, Shirvane, Sector 1, Nerul, Navi Mumbai 400 706
  E-mail: vivamumbai@vivagroupindia.net, Tel. 27721273, 27721274
- Megh Tower, Old No. 307, New No. 165, Poonamallee High Road, Maduravoyal, Chennai 600 095
  E-mail: vivachennai@vivagroupindia.net, Tel. 28290304, 28294241
- B-103, Jindal Towers, 21/1A/3 Darga Road, Kolkata 700 017
  E-mail: vivakolkata@vivagroupindia.net, Tel. 22836381, 22816713
- 7, Sovereign Park Apartments, 56-58, K. R. Road, Basavanagudi, Bangalore 560 004
  E-mail: vivabangalore@vivagroupindia.net, Tel. 26607409, 26607410
- 101-102 Mughal Marc Apartments, 3-4-637 to 641, Narayanguda, Hyderabad 500 029
  E-mail: vivahyderabad@vivagroupindia.net, Tel. 27564481, 27564482
- Beevi Towers, First Floor, SRM Road, Kaloor, Kochi 682 018, Kerala
  E-mail: vivakochi@vivagroupindia.net, Tel: 0484-2403055, 2403056
- 232, GNB Road, Beside UCO Bank, Silpukhuri, Guwahati 781 003
  E-mail: vivaguwahati@vivagroupindia.net, Tel: 0361-2666386

www.vivagroupindia.com

ISBN : 978-81-309-1215-8

Published by Vinod Vasishtha for Viva Books Private Limited, 4737/23, Ansari Road, Daryaganj, New Delhi-110 002. Printed and bound by Raj Press, R-3, Inderpuri, New Delhi - 110 012.

CONTENTS

PART – B

PREFACE TO THE FIRST EDITION

Engineering Drawing is an essential subject for all engineering curricula and this book is written to serve as a text for all engineering institutions — degree or diploma levels and also for the examinations conducted by different bodies. This book, which explains the fundamentals and essentials of Drawing in a concise and self-study form, will nevertheless be of interest to the practising engineers as well.

The chapters have been arranged in a logical sequence that permits the students to grasp each topic before moving on to the next. Chapter 6 includes Essential Fundamentals of Descriptive Geometry to promote imaginative power and develop better visualisation of the orthographic projection amongst the beginners.

The Unique Feature of this book is the adoption of Indian Standards. I.S. Conventions have all along been followed throughout the text. Various tables confirming to I.S. have been included to familiarise the beginners with I.S. Specifications. Emphasis has also been put, as would be seen on analysing the different problems given, on selection of sizes of different machine elements from available standards.

In preparing this text, all the available books and literatures on the subject inclusive of the valued drawing problem sheets prepared by our Ex-Professor Late A.K. Pal and relevant IS publications have been consulted for which heartfelt gratitude of the authors knows no bounds.

Authors are thankful to Sri A. Sengupta, their student and Sri S. Bhattacharyya of A.P.C.R Polytechnic for preparing the drawings in time.

Authors also record their sincerest thanks to Prof. R.K. Chakraborti of Mech. Engg., Jadavpur University; Prof. N.K. Das Talukdar, Head of Mech. Engg., Benaras Hindu University; Prof. A.K. Sen of Mech. Engg., R.E. College, Durgapur; Prof. G.L. Sinha of I.I.T. Kharagpur and Dr. S.K. Ghosh, Pool Officer, Jadavpur University, for their valuable and inspiring suggestions.

Any criticism/suggestion from the learned users of the book is welcome and will be taken cognizance of in future edition.

Authors would consider their labour amply rewarded if this book proves to be of help to its users.

Dr. Subrata Pal
Dr. Madhusudan Bhattacharyya

Kolkata
15.06.1978

PREFACE TO THE SIXTH EDITION

The authors are happy to publish the enlarged sixth edition of "Fundamentals of Engineering Drawing" renamed as *Engineering Drawing.* All the special features of this book are kept intact as in earlier editions. The standard rules as specified by Bureau of Indian Standard in SP: 46-1988 have been adopted as far as practicable, as this is very important for uniformity of teaching the subject.

In this edition the drawings were prepared using drafting software and the entire book was totally computerized painstakingly by Sri Amit Mallick and Sri S. Mandal. We thank both of them for their sincere effort.

This book is like a mini hand book packed with information on dimensioning, material specifications, tolerance, surface finish and some functional and manufacturing aspect of design. No other book on Engineering Drawing provides so much of information. This covers the major part of syllabi of all undergraduate courses in Engineering, Diploma in Engineering and AMIE courses, specially Mechanical Engineering.

This text has explained the fundamentals of engineering drawing clearly which is essential for any drafting procedure. It is sincerely believed that this edition will be received by the users with the same enthusiasm as before. Any criticism and suggestion to improve the quality of the book is most welcome; the same will be incorporated in the future edition.

Dr. Subrata Pal
Dr. Madhusudan Bhattacharyya

PREFACE TO THE SEVENTH EDITION

We are happy to publish the enlarged, seventh edition of *Engineering Drawing* revised as *Computer-Aided Engineering Drawing*. We updated this book keeping in mind the increased use of computers in various spheres of human activity. In the 15th chapter we have explained in detail how to use AutoCAD, the most popular application software in the field of engineering drawing. This software is upgraded regularly by the developer, and students will have to procure a suitable version and install it in their computer. All the special features of this book are kept intact as in earlier editions. The standard rules as specified by Bureau of Indian Standard in SP: 46-1988 have been adopted as far as practicable, as this is very important for uniformity of teaching the subject in different colleges, institutes, polytechnics.

In this edition the drawings were prepared using drafting software and the digital copy of the book was prepared by Sri Amit Mallick. He also helped in writing the new chapter. We thank him for his sincere effort.

This book is like a mini handbook packed with information on dimensioning tolerance, material specifications, surface finish and some functional and manufacturing aspect of design. No other book on engineering drawing provides so much of information. This covers the major part of the syllabi of all undergraduate courses in core engineering, diploma in engineering and AMIE courses, especially those in mechanical engineering.

This textbook has explained the fundamentals of engineering drawing clearly which is essential for any drafting procedure. We sincerely believe that this edition will be received by the users with the same enthusiasm as before. We welcome all criticism and suggestions to improve the quality of the book. They will be incorporated in the future editions.

Dr. Subrata Pal
Dr. Madhusudan Bhattacharyya
2009

FOREWORD

I am extremely happy to note that Dr. Subrata Pal, Professor of Bioengineering and Prof. M. Bhattacharyya, Ex-Head, Dept. of Mechanical Engineering, Jadavpur University have revised the text on engineering drawing and fuflly computerised it in conformity to the requirements of the modern time.

The publication of the sixth edition shows how useful it is to the students who are real beneficiaries. I sincerely believe that it will definitely improve the quality of teaching of engineering students of large number of engineering colleges and universities of the country.

I understand that they have prepared the drawing using BIS-standards and explained the fundamentals and practice from their vast experience over years. I hope, the authors will incorporate all new information also in future publication.

Prof. Shyamal Kanti Sanyal
Pro-vice-chancellor
Jadavpur University
Kolkata

PART – A

CHAPTER – 1

INSTRUMENTS AND AIDS

INTRODUCTION

Drawing is a medium of expression of ideas, thoughts and feelings. An artist expresses his thoughts, human experience through his artwork known as artistic drawing. Similarly, an engineer also creates his work, gives concrete shape to his ideas and design through technical drawing. It is a marvelous medium of expression, a powerful language. It does not change its complexion like spoken languages with the crossing of the boundary of a country. The practice and implementation is more or less the same over the entire planet. This medium of expression is unique as it can be understood interchangeably by the Japanese, American, Russian, European or Indian engineers at ease. It may also be mentioned that the drawing of the space-shuttle Columbia or the Hooghly Bridge or the Tajmahal is no less beautiful then the artwork of a great master artist. Therefore students of engineering should try to master this subject right from the beginning since it has far reaching effect in their post-educational professional practice. This subject distinguishes an engineer from a basic scientist who lacks a formal training in this subject.

INSTRUMENTS AND MATERIALS

For good quality drawing one needs good quality instruments with proper adjustments. The following instruments and materials are necessary.

1. Drawing Board

This is used to fix the paper on which drawing is made. The paper is fixed preferably with clamps, otherwise the thumbtacks (board pins). Students are advised to procure one drawing board large enough to accommodate all drawing sheets (Table 2.1). But the most common size of drawing board available in the market accommodates imperial size (594 × 841) A1 size drawing sheet.

2. Tee-square

It consists of a straight edge ruler and a cross-piece (or head) fixed at right angles at one end. While using, its head should be kept properly pressed against the left side edge of the drawing board. Careful handling is necessary to keep the T-square in good shape. To avoid warping it should be hung from a nail fixed on a wall while not in use. It can be used to draw horizontal parallel lines. Parallel lines at angles of 30°, 45° & 60° and combinations of these can be drawn with the aid of T-square and triangles (Fig. 1.1)

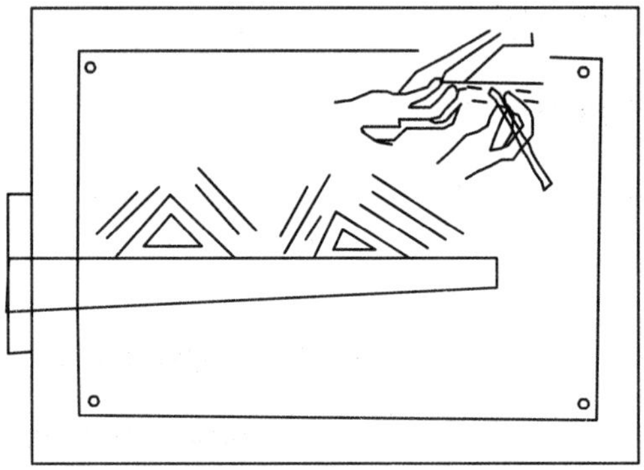

Fig. 1.1

3. Set-squares (Triangles)

Two types of transparent set-squares or triangles, *e.g.* 45°-45° and 60°-30°, as shown in Fig. 1.1 are used for drawing.

4. Diagonal Scale

This is a unique aid graduated in millimeters and inches and also graduated in degrees from 0°-180° to facilitate laying out and measuring angles. Fractions of inches are also shown. For measurement and laying out of dimensions in fractional/decimal form, diagonal scaling system is provided (details about methodology of construction and use of diagonal scale are given in the chapter on Scales and Dimensioning).

5. French Curve

French curves are used to lay out regular and smooth curves. A set of french curves numbered differently are available in different sizes. Usually no. 5 and no. 3 are mostly used for fundamental courses in drawing. The edges of the french curves must be absolutely smooth. It needs some skill to use this instrument properly and effectively to generate a smooth curve gradually merging from one section to another. Now-a-days flexi-curves are also available for drawing all types of curves by suitable manoeuvring.

6. Instrument Box

It consists of compasses, dividers, ruling pens and other instruments in a specially made case.

(i) *Compass* can be used for drawing arcs and circles both in pencil and ink. Two legs are pivoted at the top; the other end of one leg is fixed with a steel needle while the other (shorter) leg is provided with a slot or socket, Fig. 1.2 (a), for attaching various inserts like lead for drawing circles in pencil and a ruling pen for drawing circles in ink, Fig. 1.2 (c). For drawing circles of larger radii 3 lengthening bar should be used, Fig. 1.2 (d). & Fig.1.2 (g).

(ii) *Spring Compass*: This compass is used for drawing of every small circles Fig. 1.2 (e).

(iii) *Dividers*: These are used for transferring distance and occasionally for dividing spaces into equal parts, Figs. 1.3 (e). & 1.3 (f). Spring divider [Fig. 1.3 (c)] does the same thing for still smaller size with precision.

(iv) *Ruling Pen*: This is meant for inking of drawing, Fig. 1.3 (d). The thickness of the line can be adjusted by screwing the nib in or out, Fig. 1.3 (g).

(v) *Drop Pen*: Fig. 1.3 (a) is used for drawing very small circles. The pencil or pen point spins around the central axis and can be raised or lowered.

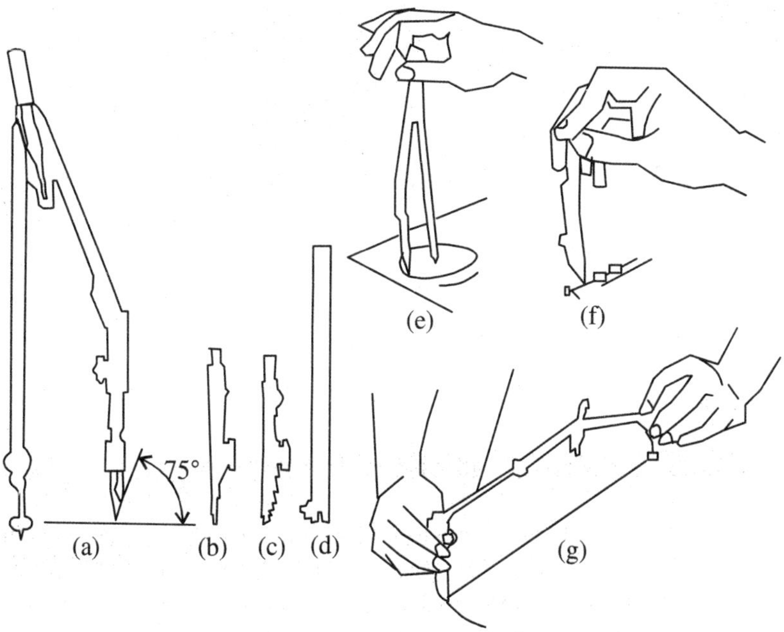

Fig. 1.2

7. Pencil

Hardness of the lead and uniform quality are the points of consideration for selection of pencil. Well sharpened good pencils are must for quality drawing. The pencils should be kept sharp by rubbing the lead on 00-emery paper. The lead may be sharpened to a chisel edge to have longer sharpened life. Depending on the hardness of the lead, pencils are graded as F, H, 2H and so on. The softer varieties are graded as B, 2B, 3B whereas medium hardness as HB. Usually 2H, H, HB and B pencils of good make will serve the purpose.

8. Drafting Machines

Modern drawing offices are equipped with drafting machines of various designs. These machines have the combined advantages of the T-square, set-square and scale. The two blades, which have simple, straight edges or the same with scales on it, can be moved parallel to their initial position. The blades are accurately set at right angles to each other. The adjustable head has a protector so that the blades may be quickly set and fixed at any desired angle.

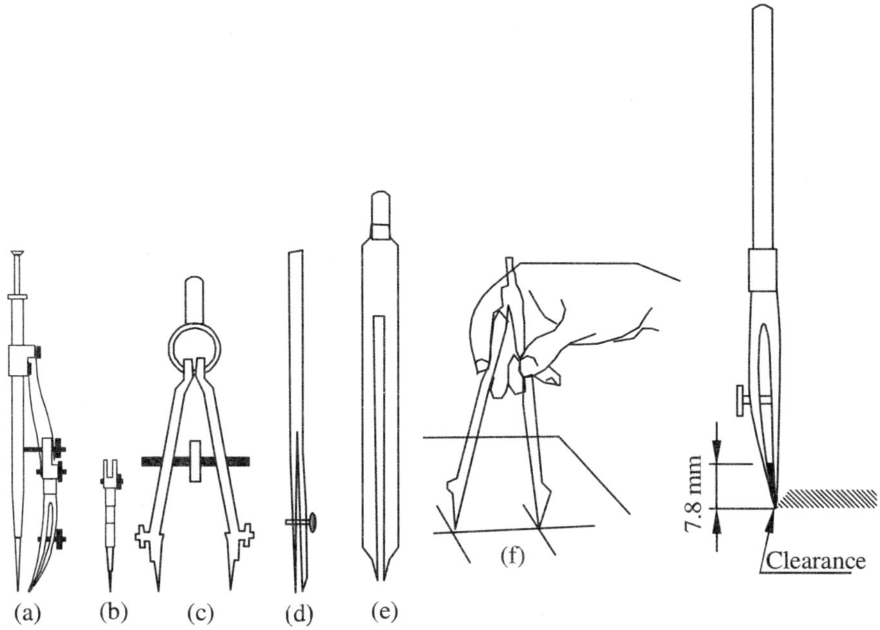

Fig. 1.3

Now to start with the learning of the fundamentals of engineering drawing one has to procure the following essential items.

(a) One good quality engineering drawing instrument box, (b) Pencils-B, HB, H and 2H grade of any good brand name, (c) One T-square (80 cm for Imperial size) or as desired, (d) A pair of medium size set squares, 45° (200mm) and 60° & 30° (250mm), (e) One drawing board (size already indicated), (f) Good quality drawing sheet (sizes are given in Table 2.2).

CHAPTER – 2

TYPES OF LINE

Different types of line shown in Fig.2.1 are used in engineering drawing. In India they are specified in SP:46-1988 which is harmonized with international standard and the former has been withdrawn.

The thickness of the lines depends on the dimensions and complexity of the part drawn, as well as on the purpose and size of the drawing. Thick continuous lines are used for drawing visible outlines or object lines and should be 0.5 to 1.2 mm thick. The chosen thickness of a particular type of line should be equal for all views of lines used for dimensioning purposes should be of one half to one third thickness of the object line. Continuous lines of similar thickness s/2 to s/3, are also used for crosshatching and breaks are indicated by wavy lines of same thickness, *i.e.* s/2 to s/3, where 's' is the thickness of object line; short-dash or dotted lines are used as centre, elsewhere lines are used to show cutting plane lines. Long chain lines (s to 1.5s) are used to indicate surfaces which are to receive additional treatment like plating, shot-penning etc..

DRAWING SHEETS

The basic principles involved in arriving at the sizes are:

(a) $x : y = 1 : \sqrt{2}$ (b) $xy = 1$

when x and y are the sides having a surface area of 1 m^2 so that

$$x = 0.841 \text{ m} \quad \text{and} \quad y = 1.189 \text{ m}.$$

DESIGNATION OF SIZES (ISO-A SERIES)

Table 2.1: Preferred sizes of the trimmed sheets

Designation	Dimensions (mm)
A0	841 × 1189
A1	594 × 841
A2	420 × 594
A3	297 × 420
A4	210 × 297

Table 2.2: Special elongated sizes (Second Choices)

Designation	Dimensions (mm)
A3 × 3	420 × 891
A3 × 4	420 × 1189
A4 × 3	297 × 630
A4 × 4	292 × 841
A4 × 5	297 × 1051

Exceptional Elongated Sizes (Third Choice) sheets are also recommended when a very large or extra elongated sheet is required e.g.,

A0 × 2 1189 mm × 1682 mm *etc.*

These sizes are obtained by extending the shorter sides of a format of the ISO-A series to lengths that are multiples of the shorter sides of the chosen basic format.

PROBLEM

Reproduce prob. 2.1 on a A4 size sheet. (First of all draw border lines determining the size of the drawing. Then draw the margin lines, which should be parallel to borderlines and spaced 5 mm away from them at the right, top, bottom and 20 mm at the left for binding. In the bottom, a space of 15 mm should be left for the block suitable for academic drawing since principal interest of the title block here is to identify the student).

Types of line	Illustrations	Approx thickness of lines	Application	Remark
Continuous thick		S	Visible outlines and edges	Outline to be bold and dark for hatching lines, should be spaced evenly to give a shaded effect.
Continuous thin		S/2	Dimension lines, leader lines, extension lines, construction lines, outlines of adjecent parts, hatching and revolve sections. Short central lines and imaginary section of intersection.	
Continuous thin-wavy		do	Irregular, boundary lines, short break lines.	Drawn free hand.
Short dashes–medium		S/2 or S	Hidden outlines and edges	
Long chain thin		do	Centre lines, trajectories, pitch circles and lines of symmetry.	Alternate long and short dashes.
Long chain thick at ends (S to 1.5 S) and thin elsewhere S/3 to S/2	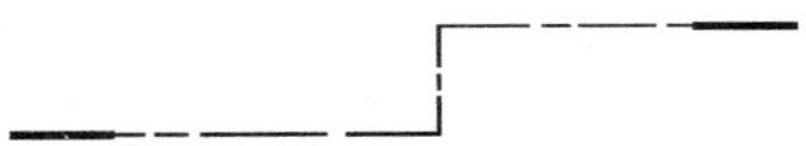	cutting plane lines		
Long chain-thick		S to 1.5S	To indicate surfaces which are to receive additional treatment.	
Ruled line to short zigzag-thin		S/2	Long break lines currently recommended for use in interrupted views.	
Chain thin double dashed		do	Outlines of adjacent part, alternative and extreme position of movable parts, initial outlines prior to forming, parts situated in front of the cutting plane.	

NB: The thickness (S) of lines should be chosen according to the size and the type of the form. The following are the ranges: (SP:46-1988) 0.18, 0.25, 0.35, 0.5, 0.7, 1 , 1.4 and 2 mm.

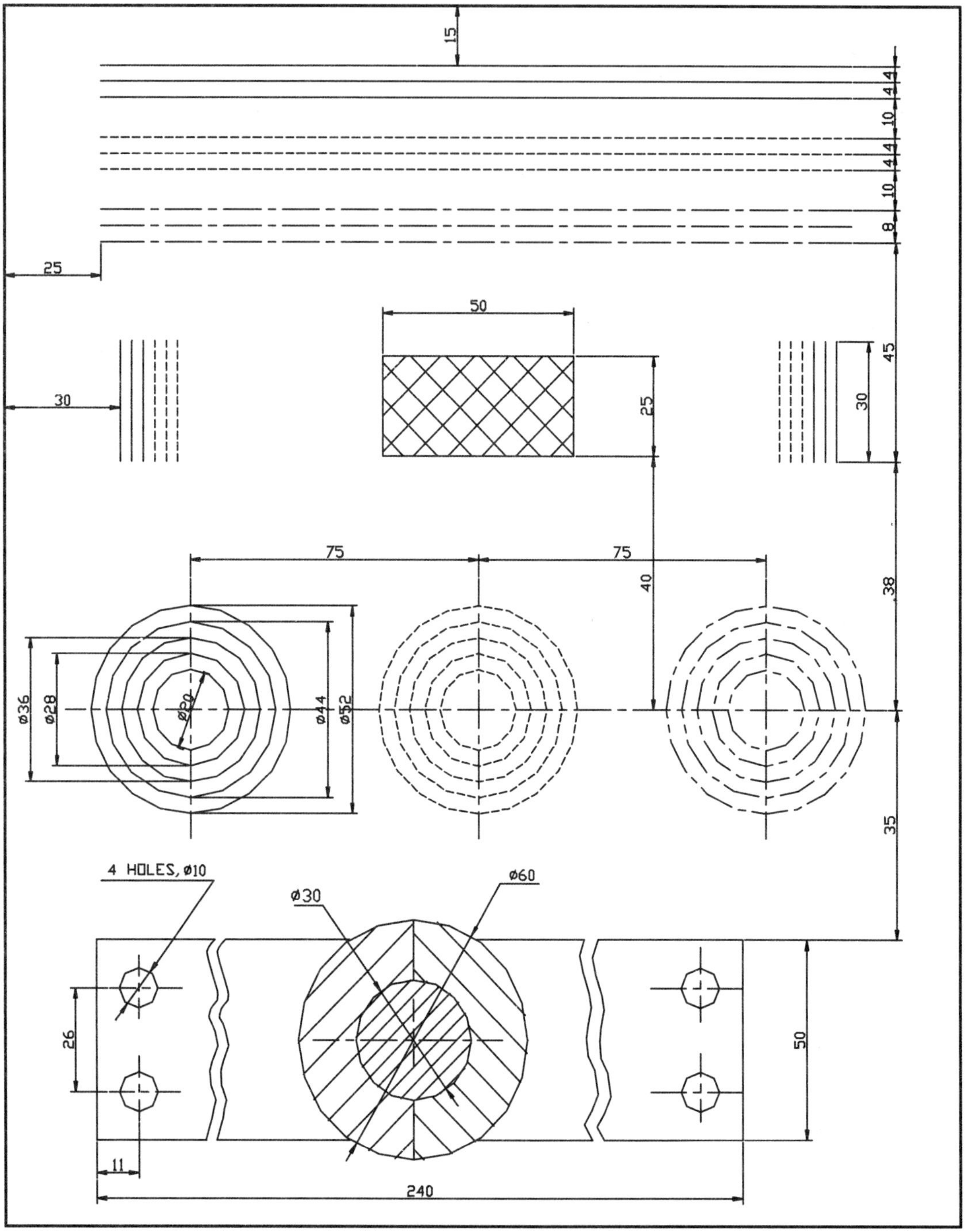
15
4
4
10
4
4
10
8
25
50
30
25
30
45
75
75
40
38
Ø36
Ø28
Ø20
Ø44
Ø52
35
4 HOLES, Ø10
Ø30
Ø60
26
50
11
240

Prob. 2.1

CHAPTER – 3

ENGINEERING LETTERING

All engineering drawings should be carefully prepared and appropriately lettered as per the requirements. The uniformity and depth of line-work and proper and pleasing lettering are very important for effective draftsmanship.

The styles of alphabets and numerals to be used for engineering drawings are standardized (Figs. 3.1 & 3.2). The BIS: SP 46-1988 gives the recommended specimens of the capital and

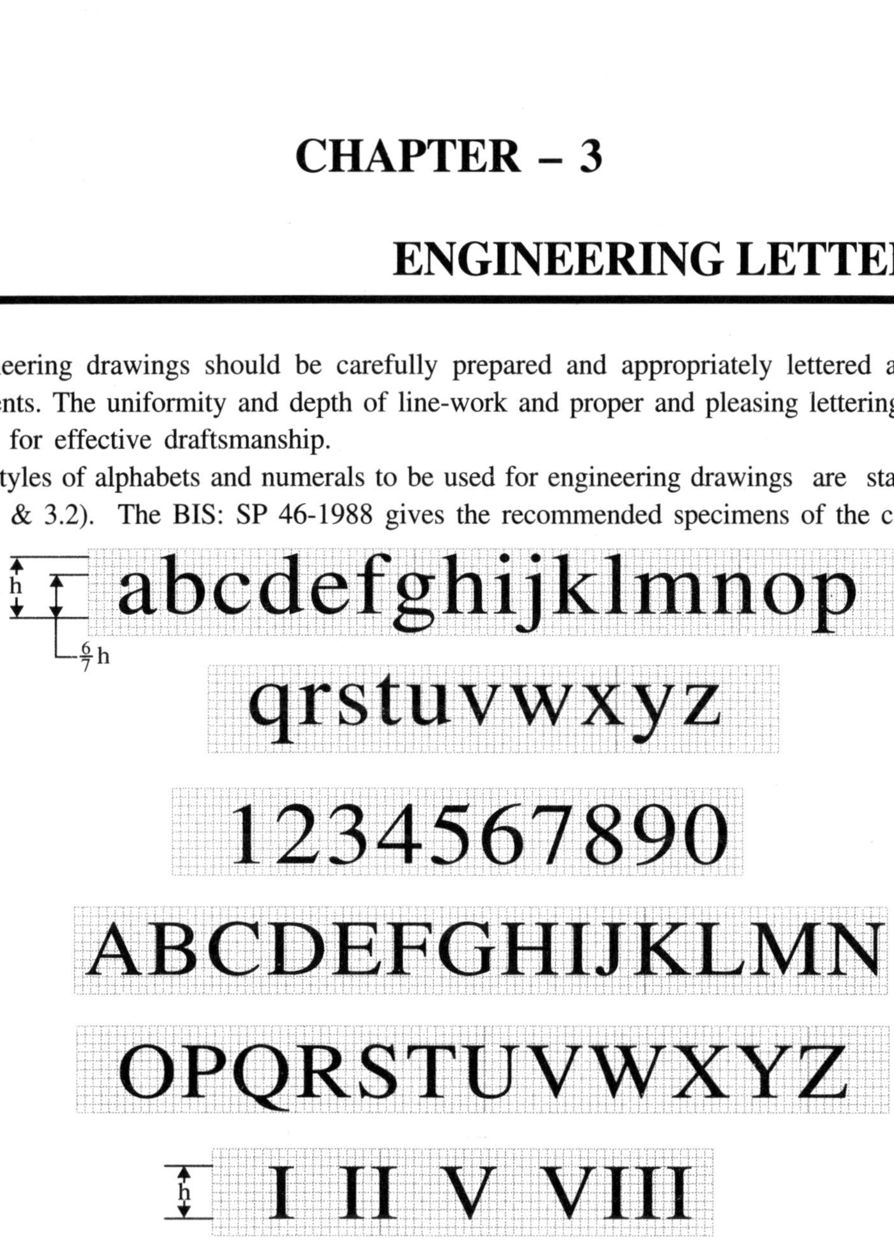

Fig. 3.1

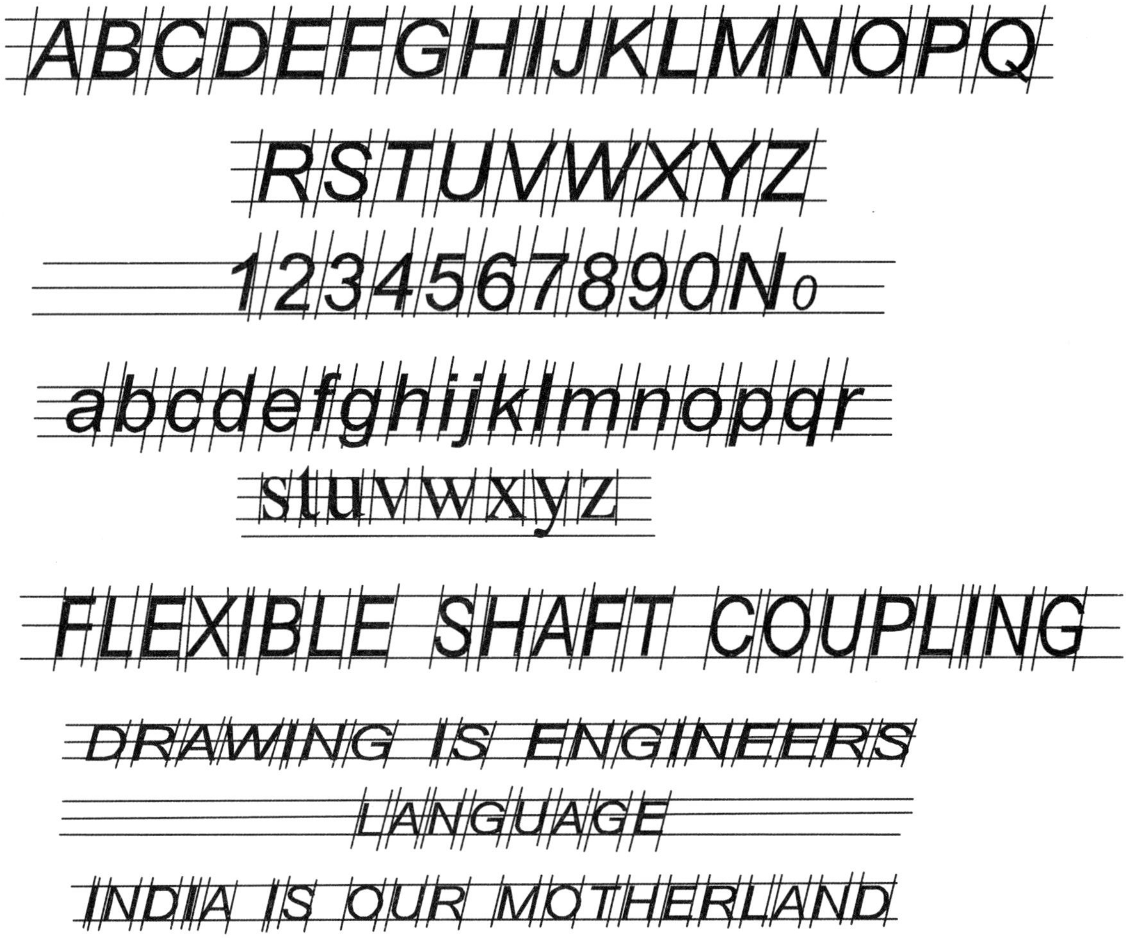

Fig. 3.2

lower-case letters of the Latin and Greek alphabets and numerals. To learn lettering one must study carefully the sizes and elements of the alphabets and numerals.

Alphabets and numerals are designated by their height h and width b. The standard proportions of Latin alphabets are given in Table 3.1.

Recommended sizes of alphabets and numerals to suit different purposes are given in Table 3.2. It must be remembered that the actual size will depend on the size of the drawing and the purpose for which it is intended. Lettering should be done so that it is readable from the bottom edge of the drawing sheet.

The beginners should use guide lines forming a grid. The inclined lines are drawn with the help of a T-square and a pair of 45° and 30° set-squares. It is to be noted that all the sizes of the characters are proportional to the quantity h/7; therefore the grid should be drawn by eight

Table 3.1 : Specification of letters (for d/h = 1/14)

Specifications	Designation	Specifications	Designation
Height of capital alphabets and numerals	h	Width of Lower Case Alphabets a	$b_6 = (4.5/7)$ h.
Width of alphabets and numerals	$b_1 b_2 b_3$ ....etc.	b,c,d,e,g,h,k,n,o,p,q,s,u, v,x,y and z	$b_7 = (4/7)$ h.
Width of capital alphabets A & M	$b_1 = (6/7)$h.	t,r & t i	$b_8 = (3/7)$ h. $b = (1/7)$ h.
B,C,D,E,F,G,H,K,N,O,P,Q,R, S,T,U,V,X,Y and Z	$b_2 = (5/7)$h. m & w	j & t b_{11} = h.	$b_{10} = (2/7)$ h.
I	$b_3 = (1/7)$h.	Width of Numerals 1	$b_{12} = (2/7)$ h.
J & L	$b_4 = (4/7)$h.	2, 3, 4, 5, 6, 7, 8, 9, and 0	$b_{13} = (2/7)$ h.
W	$b_5 = (8/7)$h.	Line Thickness of Alphabets and Numerals	= (1/7) h.
Height of lower case Alphabets without stem or tail		Minimum Space between words	= (3/7) h.
a,c,e,m,n,or,s,u,v,w,x, and z	=(5/7)h.	Space between characters	= (1/7) h
b,d,f,g,h,i,k,l,p,q,t,and y	h.	Space between Words and between Numbers	= not less than 'h'
j	=(9/7)h.	Space between Lines of Lattering	= Not less than (10/7) h.

Table 3.2 Recommended sizes of alphabets and numerals

Item	Size h,mm
Drawing number in Title block & alphabets denoting cutting plane section	10,14,26
Title of drawing	5,7
Subtitles & headings	3,5,5,7
Notes such as legends, schedules, material list, dimensioning	3,5,5,7
Alterations entries & the tolerances	2,5,3,5

horizontal lines intersected by series of parallel lines inclined at an angle of 75° to the horizontal. In case of vertical characters, vertical guide-lines should be used.

Lettering may also be made with the help of simplified grid. The capital alphabets can be grouped into two. The first group comprises figures of simple shape, such as T, I, N, L etc. that can be constructed without drawing auxiliary guide lines. Then the alphabets E, F, H, P, R etc. requiring an additional horizontal guide line at half the height comprise the second group. The middle elements are either on the guide-line so drawn or above in general and should be straight. For numerals 3 and 8, the upper part should be somewhat smaller than the lower one.

Lettering should be done with soft pencils. Large and medium size alphabets and numerals are finished with blunt lead and small sizes with well sharpened lead.

Initially the students will have to use guide lines. When they have mastered this operation of lettering they should try free-hand lettering without using any grid; only the main guide lines for heights may be drawn.

Lettering should be done on the drawing in such a manner that it may be read when the drawing is viewed from the bottom edges except where it is used for dimensioning purpose. Last but not the least, it may be mentioned that the presently used softwares have modernized these systems making drafting very much user-friendly.

PROBLEMS

Prob. 3.1 : Reproduce Figs 3.1 & 3.2

Prob. 3.2 : Fig.3.3 shows a typical Title Block in conformity with B.I.S. Reproduce the Title Block.

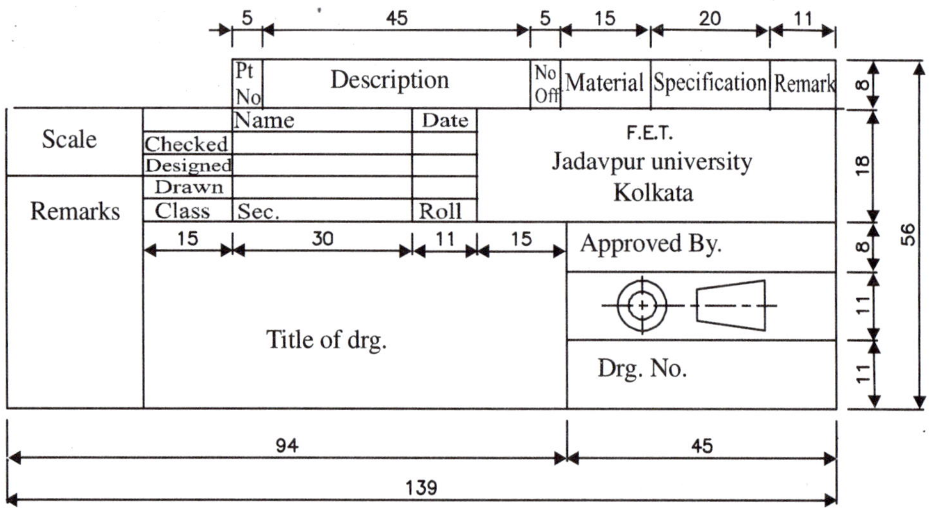

Fig.3.3

CHAPTER – 4

GEOMETRICAL DRAWINGS AND CURVES

COMMON GEOMETRICAL CONSTRUCTION

1. To Draw a Regular Pentagon, Circumscribing Circle Being Given.

Construction: Draw two diameters at right angles (intersecting at the centre 0) (Fig.4.1). Bisect the radius AO at B. With B as centre and radius BC, draw an arc CD. With centre C and radius CD draw an arc intersecting the circle at E. Join EC, which is one side of the regular pentagon. Then divide the circle with this chord length into five divisions and finally draw the regular pentagon.

2. To Inscribe a Regular Hexagon in a Circle.

Construction: Locate any point A on the circle (Fig.4.2). With A as centre and radius equal to that of the circle draw an arc cutting the circle at B. Joint AB. AB is a side of the required hexagon.

3. To Circumscribe a Regular Hexagon About a Circle.

Construction: Draw the hexagon with T-square and 30°-60° set square as shown (Fig.4.3). Note that the diameter AB is the (Flat to flat) distance between parallel faces.

4. To Construct any Regular Polygon on a Given Line, AB.

Construction: Produce BA and with centre A and radius AB describe a semicolon. Divide the semicircle into as many equal parts as the describe figure has sides with a divider (in this case five) and number it from C. Join A with 2 giving another side of the polygon. (Always join A with the second division for any polygon). Bisect the sides AB and A2 by lines meeting at O, giving the centre of the polygon. With centre O and OB as radius describe a circle. Cut off BE and ED each equal to AB and then draw the required pentagon, Fig.4.4.

5. To Inscribe any Regular Polygon (approximate) in a Circle, e.g., a Heptagon.

Construction: Draw any diameter AB and divide it into 7 equal parts (Fig.4.5). With centers A and B and radius AB draw arcs intersecting at C. From C through the second part, rule CD (applicable

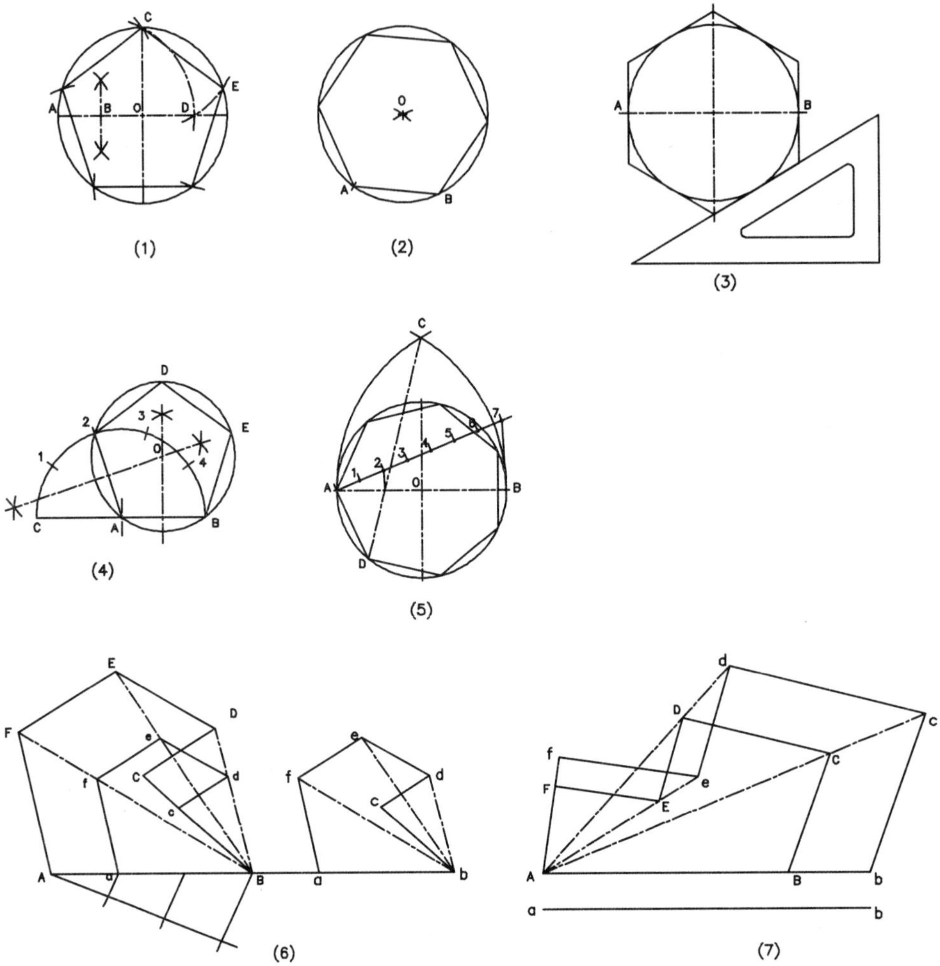

Fig. 4.1 to 4.7. Geometric construction.

for any polygon) and join AD. Then AD is one of the sides of the heptagon. Set off AD round the circle and join the points to get the polygon.

6. To Make a Reduced Drawing of a Given Figure.

Let ABCDEF be the given figure (Fig.4.6). It is required to reduce it so that each side shall be $2/3^{rd}$ of its length in the given figure.

Method 1: Join BF, BE, BD. Divide AB into 3 equal parts. Ba will be one of the required sides. Draw af, fe, ed, dc parallel to the corresponding sides of the figure.

Method 2 : Divide AB into 3 equal parts as before. Take Ba and place it on the same straight line or on a line parallel to AB. Draw at, bf parallel to AF, BF respectively thus getting the point f. From f draw a line parallel to FE and from b draw a line parallel to BE giving the point e. Extend this procedure in a manner for the other sides.

7. To Make an Enlarged Drawing of a Given Figure.

Let ABCDEF be the given figure which is required to be enlarged so that AB may be equal to ab (Fig.4.7). Draw AC, AE, AD and extend them. Set off ab equal to Ab on the line AB extended. From b draw bc parallel to BC and proceed as in the previous problem.

8. To Divide a Straight Line AB into Proportional Parts.

Construction : (a) Draw the line AB (Fig. 4.8) which is to be divided into three parts, say, proportional to 2, 3 and 4. Draw any line AC; take 9 = (2 + 3 + 4) equal parts on this line starting from A. Join 9B. From 5 and 2 draw line 55 and 22 parallel to 9B.

Then A2 : 2-5 : 5B = 2 : 3 : 4

(b) Draw any length of line parallel to AB (Fig.4.8). Divide this length into 9 equal parts. Join OA and 9B and produce them to meet at 0. Join 05 and 02 which cut AB at 2 and 5.

Then A2 : 2-5 : 5B = 2 : 3 : 4

9. To Draw the Fourth Proportional to Three Given Lines AB, CD and EF.

Construction : Draw GH = AB and GL = CD making any angle with GH. Join HL. Measure GN = EF. Draw NM parallel to HL (Fig.4.11).

$$\text{Then } \frac{GH}{GL} = \frac{GN}{GM}$$

GM is the fourth proportional line.

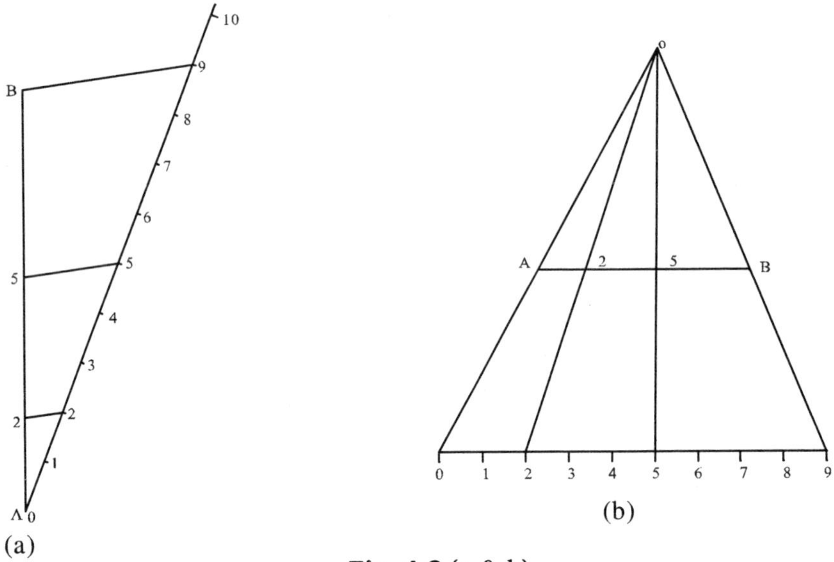

Fig. 4.8 (a & b)

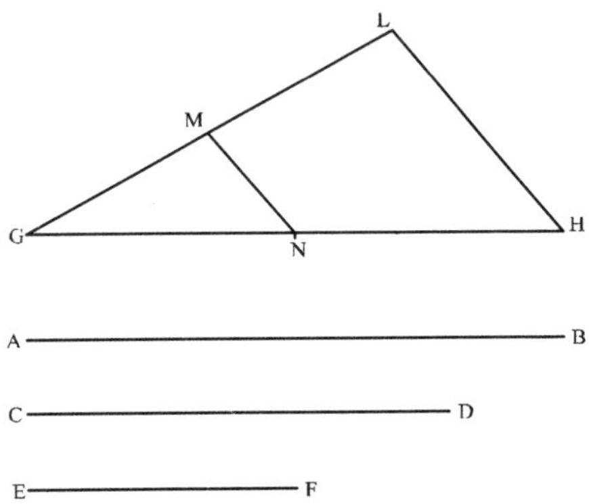

Fig. 4.9

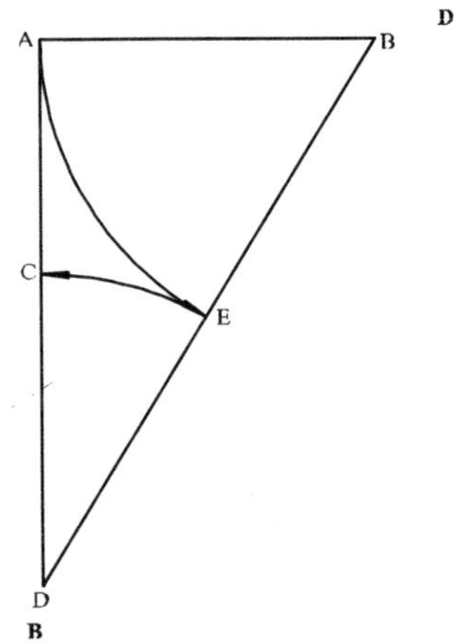

Fig. 4.10

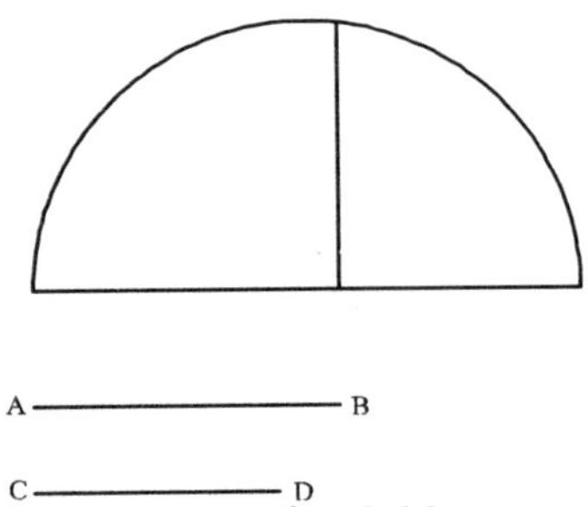

Fig. 4.11

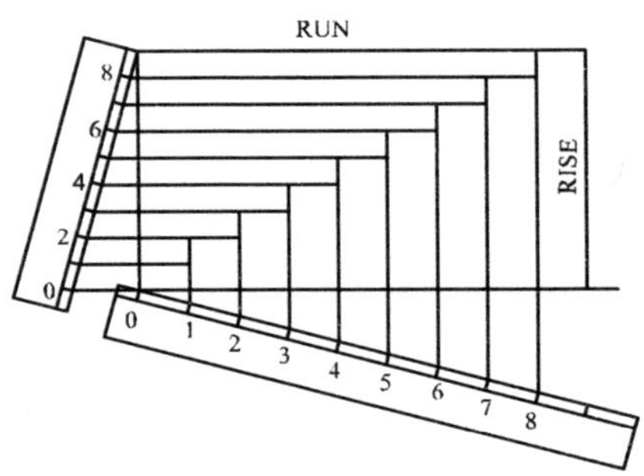

Fig. 4.12

10. To Draw a Line AB into Extreme and Mean Ratio (Golden Section).

Construction: Draw AD perpendicular to AB (Fig.4.12) and make AD = AB/2. Draw an arc with D as centre and DA as radius intersecting BD at E. With B as centre and BE as radius draw an arc intersecting AB at C.

Then AC/CB = CB/BA

11. To Find the Mean Proportional of Lines AB and CD.

Construction: Draw EF = AB. Extend EF to G such that FG = CD. On EG as diameter draw the semicircle EHG (Fig.4.11). Draw FH perpendicular to EG. Draw straight lines EH and HG.

$$\text{Then } \frac{EF}{FH} = \frac{FH}{FG}$$

FH is the mean proportional.

Also note that,

EH^2 = GE.. GF.

12. To Draw the Steps of a Stair, Run and Rise being Given.

Construction: Generally the Rise (the number of steps) should be divided into one more than the Run. In the construction (Fig.4.12), the Rise is divided into 9 equal divisions and the Run into 8. Generally, the Riser *i.e.*, the height of individual stair varies between 150 mm to 200 mm and the tread *i.e.*, the width of individual stair between 250 mm to 300 mm.

CONIC SECTIONS

The ellipse, parabola and hyperbola are called conic sections since they may be obtained by taking various sections of a right circular cone (Fig.4.15).

An ellipse is a section cut from a cone by a plane making an acute angle with the axis and cutting the entire curved surface of all the elements.

A **parabola** is a section cut from a cone by a plane parallel to one of the elements.

A **hyperbola** is a section cut from a cone by a plane making a smaller angle with the axis than the element does. Usually the section is taken parallel to the axis which generates equilateral hyperbola.

To Construct an Ellipse, Major and Minor Axes being Given (Fig.4.13)

(i) *Two circle method:* Draw two circles on two axes as diameters on a common center 0. Through center 0 draw radial lines 11', 22', etc. at some suitable interval say 15° or 30°. From points 1', 2', 3' *etc.* draw vertical lines intersecting horizontal lines drawn from the corresponding points 1, 2, 3 *etc.*

These points of intersection will be the points on the ellipse. Join the points to a smooth curve with the help of a french curve.

(ii) *Four-center method:* Join points C and A of given major axis CD and minor axis AB respectively. Lay off the distance AE equal to the difference of CO and AO. Bisect CE. The points at the intersections of the bisector with the two given axes locate the centers R and S for the large and small icons forming the ellipse. Similarly the other two centers P and Q are obtained. The short axis may be extended if required to find points P and R.

To Construct a Parabola (Fig.4.13)

(i) *Tangent method:* The given lines of tangency AB and BC are divided into an equal number of parts and the divisions are numbered as shown. Draw lines connecting similarly numbered points 1 and 1', 2 and 2', *etc.* These lines are tangent to the required parabola.

(ii) *Through two given points (i.e.) origin and another point:* Let OX and OY be the coordinate axes and P be the given point. Divide OY and YP separately into same number of equal parts 1, 2, 3 and 1', 2' *etc.* Join O with 1', 2', 3' *etc.* Draw lines parallel to OX from points, 1, 2, *etc.* intersecting the corresponding lines drawn *e.g.* 01', 02' *etc.* giving the points on the parabola. Draw smooth curve as shown.

To Construct a Hyperbola (Fig. 4.13)

(i) Draw an *equilateral hyperbola* with asymptotes (OX and OY) and one point P on the given curve. Draw axes OX and OY, locate P. Draw PA and PB parallel to the axes. Draw straight lines arbitrarily from origin O intersecting PB and PA at 1, 2, 3 *etc.* From the corresponding points of intersection draw lines parallel to OX and OY which intersect to provide the points on the hyperbola.

(ii) Construct a *rectangular hyperbola*, asymptotes and length of transverse axis being given. Draw transverse axis AB and asymptotes making 45° with AB. Mark arbitrarily some points p, q, r on one branch of the asymptotes, join these points with A and produce them to meet the other asymptote at 1, 2, 3 *etc.* Mark the points 1', 2', 3' *etc.* such that P1' = A1' = q2' - A2.... *etc.* These points 1', 2', 3' are on the required curve. Similarly complete the other half to the left of A and the conjugate part.

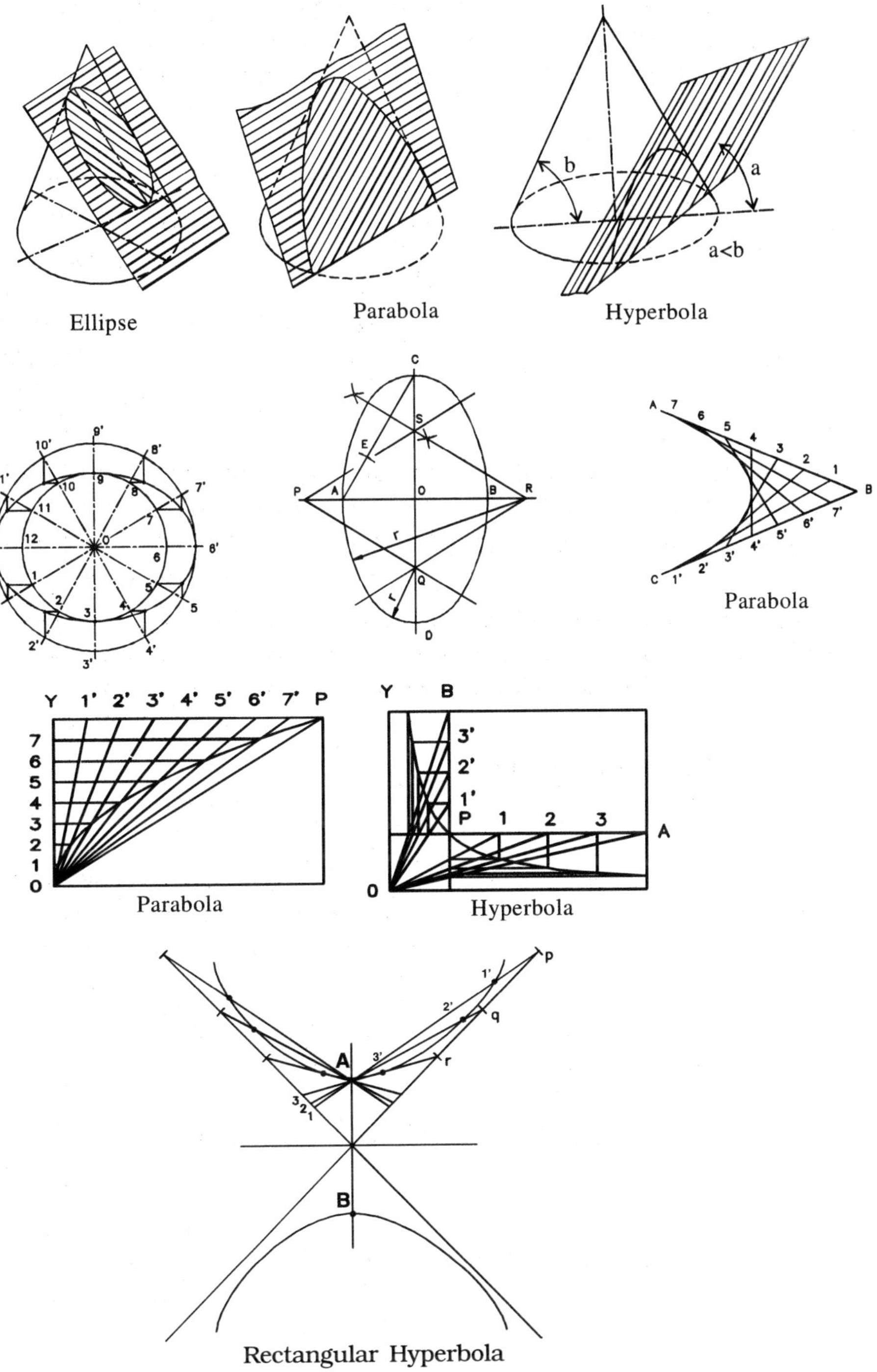
b
a
a<b
Ellipse
Parabola
Hyperbola
C
S
E
P
A
O
B
R
r
Q
D
A
B
C
Parabola
Y 1' 2' 3' 4' 5' 6' 7' P
7
6
5
4
3
2
1
0
Parabola
Y
B
3'
2'
1'
P 1 2 3
A
0
Hyperbola
p
1'
2'
q
A
3'
r
B
Rectangular Hyperbola

Fig. 4.13

SINUSOID

The sine curve is defined as a plane curve depicting the law of variation of the sine of an angle varied from 0° to 360°.

D/2 in Fig. 4.16 is called the amplitude of the sinusoid and L the wave length. The wave length of a normal sinusoid is πD. In case of a compressed sinusoid L is less then πD (*e.g.* in worm). L is more than πD (*e.g.* in drill) when we have an extended sinusoid.

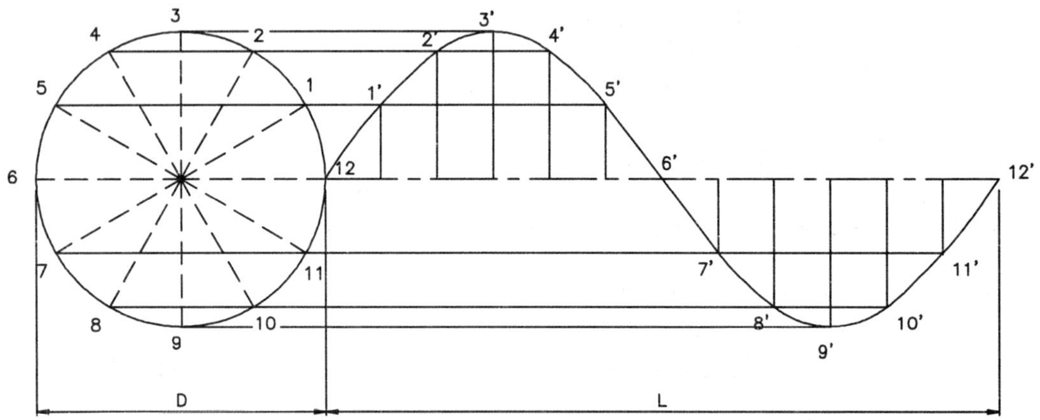

Fig. 4.14 Sinusoid

To Draw a Sinusoid

Draw a horizontal line, the axis of the sinusoid and mark off the given wave length L which is then divided into any number of equal parts, say 12 (Fig.4.16). To the left, a circle of diameter D (twice the amplitude) is drawn as shown. Divide the circle also into 12 equal parts and number them as 1, 2, ... 12. Draw horizontal lines from these points to intersect the perpendiculars drawn on the wave length line at 1', 2', 3', ... 12' as shown. A smooth curve through 1', 2', 3', ... 12' is obtained using a French curve.

The sinusoid is commonly used in drawing screws, worms, drills, displacement diagram for cams *etc.*

SPIRALS

Archimedean Spiral is a plane curve traced by a point which, starting from a fixed point called pole, 0, moves with uniform velocity along radial line (the radius vector) which itself revolves with uniform angular velocity. The equation of the curve in polar form is

$$r = a\theta, \text{ where 'a' is the pitch.}$$

The spiral of Archimedes is widely used in cam design and also in chucking device used on lathes.

To Construct an Archimedean Spiral

Let 'a' be the pitch given (Fig. 4.17). With center O and a radius draw a circle and divide it into some suitable equal number of parts by drawing radial lines, say twelve. Divide the radius into the same number of equals parts at 1', 2', 3' etc. With center 0 and radius equal to 01', 02', 03' *etc.* draw arcs intersecting the corresponding radial lines 01, 02, 03 *etc.* The points of intersection are the points on the spiral.

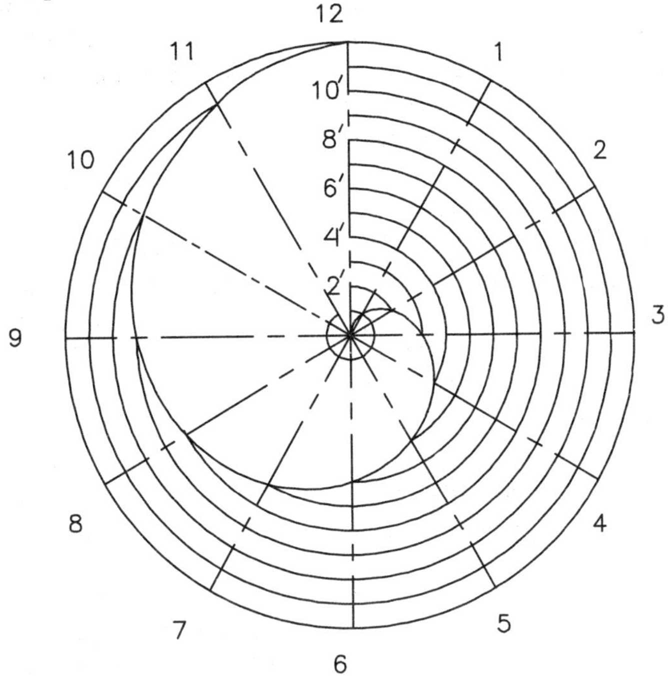

Fig.4.15 Archimedean spiral

Logarithmic Spiral

It is a plane curve which cuts the radius vectors from a fixed point (pole) at angles whose cotangent is constant. Its polar equation is given by

$$r = ae^{b\theta}$$

where 'a' is a constant which is the value of r when θ is zero.

$$b = \cot\alpha,$$

where α is the constant angle which any radius vector makes with the tangent to the curve and

e = the base of the Napierian logarithms.

To Construct a Logarithmic Spiral

If it is desired to draw a logarithmic spiral through points on radiants at a given angle apart, it is to be noted from the equation of the curve that, if successive values of q are taken with a uniform increase, the lengths of the corresponding radius vectors will be in geometrical progression.

If (r_1, θ_1) and (r_2, θ_2) are two points on the curve then

$$r_1 = ae^{b\theta_1} \quad \text{and} \quad r_2 = a_e^{b\theta_2}$$

Multiplying r_1 and r_2 and taking square root of both sides we get,

$$(r_{1.} r_2)^{1/2} = ae^{b_1(\theta_1 + \theta_2)/2}$$

or, $$r' = ae^{b\theta'}$$

where (r', θ') is a point obtained by taking the geometric mean of r_1 and r_2 and locate on the radius vector at the bisector of the angle between r_1 and r_2.

In Fig. 4.16, OA and OE are the radius vectors making an angle θ between them. The objective is to draw a logarithmic spiral through the points A and E.

Rotate OA to OA' on EO produced. Draw the semicircle A'B'E on diameter A'E and OB which bisects the angle AOE. Perpendicular OB' at O on A'E which intersects the previous semicircle at B' is the length of the new radiant. Make OB' = OB. B is a point on the spiral.

Draw the semicircle OC'E on diameter OE. Make OB" = OB. Erect perpendicular at B" on OE which intersects the semicircle OC'E at C'. Join OC' which is the length of another radiant (radius vector). Measure OC' = OC, the bisector of the angle BOE. Then C is another point on the spiral. Similarly, point D can be found out. Join the points A, B, C, D and E by a smooth curve which is the required logarithmic spiral.

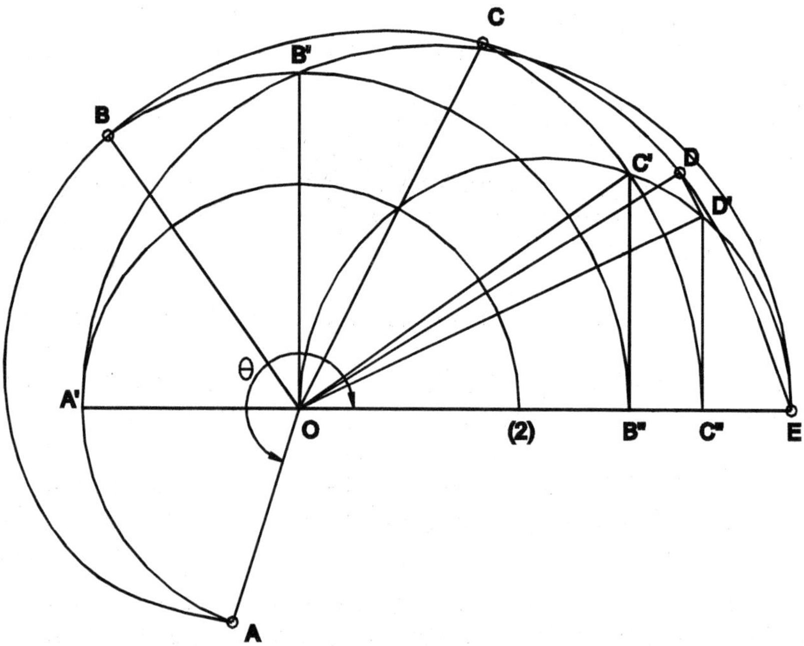

Fig. 4.16 Logarithmic spiral

Note that the point B could also be found out by this method using the semicircle OC'E.
A circle is a special case of the logarithmic spiral.

A logarithmic spiral sector is sometimes used to obtain varying translatory motion of a slide whose surface coincides with the tangent to the spiral. The linear speed of the slide will be equal to the angular speed of the spiral sector multiplied by the length of the radiant in contact.

Involute

It is spiral curve traced by a fixed point on a straight line which rolls on a closed curve *e.g.* circle, ellipse *etc.* The involute is named after the closed curve. Suppose a cord of length πD is wound on a fixed disc of diameter D. One end of the cord is kept fixed, then the other end will describe the involute of the circle if it is unknown with the cord kept taut.

The involute of a circle is a basic curve for gear teeth and gear-cutting tools.

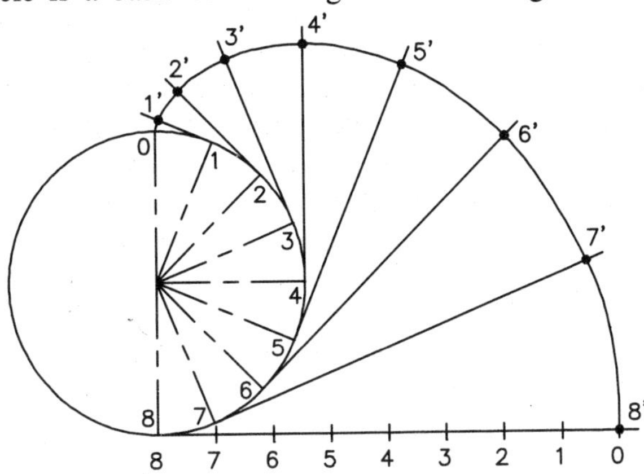

Fig. 4.17 Involute of a circle

Construction

Draw the circle of given diameter D, then divide the semicircle into any convenient number of equal parts and mark the points of division by figures 1 to 8 (Fig. 4.17). Draw a line of length equal to πD/2 and tangent to the circle at point 8. Divide this length into the same number of equal parts. Draw tangents at each points from 1 to 7 and lay off on them the length equal to 8-1, 8-2, 8-3 etc. respectively and mark them as 1', 2', 3' ..... 8'. These points are joined by a smooth curve to get the required involute of a circle.

Cycloidal Curves

Cyloidal curves are a group of curves generated by the path of a fixed point on the circumference of a rolling circle. When the circle rolls on a straight line, the locus of the point is called a cycloid. When the rolling circle rolls on the outer periphery of a circle, then the tracing point generates an Epilcycloid; if on the inside periphery then a Hypocycloid is described.

To Draw a Cycloid (Fig.4.18)

Draw the rolling circle of diameter D, then draw a straight line AA_{12} and mark off the perimeter of the circle equal to πD. Divide the line AA_{12} and the circle into same number (say 12) of equal parts and number them as 1', 2', 3' . . . . 12' and 1, 2, 3, ... 12, respectively. Perpendiculars at points 1', 2' . . . . 12' are drawn to intersect extended horizontal axis of the rolling circle at points O_1, O_2, O_3, ... O_{12}. Then draw horizontal lines passing through the points 1, 2, 3, *etc.* on the circle. Now O_1, O_2, O_3, ... O_{12} as centres and radius equal to the radius of the rolling circle draw arcs to cut the corresponding horizontal lines at the points A_1, A_2, A_3, ... A_{12} of the required cycloid. A smooth curve drawn through these points is the required cycloid.

To Draw an Epicycloid (Fig.4.18)

Draw the rolling circle of diameter D and the director circle of radius R tangent to it at point A (*i.e.* 12). Divide the rolling circle into 12 equal parts. With O as centre and (R+D/2) as radius, draw an arc indicating path of centres of the rolling circle. From the point A layout on the director circle 12 divisions each equal to $\pi D/12$ marked 1', 2', 3' . . . . 12'. Evidently arc AA_{12} (i.e. 12-12') will be equal to πD. [To do this calculate the angle a = (180 D/R)°. The angle a is then divided into 12 equal parts by lines $O_0$1', $O_0$2, ... $O_0$12]. From O_0 draw radial lines through these points and extend them to intersect the path of centres at points O_1, O_2, O_3 . . . . O_{12}. With O_0 as centre draw series of auxiliary arcs through the points of division of the rolling circle. Now starting from O_1 to O_{12} as centres and radius equal to D/2 draw arcs to intersect the corresponding auxiliary arcs at points A_1 to A_{12} of the desired epicycloid.

To Draw a Hypocycloid (Fig.4.18)

Draw the rolling circle of diameter D tangent to the director circle of radius R, at A (*i.e.* 12). Divide the angle a {which is equal to (180 D/R)°} into 12 equal parts and the director circle in the same number of parts. From O_0 and through the points of divisions on the director circle, draw radial lines which intersect the locus of centre circle radius (R = D/2) at points O_1, O_2, O_3 . . . . O_{12}. Then with O_0 as centre, draw auxiliary arcs passing through the points of division of the rolling circle. With O_1, O_2, O_3 . . . . O_{12} as centres and D/2 as radius, draw arcs to intersect the corresponding auxiliary arcs at the points A_1, A_2, A_3, ... A_{12} of the desired hypocycloid.

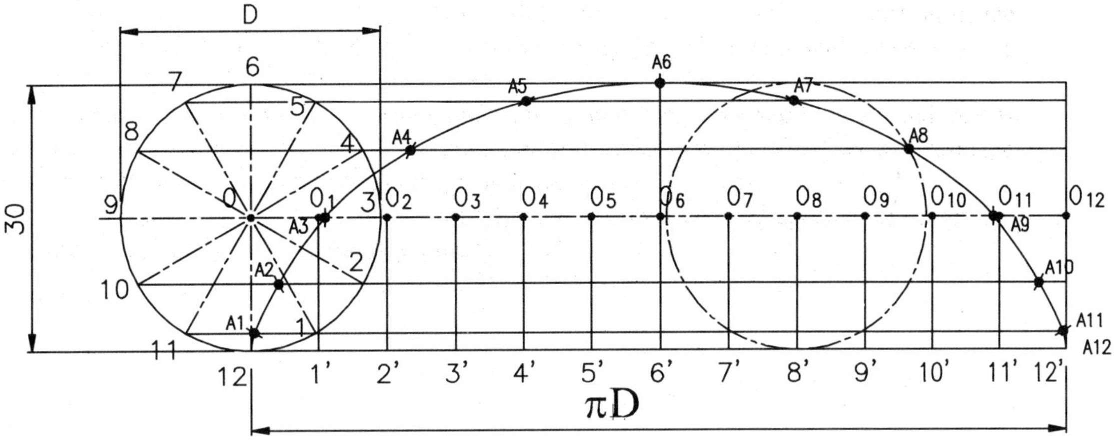

(a) Cycloid

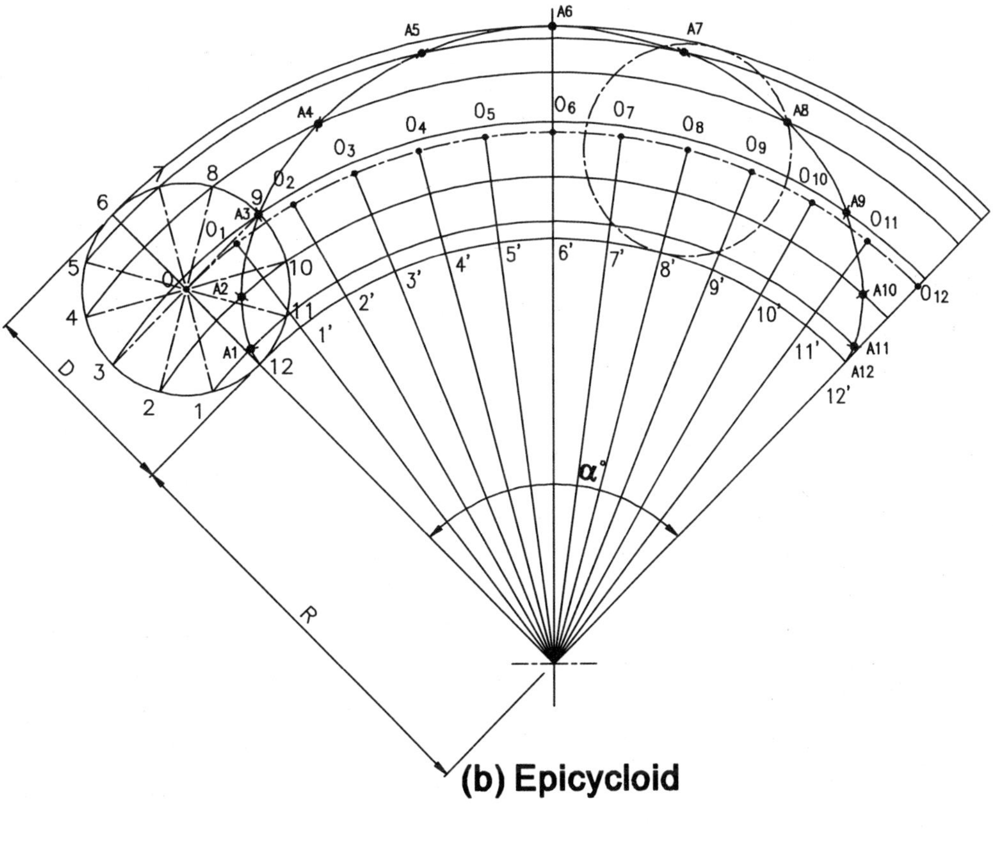

(b) Epicycloid

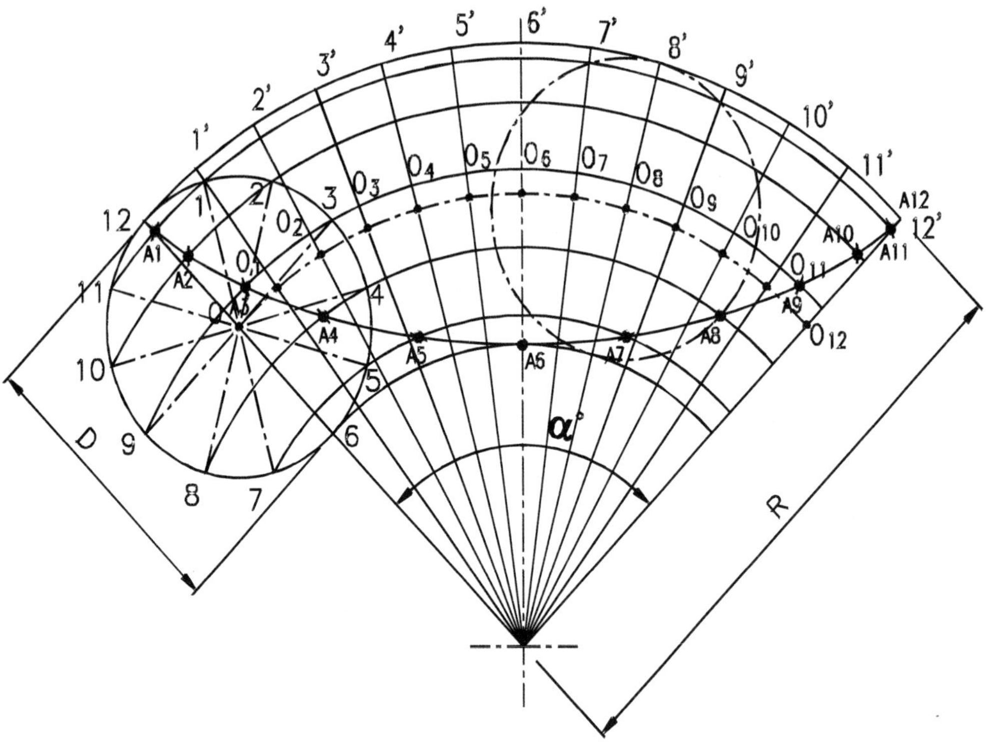

(c) Hypocycloid

Fig 4.18. (a) Cycloid (b) Epicycloid (c) Hypocycloid

PROBLEMS

4.1 (i) Draw the following, the circumscribing circle being 50 mm in diameter.
(a) Equilateral (b) Regular pentagon (c) Regular hexagon
(d) Regular heptagon (e) Regular octagon.

(ii) Draw a regular octagon circumscribing square being ABCD of area 25 cm^2.

(iii) Draw a regular octagon, one side AB being given.

4.2. Draw the locus of the point in the following cases:

(i) The point P moving in a plane about another point in such a way that its distance from it is constant.

(ii) The point Q moving in a plane in such a way that its distance from the fixed line AB is constant.

(iii) The point R is always equidistant from two fixed points A and B.

(iv) The point S equidistant from two fixed circle of 100 mm and a fixed point 200 mm away from the centre of the circle.

4.3 Construct an ellipse whose major and minor axes are 80 mm and 60 mm respectively. [Hint: Construction by radius of curvature method given.]

4.4 Two lines AB and BC of length 60 mm each, make an angle of 40°. Construct a parabola tangent to the lines.

4.5 Draw a parabola through point P (60,50) and the origin.

4.6 According to Boyle's law, PV = Const. At a particular moment the volume of an enclosed gas is $4m^3$, the pressure is 0.20 kgf/cm^2. Draw the isothermal expansion curve.

4.7 A fixed point is 75 mm away from a fixed line. Draw the locus of a point P moving in such a way that its distance from the fixed line is twice its distance from the fixed point.

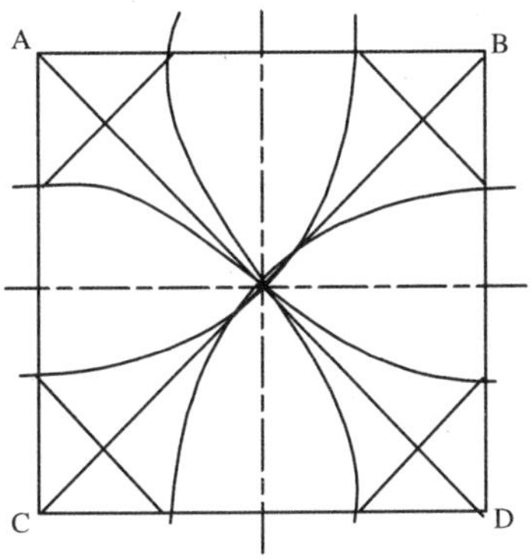

Hint on Prob.4.1(ii)

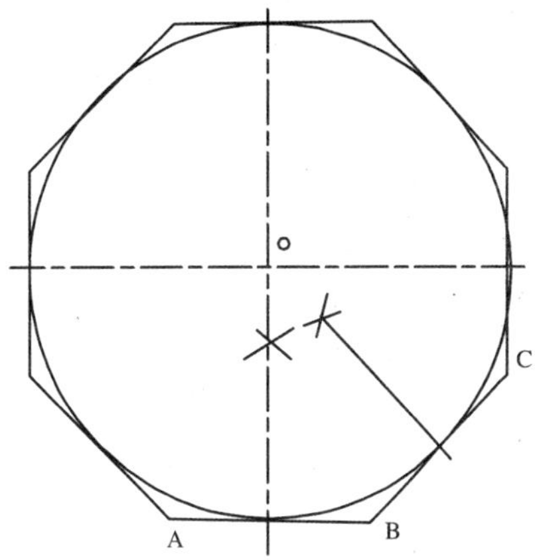

Hint on Prob.4.1(iii)
($\angle ABC = 135°$)

4.8 Draw the sine curve having 30 mm amplitude and 60 mm wave length, (Note the difference in drawing of a sine curve and that of a cylindrical helix. In case of a helix wavelength is known as the lead.)

4.9 Draw an Archimedean Spiral of 1.5 convolution, having the initial and final radii as 25 mm and 75 mm respectively.

4.10 Draw the Logarithmic Spiral having the following:
r_1 = 25 mm, r_2 = 80 mm, angle between r_1 and r_2 = 240°.

4.11 Draw one convolution of a logarithmic spiral having the shortest radius as 16 mm and ratio of lengths of radius vectors enclosing an angle of 30^0 as 9:8. Find the value of the angle which the tangent at any point makes with the radius vector.

4.12 Construct the involute of a 50 mm diameter circle for one revolution.

4.13 Draw the involute of a regular hexagon of 40 mm side for half a turn.

4.14 A roller of 40 mm diameter rolls without slip. Draw the locus of any point on the periphery of the roller for a complete turn.

4.15 ABCD is a square. A circle of 5 cm diameter rolls without slip on AB for half the revolution and then on BC to complete the revolution. The circle rolls inside the perimeter of the square. Trace the path of a point P situated on the circumference of the circle. Assume the point at the bottom-most position initially. Also trace the path traversed by another point Q situated at a distance of 10 mm from the centre of the rolling circle.

4.16 A 40 mm diameter roller rolls without slip (i) inside (ii) outside a 240 mm diameter cylinder. Draw the locii of any fixed point on the circumference on the roller and identify the curves.

4.17 Typical cross-sectional drawing for I-Beams, Channels, Tees and High Tees are shown. From Indian Standards find out an available section of each and draw the shapes.

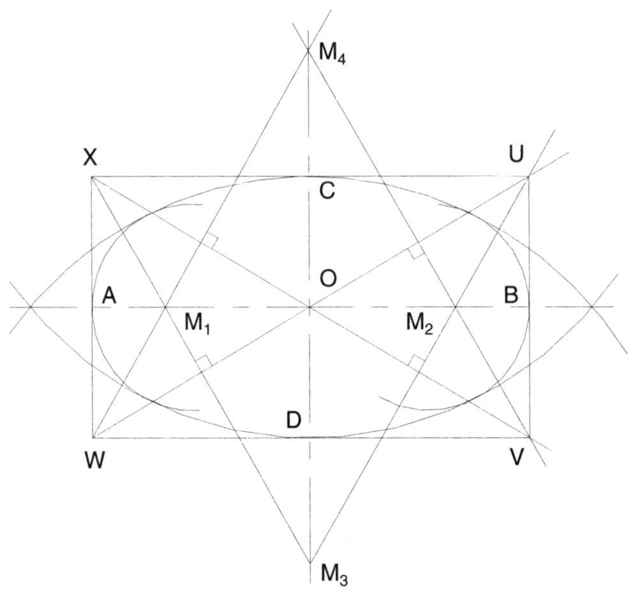

Hint on **Prob.4.3**

Let O be the centre of the ellipse. Draw two lines AB and CD perpendicular to each other representing the major and minor axes respectively and extend them. Complete the rectangle UVWX and drop perpendiculars from W & V and X & U on the diagonals and extend them and get the points M_1 & M_2 and M_3 & M_4. With these points as centers, draw circles to get the required ellipse.

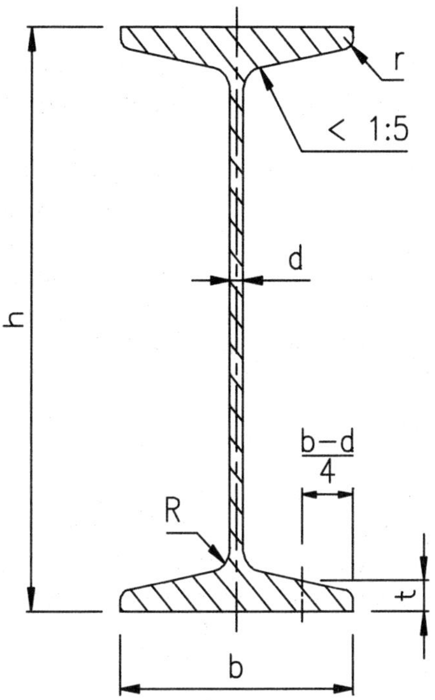

I– Beam

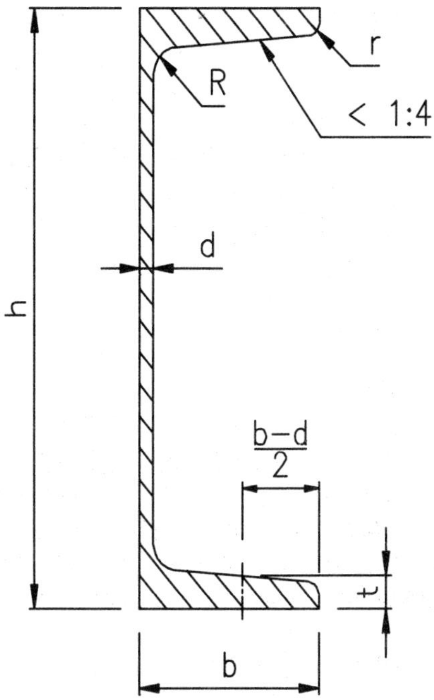

Channel

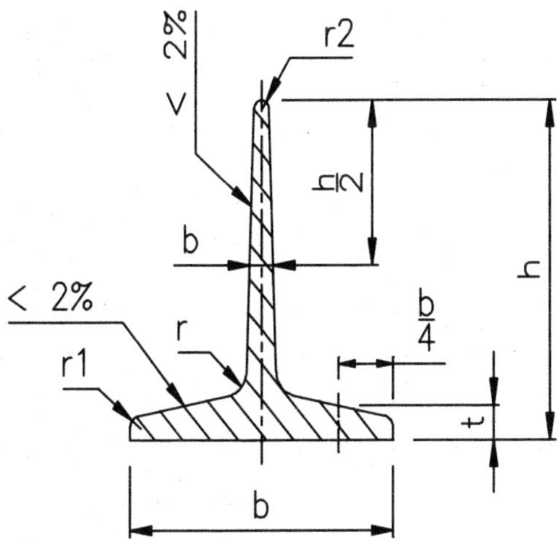

High Tee

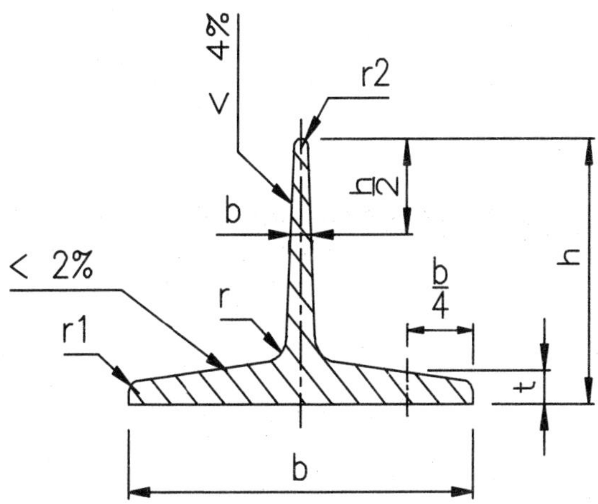

Tee

Prob. 4.17

T E E †						
Serial No.	b	h	d = t	r	r_1	r_2
1	60	30	5.5	5.5	3.0	1.5
2	70	35	6.0	6.0	3.0	1.5
3	80	40	7.0	7.0	3.5	2.0
4	90	45	8.0	8.0	4.0	2.0

CHANNEL*

Designation	h	b	d	t
ISLC 75	75	40	3.7	6.0
ISLC 100	100	50	4.0	6.4
ISLC 125	125	5.5	4.4	6.6

I – BEAM*					HIGH T E E*		
Designation	h	b	d	t	Designation	h = b	d = t
ISLB 75	75	50	3.7	5.0	ISNT 60	60	6.0
ISLB 100	100	50	4.0	6.4	ISNT 80	80	8.0
ISLB 125	125	75	4.4	6.5	ISNT 100	100	10.0
ISLB 150	150	50	3.0	4.6	ISNT 150	150	10.0

All demensions are in mm

† Not at par with BIS

*Based on BIS Hand Book for Structural Engineers

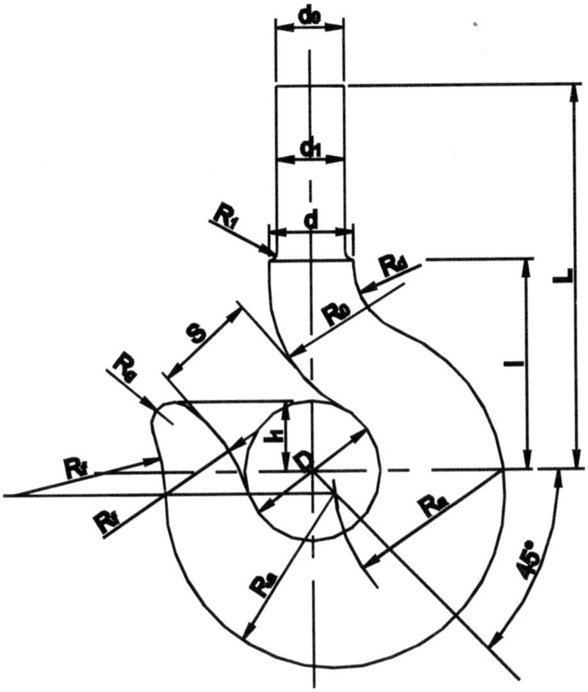

HOOK

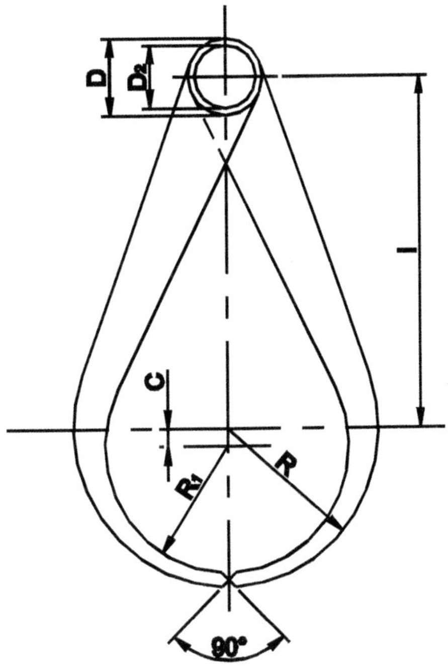

CALIPER

Prob. 4.18

Prob. 4.18 Figures above show a forged Crane Hook and an Outside Caliper. Draw the shapes and dimensions, to be taken from the figures proportionately.

CALIPER †						
Serial No.	I	d	d_2	R	R_1	C
1	120	24	20	48	38	6
2	100	22	18	45	36	5
3	90	20	16	40	32	4
4	85	20	16	30	22	4

HOOK †

Serial No.	D	S	b	d	d_1	d_0	L	l	l_1	R_a	R_d	R_o	R_t	R_1	R_g
1	20	14	18	15	12	12	60	30	10	25	8	22	28	1	4.5
2	25	18	24	18	15	14	70	35	12	32	9	26	30	1.5	5
3	30	22	26	20	17	16	80	45	15	37	10	30	35	1.5	5.5
4	32	22	28	20	17	16	90	45	16	40	11	32	38	1.5	6
5	36	26	32	25	20	20	100	50	18	45	13	36	40	2.5	6.5
6	40	30	36	25	20	20	110	60	20	50	15	40	45	2.5	6.5

All dimensionsare in mm † Not at par with IS

CHAPTER – 5

SCALES AND DIMENSIONING

Drawings are usually prepared smaller than the object they represent though it is desirable to make full-size drawings of machine parts and assemblies to represent their true shape and size. Large objects of simpler shapes are drawn smaller than full-size, whereas small objects (instrument part, watch-components *etc.*) are drawn larger. The scale is the ratio between the dimensions of the drawing of an object and its actual dimensions. Scales should be selected as per BIS:SP-46-1988 the relevant recommendation being:

Full Scale	1:1
Reduced Scale	1:2; 1:5; 1:10; 1:20; 1:50; 1:100; 1:200; 1:1000; 1:2000; 1:10,000.
Enlarged Scale	50:1; 20:1; 10:1; 5:1; 2:1

All drawings should be made to scale which must be stated as follows:

Scale 1:1 (for full size), Scale 1:2 (for half size)

Scale 2:1 (for twice full size) and so on.

Formally the above ratios were known as Representative Fraction (R.F) of the scale.
At present, however, the word "Scale" is used to mean the term.

Plain Scales may be drawn to show two dimensions, *e.g.*, mm and cm, tens of meters (decameter), meter and so on. When three dimensions are to be shown, *e.g.* centimeter, tens of centimeter, hundreds of centimeter *i.e.* meter, decimeter, centimeter or likewise, we have to construct **Diagonal or Vernier Scale**. It should be particularly noted that unless stated otherwise all scales should be made 150 mm in length or as near as 150 mm.

CONSTRUCTION OF SCALES

Plain Scale

Let us explain the procedure with an example. Construct a scale of m and dm when 2.5 mm represent 10 cm. This scale, of course is not recommended in BIS.

$$\text{Scale} = \frac{\text{distance on drawing}}{\text{distance on object}} = \frac{2.5 \text{ mm}}{100 \text{ mm}} = \frac{1}{40} = 1{:}40$$

Now 2.5 mm represents 10 cm.

Therefore, 150 mm represents [(10/2.5) × (150/100) = 6 meter]
[150 mm because we want scale as near 150 mm as possible.]

Rule three parallel lines AA, BB, CC as in Fig. 5.1. AA being 6 mm above BB and BB about 2 mm above CC. Draw the scales as shown. Divide 150 mm into 6 equal parts, each of which represents 1 m. Divide the first meter on the left into ten equal parts, each of which represents 1 decimeter. Number the main division marks (counting from left) O_2, and rest as shown. 1.6 m is shown on the drawing.

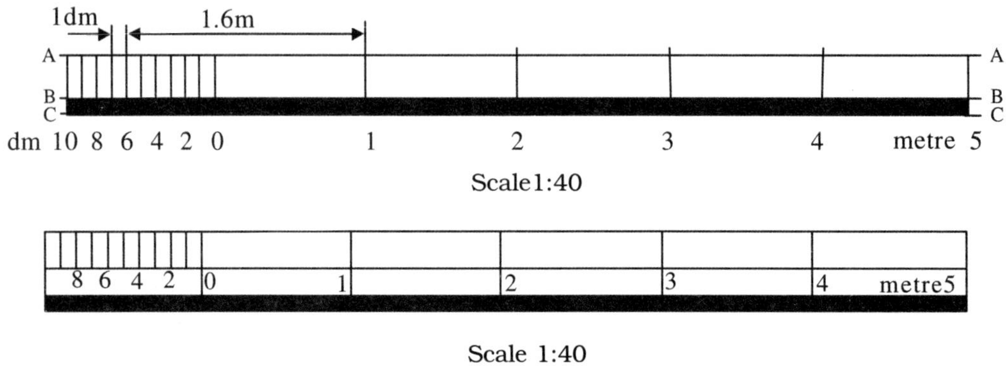

Fig. 5.1 Plain Scale

Diagonal Scale

It is already mentioned that when we require dimensions as m, dm and cm or in other words hundredth of a dimension we use diagonal scale.

PQRS is a rectangle divided into five equal parts by parallel lines drawn equidistant from each other (Fig. 5.2a). PQ is bisected at X. S and X are joined and YQ is drawn parallel to SX. Now on drawing perpendiculars, a series of triangles or steps are produced of the same number as the parallel lines above PQ. It is clear that as one moves along diagonal Y Q, every time a fresh line is intersected we have moved (1/5) XQ farther to the left.

Thus at BB, Distance moved = (3/5) XQ.

When Y is reached, we have moved as far to the left as we had gone to X, thereafter return to X and climb up the diagonal XS in the same way. On reaching at A, we find that

$$AA = (4/5)\ PX + XQ, \text{ but } PX = XQ,$$

$$\therefore \quad AA = (9/5)\ XQ.$$

It may be recalled in this case that five lines were drawn above PQ, and as a result 5 steps were obtained. Had we drawn 10 lines we would have 10 steps, 15 lines 15 steps and so on.

In drawing a diagonal scale, therefore, it must be remembered that the same number of parallel lines (equidistant) must be drawn above the line PQ, as there are units of the third dimension in the unit of the second dimension (usually 10 in metric system).

Example

Draw a scale of 1:40, showing meter, decimeter, centimeter.

Now, Scale = 1:40 (1 mm represents 40mm)

150 mm will represent 40 × 150mm

= 6,000 mm = 6 m.

Take a line 150 mm long, divide into 6 equal divisions. Mark it as shown in Fig. 5.2. Divide the extreme left portion into 10 equal divisions. Rule 10 equidistant parallel lines along the length of the scale. The distance between the lines is optional, but 3.0 mm gap is a fair one. Join the diagonals as shown.

The line AA measures 4.57 m. If the scale is required to be manufactured then all the inscriptions should be placed inside the border of the scale as shown in Fig. 5.2

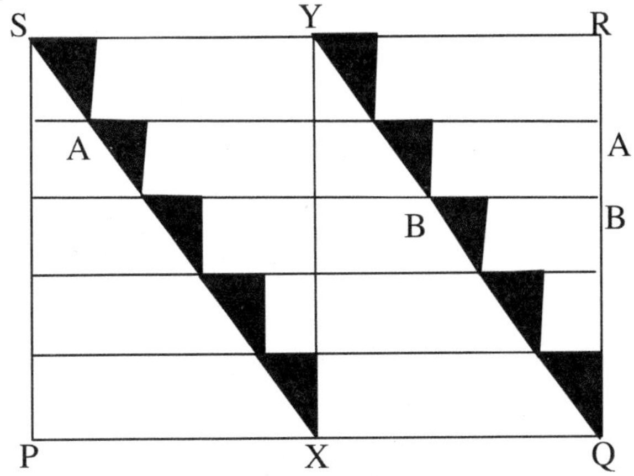

Fig. 5.2 (a)

The scale of chords: This scale marked as C or CHO on instruments is used to measure angles. This can be constructed as follows. Describe a quadrant of a circle (Fig.5.3a). Trisect the arc and subdivide each again into three, thus dividing the arc into 9 equal parts of 10° each.

With centre O and distance Oa, Ob, Oc and so on, describe arcs cutting the base line at 90, 80 *etc.* Complete the scale as shown. It may be observed that

(i) a series of isosceles triangles can be drawn, having their apices at the base of the perpendicular.

(ii) while the length of the equal sides is always the same, *viz.*, the distance between 0 and 60°, the bases increase as the apexangle increases. The bases are the chords of the arcs of the various angles.

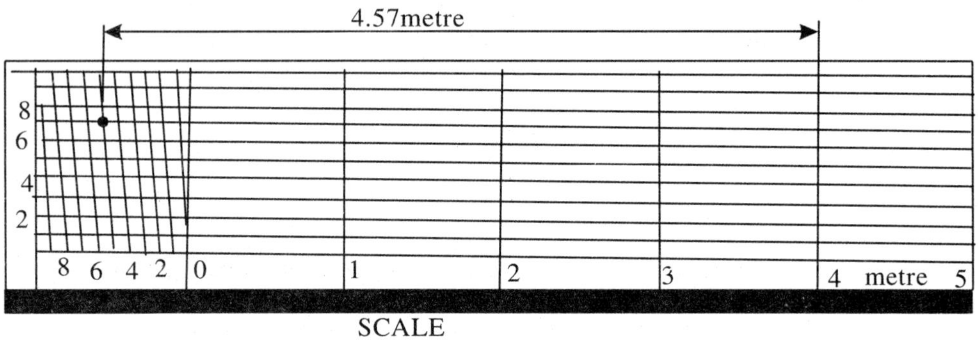

Fig. 5.2 (b)

Example

Draw any angle, say 50° by means of the scales of chords.

Let XY be any line and X be the point at which it is required to make the angle (Fig. 5.3b). With X as centre and radius equal to the distance between 0 and 60, draw arc ZY cutting line XY at Y (*i.e.* XY = 0 – 60). With Y as centre and radius equal to distance between 0 and 50 (the required number degrees) swing an arc cutting the arc YZ at Z. Join XZ. The angle ZYX is the required angle.

Dimensioning

Uptill now in this treatment of engineering drawing we were concerned mainly with representation of objects. It is needless to mention that the description of shape is not meaningful if not associated with proper and complete indication of size. The shop-worker must be able to get all the dimensions *e.g.* length, breadth, width, diameter, and location of holes, grooves and any other feature and tolerance that can be allowed during manufacturing. These information should be supplied by the draftsman on the drawing and is called dimensioning.

General Principles

All dimensions, tolerances, *etc.* required for correct functioning of the parts should be placed directly on the drawing to describe the part in its finished form with due consideration to manufacturing and inspection requirements. Dimensions should be placed on drawings so that they may be easily read.

There are two recommended systems of placing the dimensions:

(a) *Aligned system:* In this system, all dimensions are so placed that they may be read from the bottom or right hand edges of the drawing sheet and dimensions should be placed above the dimension line which should be continuous.

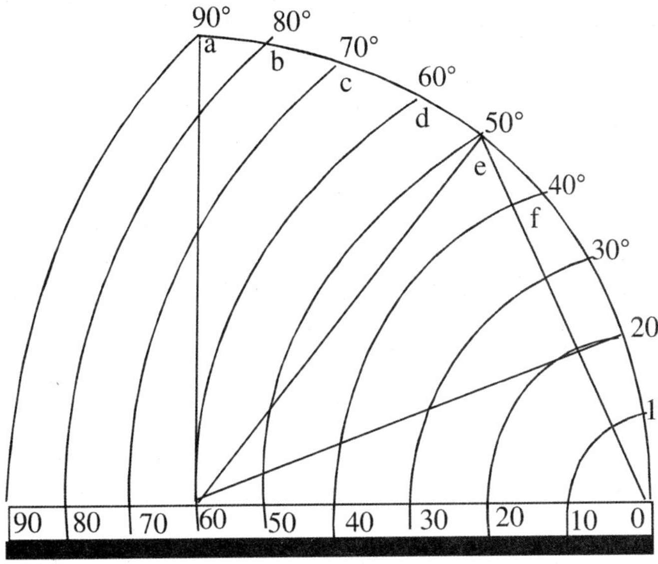

Fig. 5.3 (a)

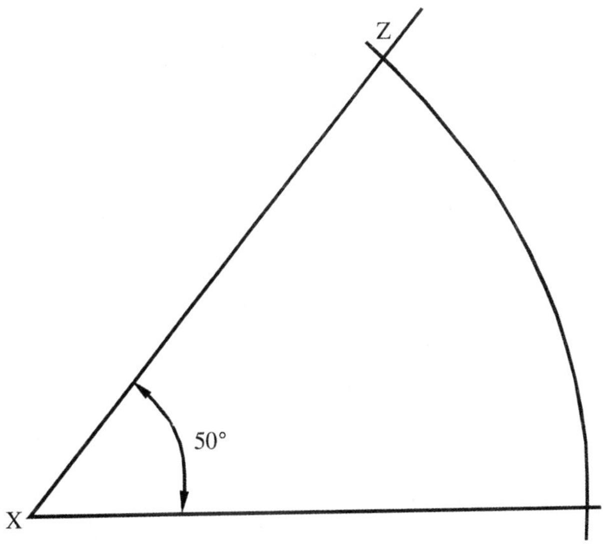

Fig. 5.3 (b) Scale differs from that used in Fig. 5.3 (a)

(b) *Unidirectional system:* All dimensions are placed so that they are readable from the bottom edge of the drawing sheet. The figures indicating dimensions are always placed upright irrespective of the orientation of the dimension line which is provided with a central gap to house the dimension (Fig. 5.4).

In this text *Aligned system* of dimensioning is used throughout.

Some General Rules

(1) Dimensions for one particular operation (*e.g.* drilling, threading) should be given in one view only.

(2) Dimensions should be placed outside the view in general unless the placing dimension within the view makes the drawing more clear.

(3) Dimensions are to be given from visible outlines, a base line, a centre line of a hole or of a cylindrical part or a finished surface.

(4) The intersection of dimension lines should be avoided as far as practicable.

(5) An axis or contour line should not be used as a dimension line but such a line may be used as an extension line.

(6) Overall dimensions should be placed outside the intermediate dimensions. Where an overall

dimension is shown, an intermediate dimension, normally the unimportant ones, should be omitted.

(7) Production processes or inspection methods should not be specified unless othey are essential.

(8) Neither excess dimensions that are required to describe the finished product should be given nor should any feature be located by more than one dimension in any one direction.

(9) It should not be necessary for dimensions to be derived by a craftsman.

Method of Execution

(i) Dimension lines and extension lines are drawn as continuous thin lines and extension lines should extend slightly beyond the dimension lines as shown in Fig. 5.4 in different examples.

(ii) In aligned system of dimensioning, except where unavoidable, an angular space of 30° arc should be left out as shown in Fig. 5.5(1).

(iii) Leaders (pointer lines) are continuous thin lines drawn from notes or figures to the spot or area where they apply. Leaders are terminated with arrow heads or dots. Leader lines should not be curved or drawn free-hand. Leaders should make an angle greater than 30° with surface line they touch [Fig. 5.5(2)].

(iv) Each end of the dimension line is defined by an arrow head which should be drawn within the limits of the relevant extension lines; when enough space is not available, they should be placed outside. Adjacent arrow-heads may be replaced by small dots.

(v) Arrows: The length of the arrow-head may be taken as about three times the depth and space in the arrow-head should be filled in. The size of arrows should be proportionate to the thickness of the lines [Fig. 5.5(3)].

(vi) Dimensions of parts not drawn to scale should be underlined *e.g.* 12.5. However, when it is required to draw the entire drawing not to scale the abbreviation, NTS may be used and the space for scale scored off.

(vii) Put before a dimension for a diameter the symbol f, for a radius the letter 'R', for a square the letters 'SQ', for hexagon 'HEX' and the radius or diameter dimension of a spherical surface should be preceded by the word SPHERE [Fig. 5.5(4)].

(viii) Decimal points should be placed at the bottom *e.g.* 5.7, 12.3.

(ix) A leader line should terminate:

(a) with a dot, if it ends within outlines of an object.

(b) with an arrow head, if it ends on the outline of an object.

(c) without a dot or an arrowhead, if it ends on a dimension line.

Dimensioning of Some Common Features

Fig. 5.6 shows the methods of progressive dimensioning, angular dimensioning and the features like diameter, radiusing and filleting.

Fig. 5.7 indicates the use of co-ordinates in location of holes, dimensioning of curved contour & bent up components and the method of dimensioning chamfers, external and internal.

Fig. 5.8 illustrates how to dimension countersinks, counter bores and spotfaces.

Fig. 5.9 depicts the correct method of dimensioning from a flat surface and the incorrect method of dimensioning with reference to the vertical centre line, limiting dimensions with one intermediate dimensions omitted and the dimensioning of key & keyways.

Fig. 5.10 indicates dimensioning of tapered features.

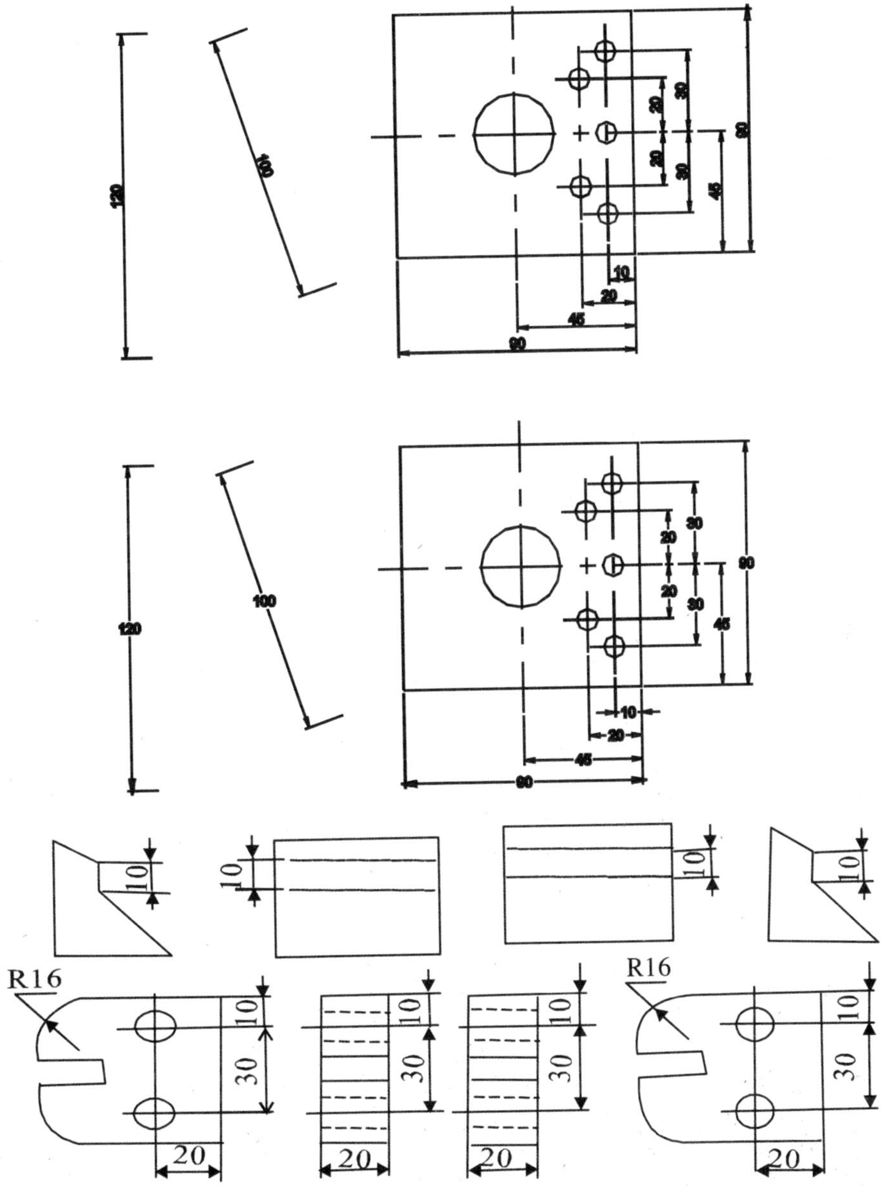

Fig. 5.4

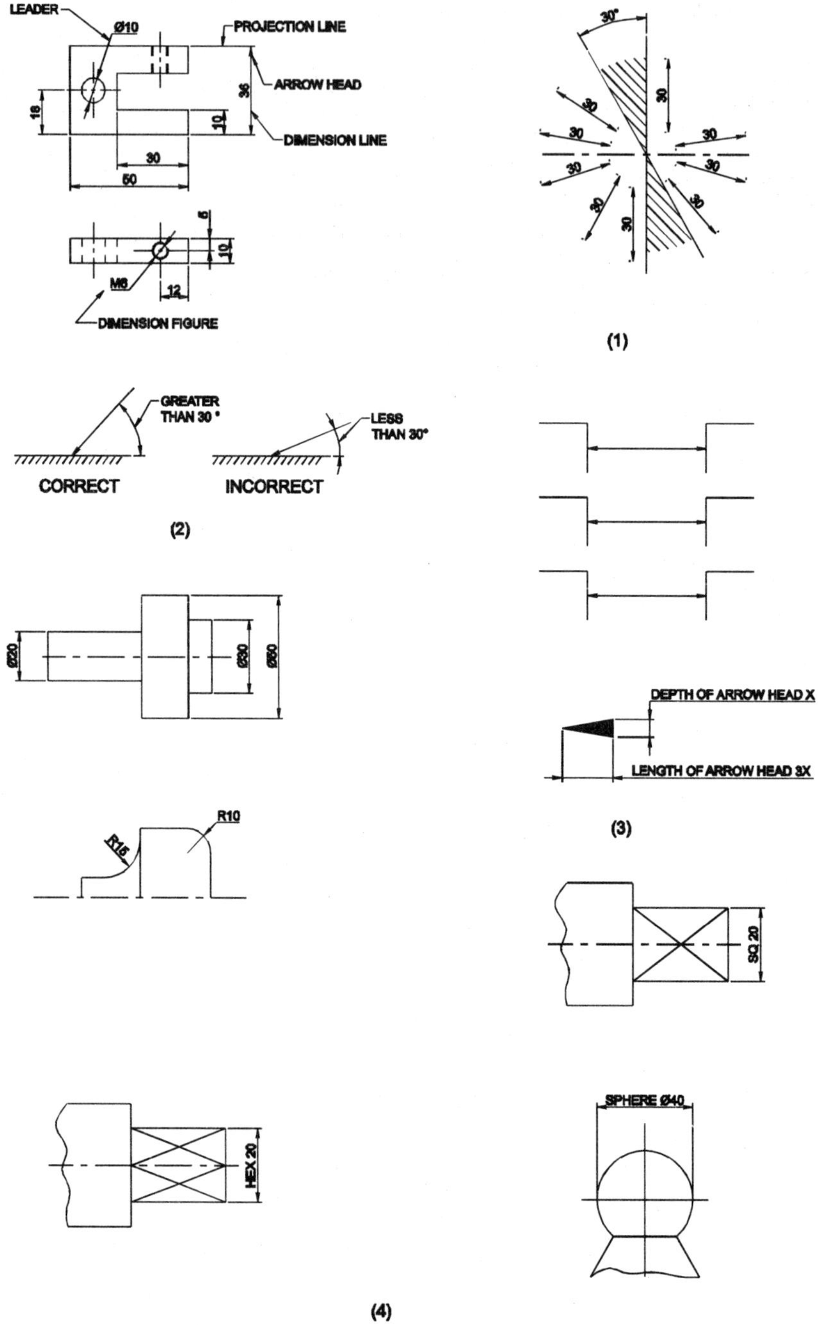
LEADER
Ø10
PROJECTION LINE
ARROW HEAD
DIMENSION LINE
36
16
10
30
50
5
10
M6
12
DIMENSION FIGURE
30°
30
(1)
GREATER THAN 30°
LESS THAN 30°
CORRECT
INCORRECT
(2)
Ø20
Ø30
Ø50
DEPTH OF ARROW HEAD X
LENGTH OF ARROW HEAD 3X
(3)
R10
R15
SQ 20
HEX 20
SPHERE Ø40
(4)

Fig 5.5

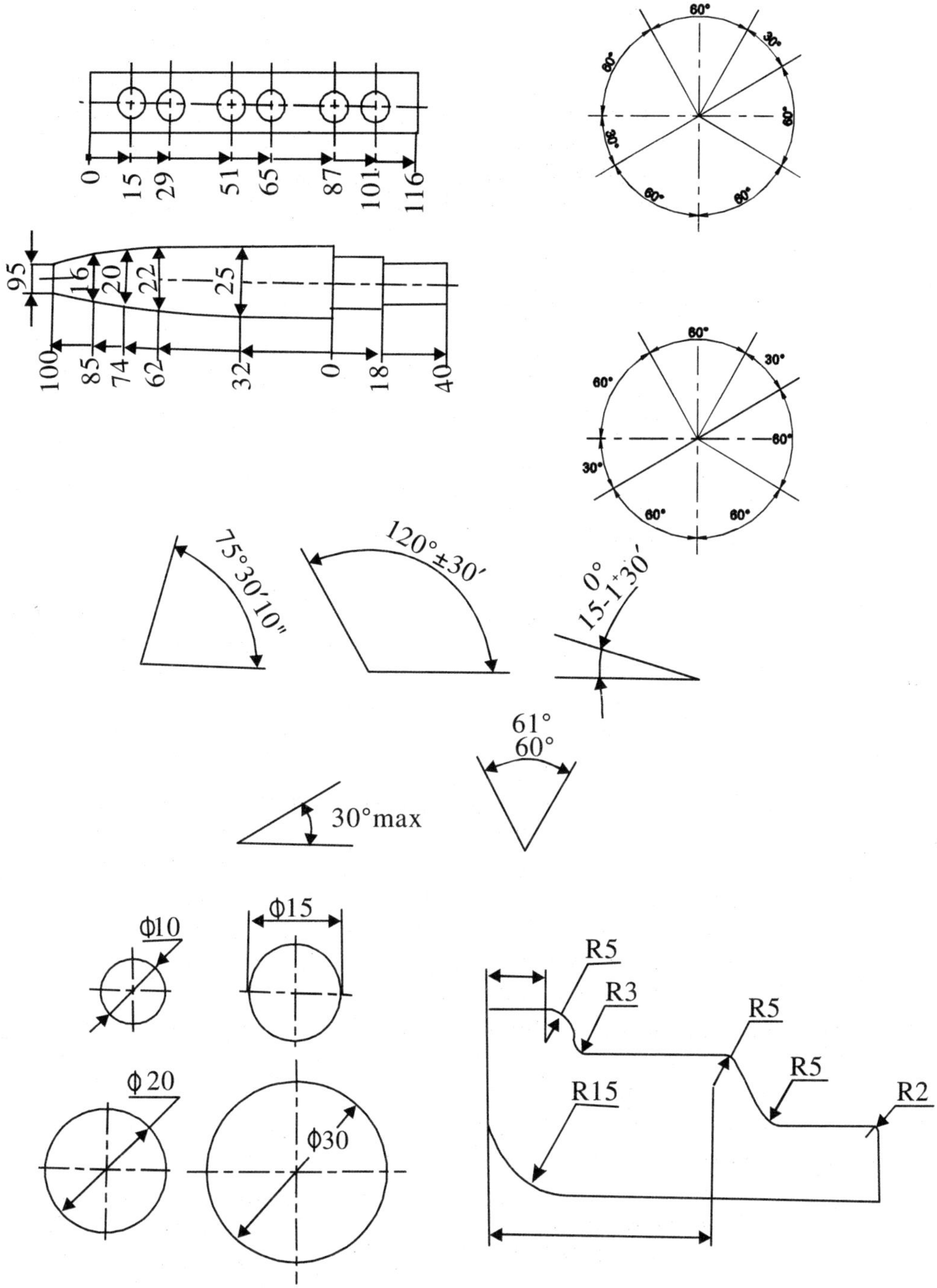
0
15
29
51
65
87
101
116
60°
30°
60°
95
16
20
22
25
100
85
74
62
32
0
18
40
75°30′10″
120°±30′
15-1°30′
61°
60°
30°max
ϕ10
ϕ15
ϕ20
ϕ30
R5
R3
R5
R5
R15
R2

Fig. 5.6

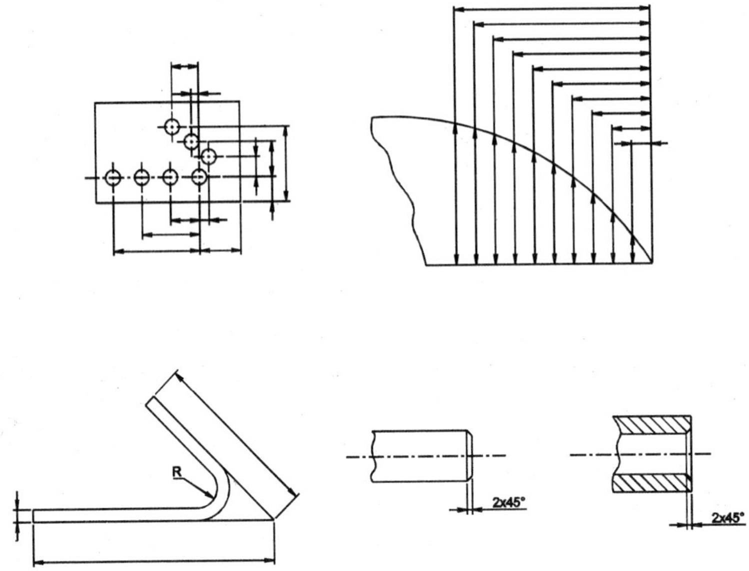
R
2x45°
2x45°

Fig 5.7

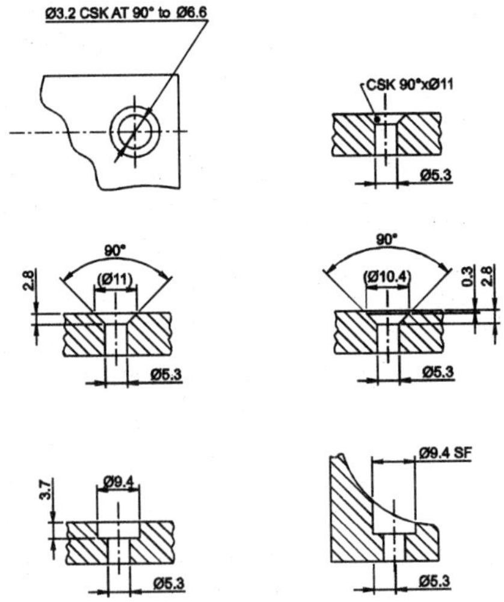
Ø3.2 CSK AT 90° to Ø6.6
CSK 90°xØ11
Ø5.3
90°
(Ø11)
2.8
Ø5.3
90°
(Ø10.4)
0.3
2.8
Ø5.3
3.7
Ø9.4
Ø5.3
Ø9.4 SF
Ø5.3

Fig 5.8

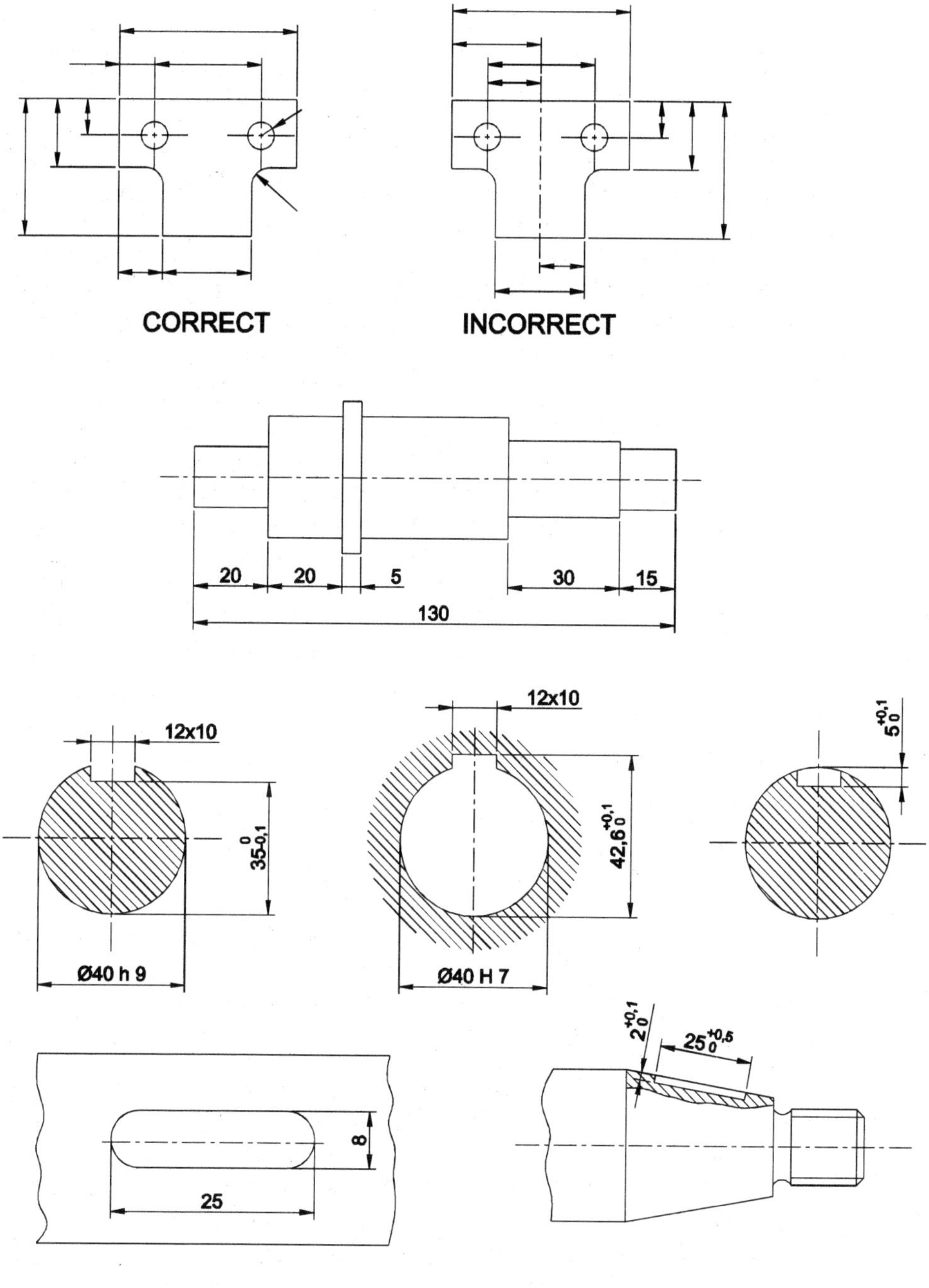
CORRECT
INCORRECT
20
20
5
30
15
130
12x10
35 0 -0,1
Ø40 h 9
12x10
42,6 +0,1 0
Ø40 H 7
5 +0,1 0
8
25
2 +0,1 0
25 +0,5 0

Fig. 5.9

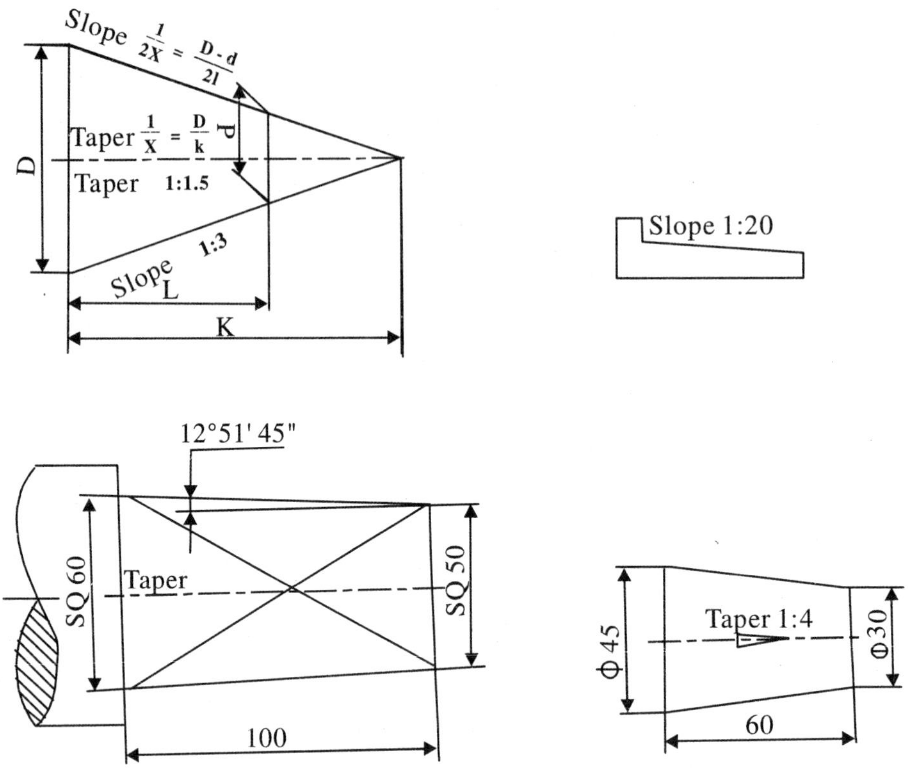

Fig. 5.10

PROBLEMS

5.1 Construct a scale to measure decimeter and centimeter and mark 2 decimeter, 4 centimeter on the scale.

5.2 Draw a scale showing kilometer, hectometer and decameter. Show 5km, 7hm, 7dm on the scale.

5.3 Draw a diagonal scale to show mm, cm, dm. Use a scale of 1:5.

5.4 A map 500 cm long × 50 cm wide represents an area of 6250 km^2. Construct a diagonal scale to measure kilometer, hectometer and decameter. Indicate on this scale a length of 5.55 km.

5.5 Take off the following angles using the scale of chords. 15°, 40° and 75°.

5.6 Reproduce the drawing given in Prob. 5.6.1 to 5.6.8 and add any missing dimensions following dimensioning rules.

5.7 Dimension each of the two views given in Probs. 5.7.1 to 5.7.8 following the rules.

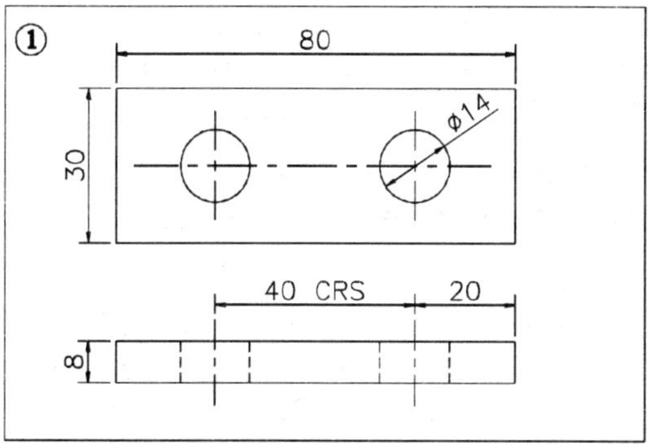

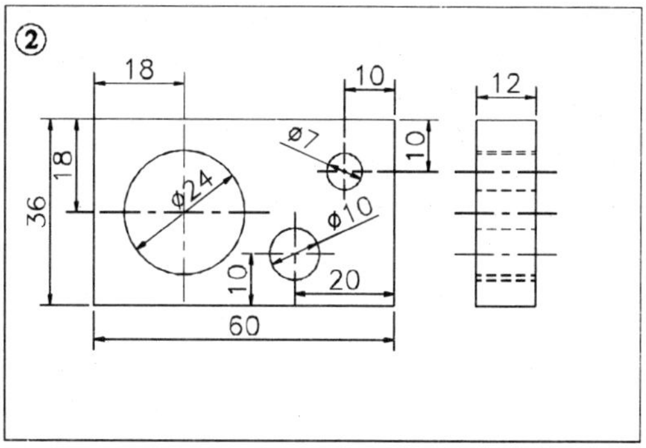

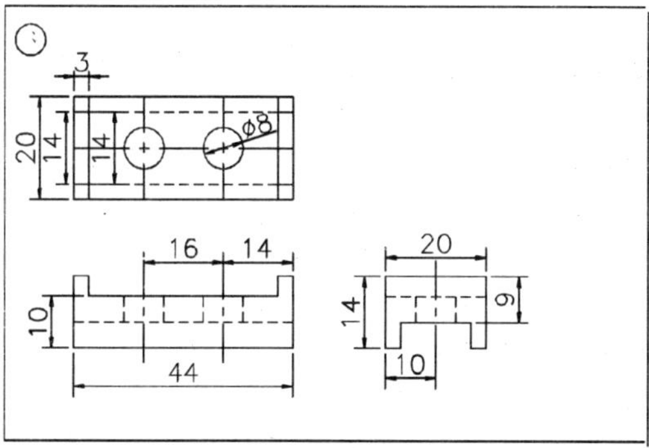

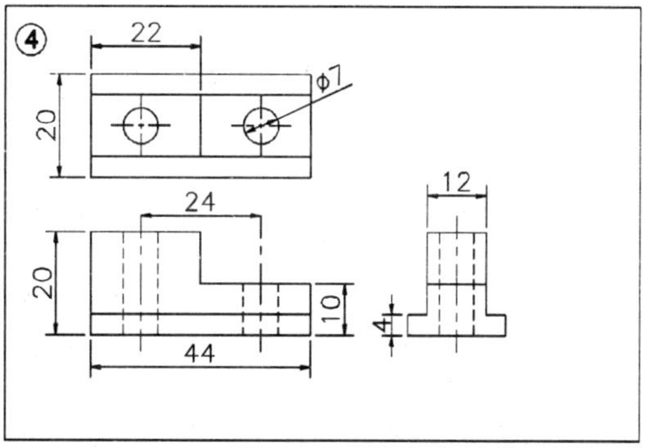

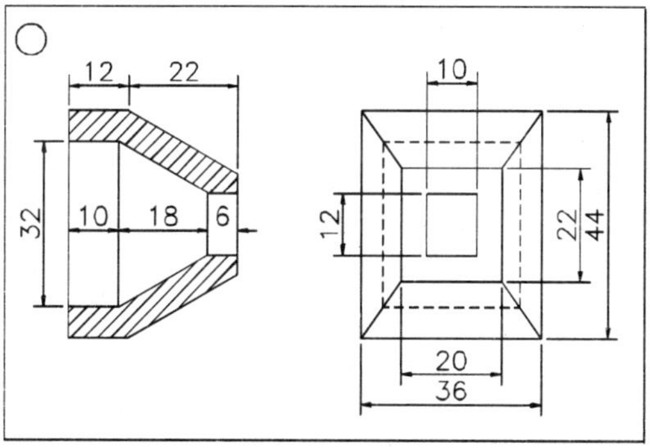

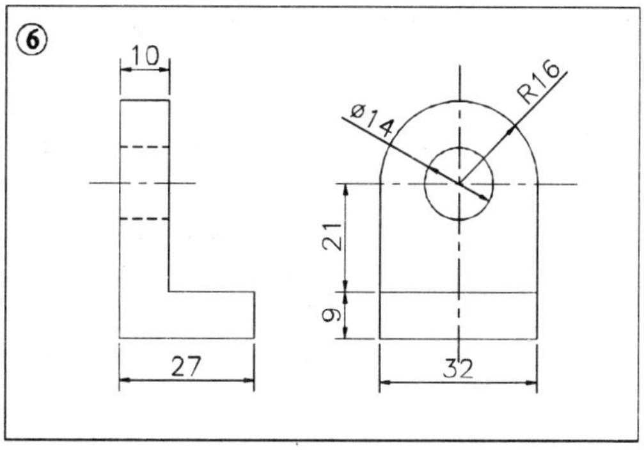

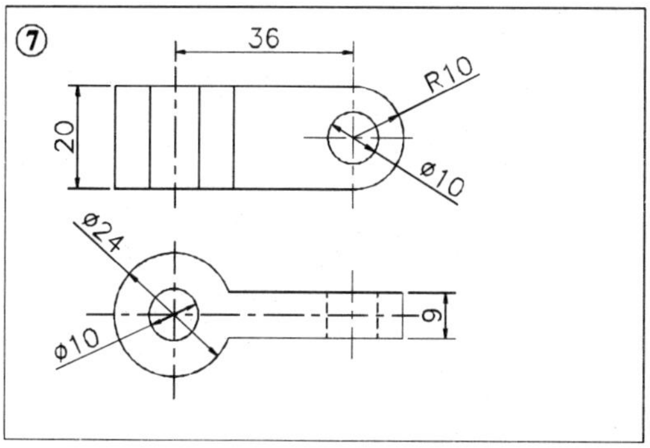

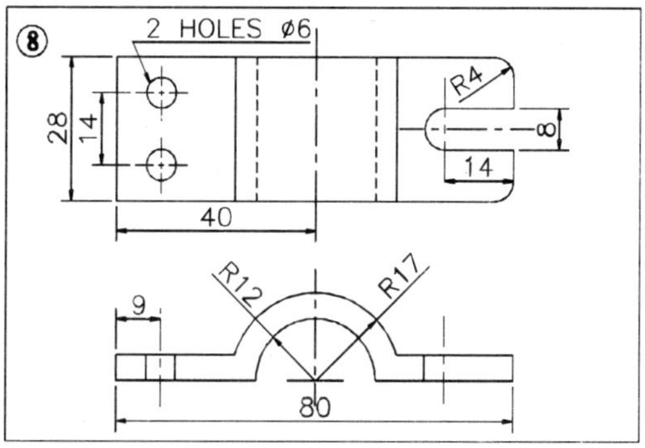

Probs. 5.6.1 to **5.6.8**

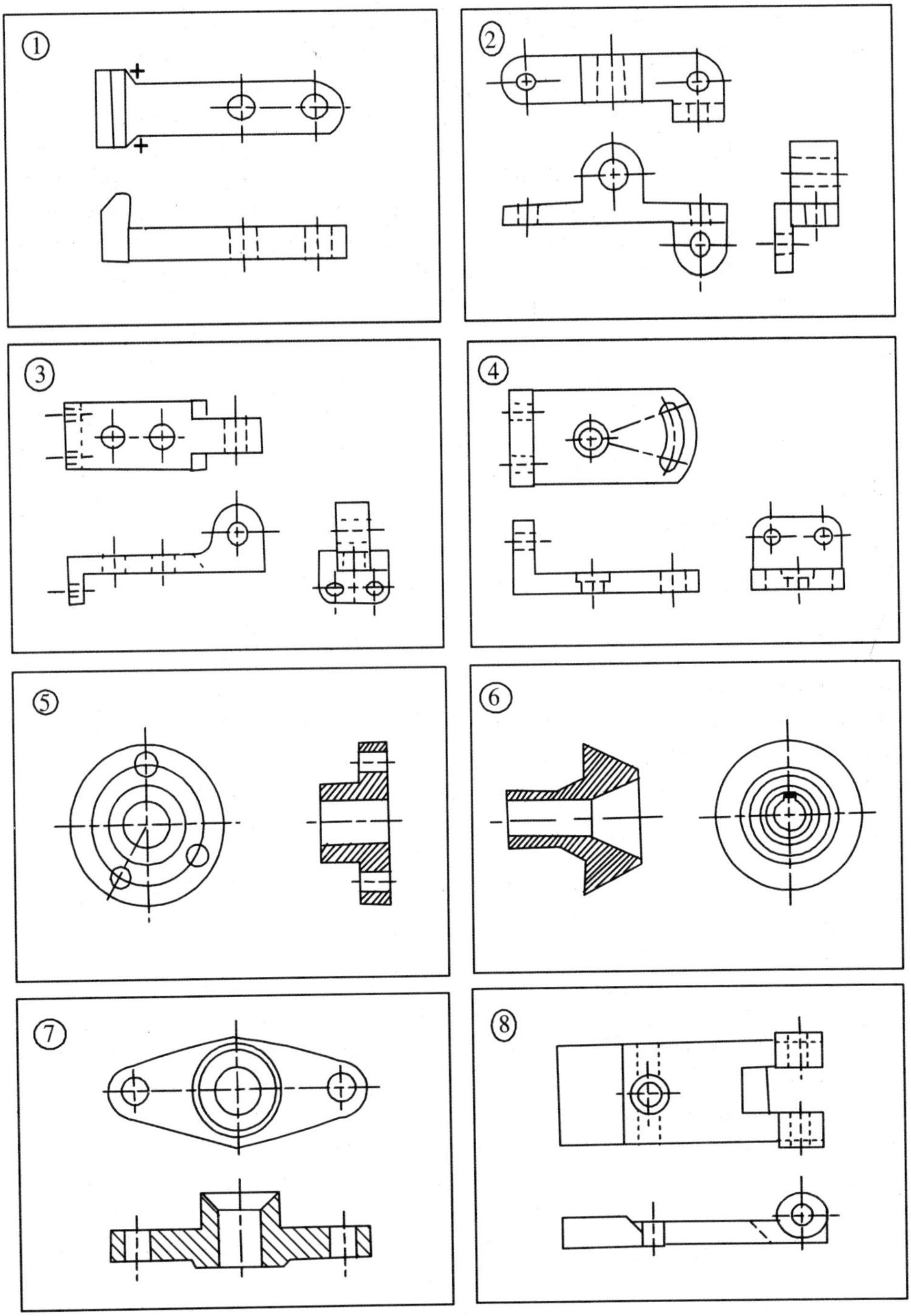

Probs. 5.7.1 to **5.7.8**

CHAPTER – 6

BASIC THEORY OF DRAWING AND ORTHOGRAPHIC PROJECTION

Theory of drawing or descriptive geometry deals with methods of projection of points, lines, planes, surfaces in space and methods of determining true lengths of lines and true shapes of surfaces. The scope of this chapter is limited to projection of points, lines and objects.

PLANES OF PROJECTION

The three mutually orthogonal planes of projection shown in Fig. 6.1a are called co-ordinate planes or reference planes. They are *Horizontal Plane* (H), *Vertical Plane* (V) and *Profile Plane* (P). These planes are assumed to be transparent and extended upto infinity. The four quadrants formed by H and V are called dihedral angles or simply angles and are numbered as shown. The position of the observer is always in front of V-plane and above H-plane. The lines of intersection of H and V planes & V and P planes are called Ground line (G.L.) & Profile Plane trace line (P.P.) respectively. The point of intersection of G.L. and P.P. may be taken as the origin (O).

ORTHOGRAPHIC PROJECTION

In this method of projection the observer is assumed to be at an infinite distance so that the projecting rays are parallel and the projecting lines for any particular plane are parallel to each other and perpendicular to the plane of projection.

REPRESENTATION OF THE VIEWS ON DRAWING PAPER

The plane of the drawing sheet represents V. To accommodate V projection (*i.e.*, **Front View** which is the view of an object from front) and H projection (*i.e.*, **Top View** which is the view of an object from above) on this sheet, H is revolved clockwise about G.L. to coincide with V (vide Fig. 6.1a). For P projections (*i.e.*, View of an object from right or left side) P is revolved about PP to coincide with V.

H and V projection lines of a point lie always on the same perpendicular to G.L. and V and P projecting lines on the same horizontal projecting line that intersects PP at right angles.

NOTATION

Points in space will be denoted by capital letters, A, B, C, *etc.* and lines by AB, BC, *etc.*, their projections by small letters, viz a^H, b^H, c^H for horizontal (H) projections (Fig. 6.1c, d) and a^V,

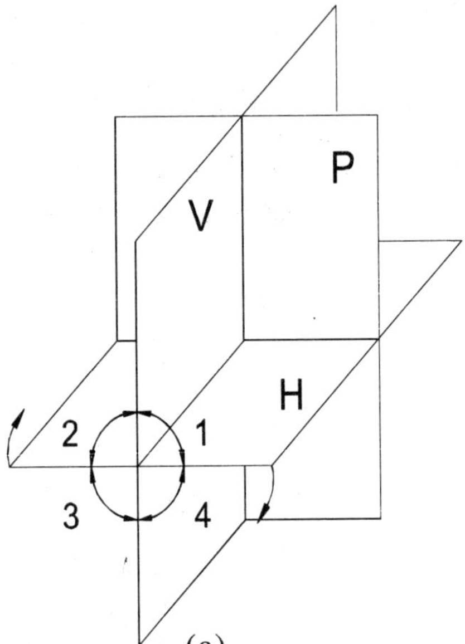

(a)

Second angle Above H Behind V	*First angle* Above H In front of V
Third angle Below H Behind V	*Fourth angle* Below H In front of V

(b)

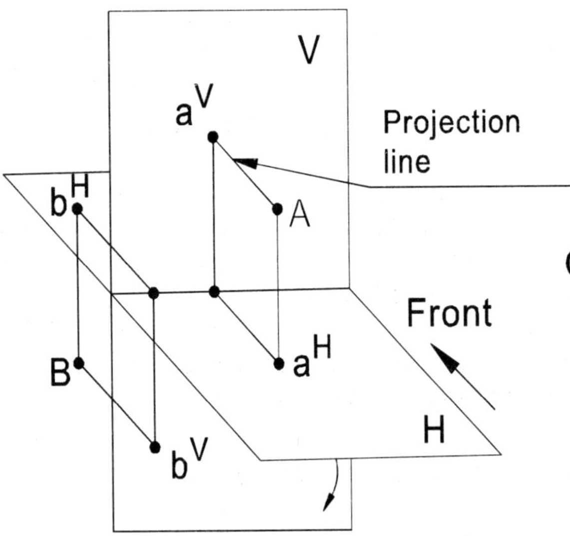

(c)

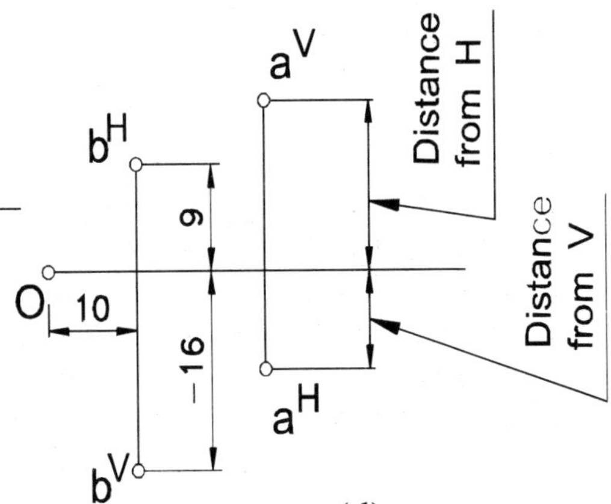

(d)

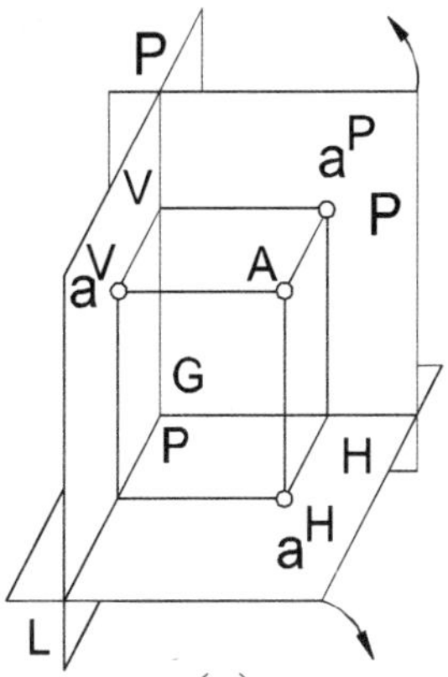

(e)

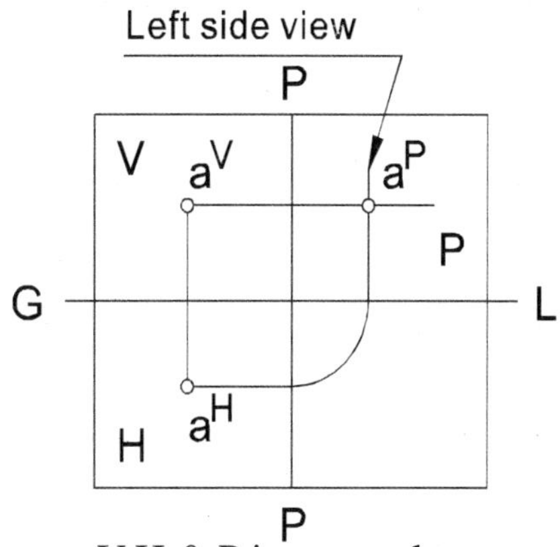

V,H & P in one plane

(f)

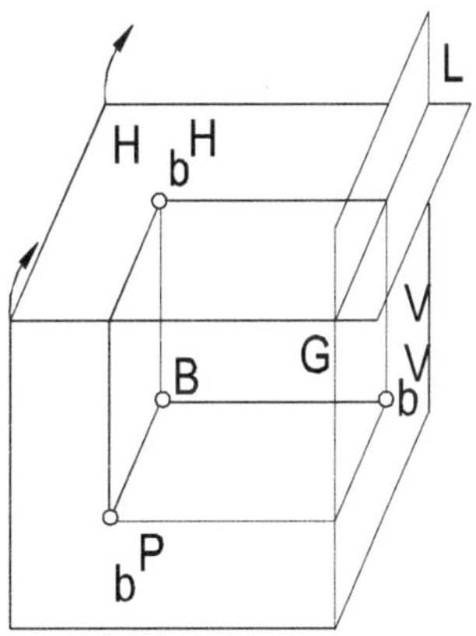

(g)

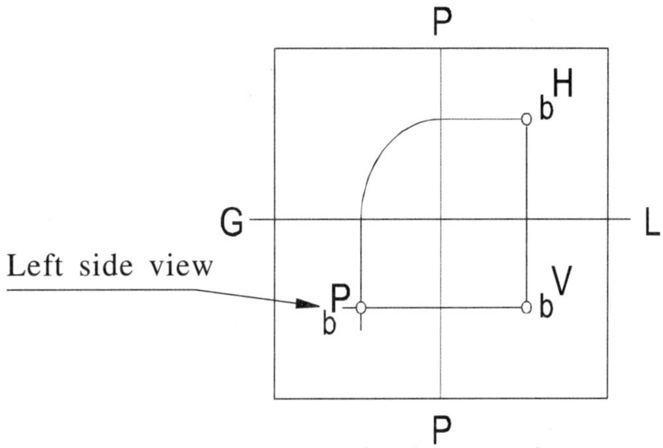

V,H & P in one plane

(h)

Fig. 6.1

b^V, c^V for projection on V-plane and a^P, b^P, c^P for P-projections (Fig. 6.1 e and f). **A** point in space may be described by its **distances from the reference planes**, H an V *viz.* **A** (3 units above H, 2 units behind V), **B** (5 units below H, 3 units in front of V).

A point may also be denoted by **co-ordinate systems** *e.g.*, (10,9, -16). The first number, 10, indicates the distance from the origin measured along G.L., the second number 9 gives the distance of H-projection from G.L. and third, -16, the distance of V-projection from G.L. Minus sign means below G.L. and plus sign above G.L. (Fig. 6.1 d).

Figs. 6.1e and f show the method of projection of points in **First angle** convention where as Figs. 6.1g and h illustrate the **Third angle** convention.

Projection of lines: Lines are obtained by joining points in space. If for a finite line AB the locations of the end points A and B and their projections are known then the projection of the line is obtained by joining the similar projections *e.g.*, $a^V b^V$, $a^H b^H$ (Fig. 6.2). Lines can be classified as under depending on their orientations with the reference planes.

Horizontal line is a line parallel to the horizontal plane, Fig. 6.2a.

Frontal line is a line parallel to the vertical plane, Fig. 6.2b.

Profile line is a line parallel to the profile plane, Fig. 6.2c.

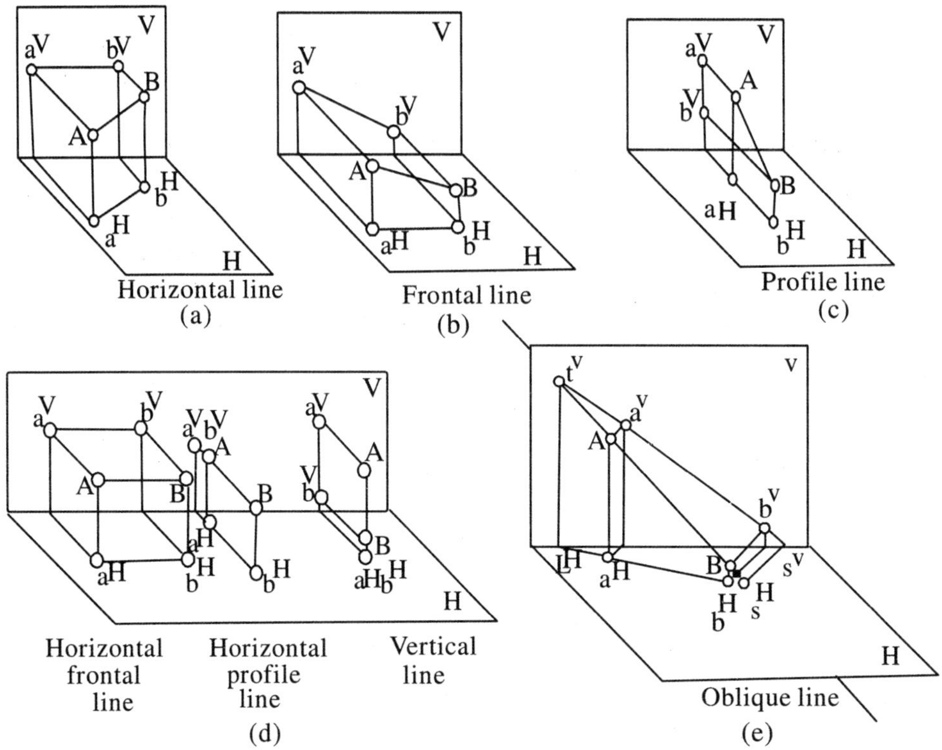

Fig. 6.2

Fig.6.2d shows a **horizontal frontal** (parallel to H and V planes) line, a horizontal profile (parallel to H and P planes *i.e.*, a^V and b^V are the same point) line and a **vertical line** (perpendicular to H-plane *i.e.*, a^H and b^H are the same point).

Fig. 6.2e illustrates the projection of an oblique line which makes angles (other than 0° and 90°) with all the reference planes. The same figure also indicates the traces of the line AB. By traces we mean the projection of the points at which the line pierces the reference planes of projection. Fig. 6.2e shows that line AB pierces the V-planes at t^V and H-plane at s^H with their projections on G.L. namely t^H s^V. It may be noted there that if H-projection of AB i.e., a^H b^H is extended it cuts the G.L. at t^H; similarly a^Vb^V when extended cuts G.L. at s^V, t^V and s^H are the required traces. The line ST lies in one of the quadrants.

It may be noted that the projection of a line represents its true length in the plane to which it is parallel.

SLOPE OF A LINE

A line may be described by its slope.

(i) Upward (U) or downward (D): Vertical projection of the line denotes this.

(ii)Forward (F) or backward (B): Horizontal projection denotes this with respect to one end of the line. Forward indicates toward the viewer whereas backward away from the observer.

TWO STRAIGHT LINES – THEIR RELATIVE POSITIONS

(a) The projections of parallel lines on any plane are parallel.

(b) In case of intersecting lines, the point of intersection is definite in position which lies on a single perpendicular to G.L. and is independent of the positions of the observer unlike the apparent intersection points for non-intersecting lines.

PROJECTION OF AN OBLIQUE LINE

Fig. 6.3 shows a line AB in space which makes an angle $\propto$ with horizontal plane and parallel to V-plane. Now, if this line is rotated about the vertical axis AY, it will describe a cone whose vertex is at A and its horizontal projection is a circle shown in the H-plane and V-projection is a circle shown in the H-plane and V-projection is a line c^Vb^V parallel to G.L. Now every generating line of this cone (say AC) makes an angle α with H. The oblique line AC is a generator of the cone.

Similarly if the line AB making an angle β with vertical plane is rotated about an axis AX perpendicular to V-plane, it generates a cone with vertex at A and its vertical projection is the circle shown in the V-plane and horizontal projection of B is on the line c^Hb^H, parallel to G.L. It is evident that an oblique line AB making an angle b with V is a generator of the cone. Therefore the line AC which makes angles α with H and β with V must be the generator common to both the cones. The sum of the angles α & β must be less than 90°, for any oblique line. When the sum of α & β equals 90°, the line is a profile line.

True length of an oblique line: Now from the above discussion it is clear that the true length of an oblique line can be obtained by considering the line to be the generator of a cone. When the generating line is at the extreme positions parallel to a reference plane, it represents the rue length in the same reference plane.

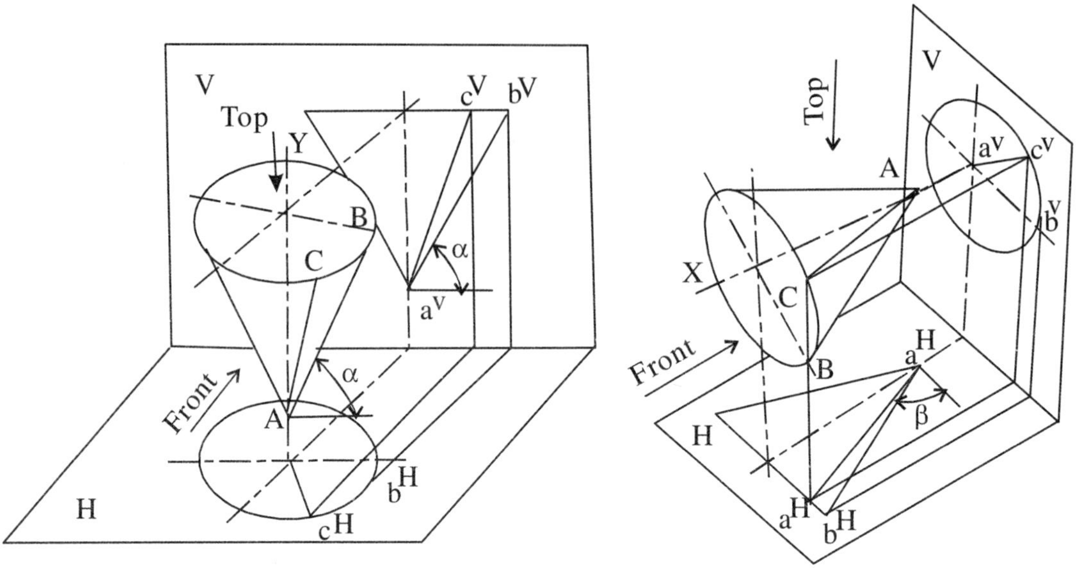

Visualisation of true length of a line

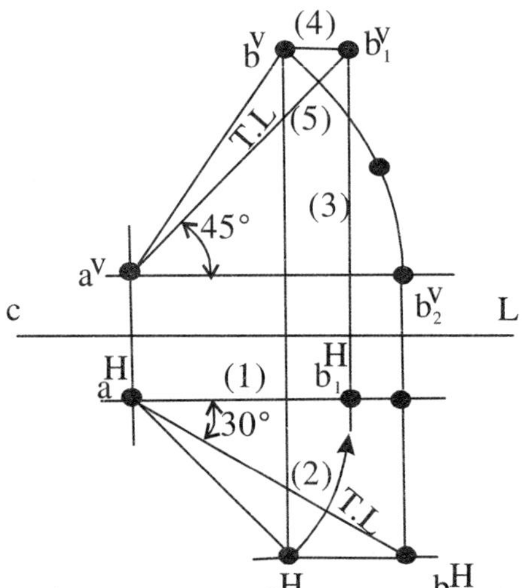

Fig. 6.3 Finding true length of an oblique line

Fig.6.3 shows the method of finding true length of an oblique line. a^Vb^V, a^Hb^H are the two projections of a line AB in space. Draw a line (I) parallel to G.L., through a^H, rotate b^H to $b_1^{\ H}$ on line (I), project line (3) perpendicular to G.L., through b^V draw a line (4) parallel to G.L. intersecting line (3) at $b_1^{\ V}$. Join $a^Vb_1^{\ V}$ which represents the true length and the angle $\alpha = 45°$ is measured. Following the same procedure for the projection a^Vb^H, the true length $a^Hb_2^{\ H}$ and angle $b = 30°$ can be obtained. Check that the true lengths thus found are equal.

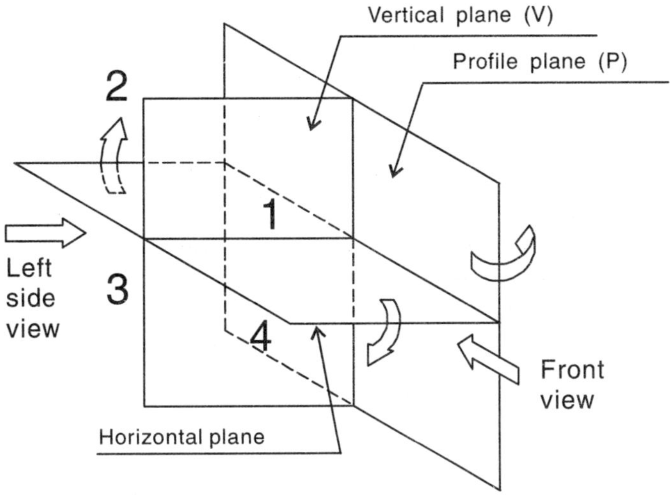

Fig. 6.4 The reference planes and the four quadrants

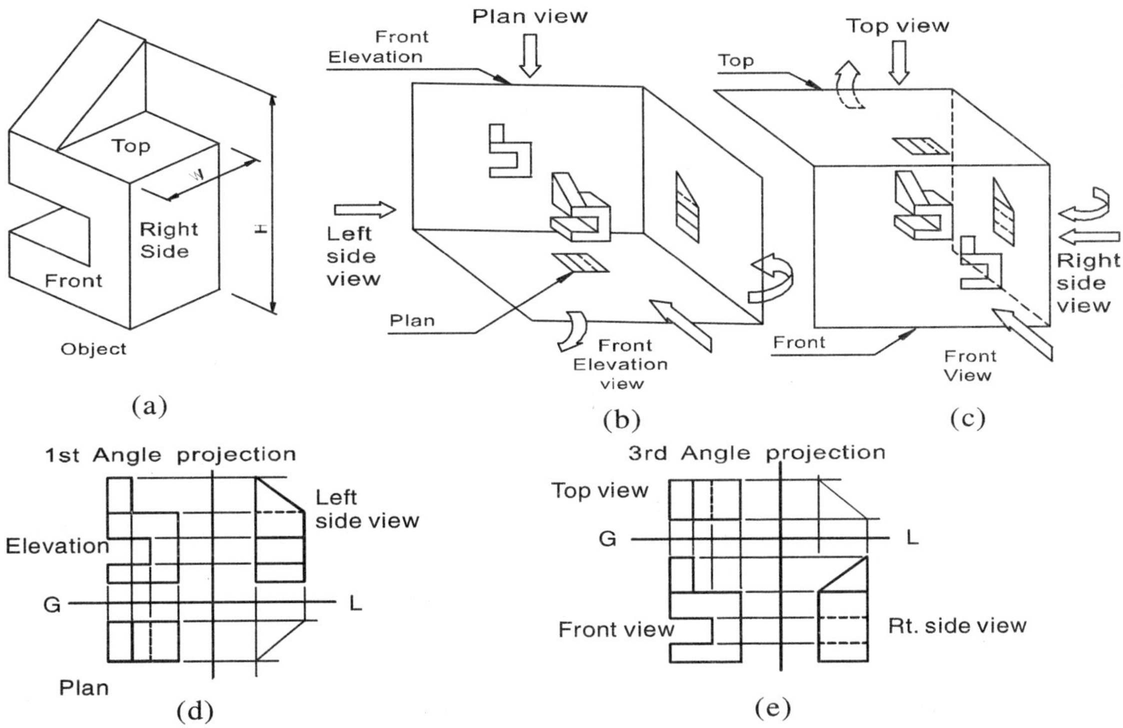

Fig. 6.5 Shows an object (a), to get the various views in Ist angle (b) and 3rd angle (c) and (d)& (e) show the projected views.

PROJECTION OF OBJECTS

Extension of the reference planes of the projections will form four quadrants or angles numbered as shown in Fig. 6.4. In theory, an object could be placed in any one of these angles and its projection drawn on the planes, which may then be rotated into the plane of the paper as shown in Fig. 6.4. In practice, first and third angle projections are used, as in other two angles views will be superimposed.

First angle projection is almost universally followed in Europe. In India both first and third angle methods are followed and both are recommended by the B.I.S. In first angle method, the object is placed between the eye of the observer and the planes of projection. The projection on the V-plane is called **elevation** and that on the H-plane is called **plan**. For side view, the object is viewed from left and is placed to the right and *vice versa*. The views are termed as left hand side view or right hand side view respectively *viz.* Fig. 6.5.

In third angle method the planes of projections are in between the object and the viewer. The H-projection is called view from top (top view) and V-projection is called view from front (front view). Side views are drawn to the left when viewed from and to the right when viewed from right *viz.* Fig. 6.5.

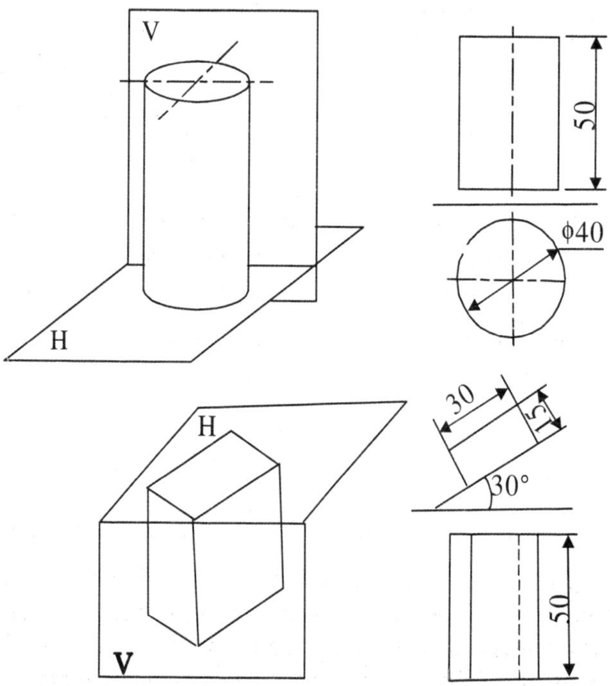

Fig. 6.6 (a)

Figure 6.6 (a) shows:

(i) A φ 40mm × 50mm long round bar placed in the 1st angle, along with its plan and elevation.

(ii) A 30mm × 15mm × 50mm long M.S. bar with its long rectangular face making 30° with V placed in third quadrant, along with its top and front views (3rd angle).

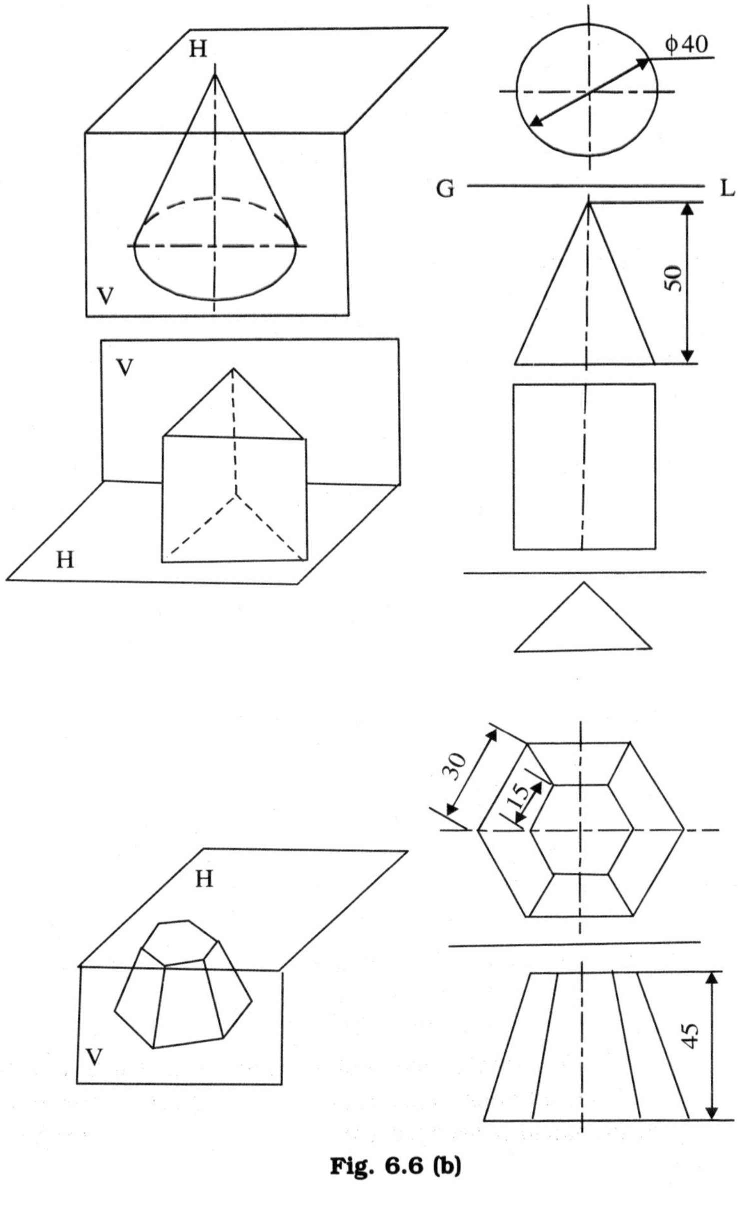

Fig. 6.6 (b)

and Figure 6.6 (b) shows

(i) Top and front view of 40mm × 50mm high circular cone (3rd angle)
(ii) Plan and elevation of a right prism (1st angle) and
(iii) Top and front views of a truncated hexagonal pyramid (3rd angle).

SKETCHING

Sketching is freehand drawing of objects in as nearly correct proportion as eyes can determine without taking measurements. After the sketch of a piece is finished, the necessary extension and dimension lines are given. The dimensions indicated on the sketch correspond to the actual dimensions of the piece.

It is preferable to make sketches in pencil (soft grades) on graph paper which facilitates freehand sketching.

While making sketches, it is recommended that lines of the types and thickness specified for drawing should be used. The views must be large enough to enable the dimensions, conventional symbols and inscriptions to be indicated.

Machined and non-machined surfaces, gradual transition from one surface to another, fillets and radii for castings, stampings and forgings *etc.* should be indicated on sketches.

In short, it can be said that there is hardly any difference between a sketch and a drawing, excepting that the former is drawn freehand and proportionate to eye estimation. A drawing drawn without any specified scale, is virtually nothing more than a neat sketch.

A sketch is a very important and useful tool to engineers who are very occasionally required to make out sketches of objects *in situ*, from which drawings would be prepared.

PROBLEMS

Descriptive Geometry

6.1 Draw the projection of the points

(a) A (20 mm above H, 25 mm in front of V)
(b) B (20 mm below H, 25 mm behind V)
(c) C (20 mm below H, 25 mm below V)
(d) D (20 mm below H, 25 mm in front of V)

6.2 Find the revolved position of the points given below and visualize them A (0, - 12,25), B (0,20, 30), C (-12, 17, -28), D (10, -21, 13)

6.3 Draw the projection of the following lines having a length of 60 mm each.

(i) Horizontal frontal line AB 15 mm over H, 20 mm in front of V.
(ii) Frontal line CD inclined 30° upward right, C (0,20, 10).
(iii) Horizontal line EF inclined 40° backward right, E (0,10, -15).
(iv) Vertical line GH, downward right, G (0, -8,15).
(v) Horizontal profile line JK, forward, J (0,-10, -10).

(vi) Line AB slopes 40° with H and 50° with V, downward backward right, co-ordinates of A (0, -45, 30).

(vii) Line CD makes 30° with H, 60° with downward forward left. Given C (0, -1 5, 25).

(viii) Line EF makes 30° with V, 45° with H and slopes upward backward left. Given C (0,20,8).

(ix) Line GH makes 25° with H, 50° with V, downward forward right. Given G (0,-10,30).

6.4 Determine by appropriate projection drawing whether the following lines AB and CD are intersecting in space. Locate the points of intersecting if, any.

(i) A (5, -10, 10), B (30, -15, 25),
C (10, 30, 25), D (50, -20, 5).

(ii) A (-50, -25, 15), B (25, -55, 60),
C (-40, -75, 5), D (40, -15, 90).

(iii) A (2, -8, 15), B (20, -14, 15),
C (10, -5, 10), D (10, -20, 25).

(iv) A (10, -30, 10), B (90, -30, 50),
C (10, -10, 30), D (90, -50, 30).

PROJECTION DRAWING

(Draw three principal views of the objects described below. Views should be placed 6 mm away from the reference planes unless stated otherwise. It is advisable to solve the even nos. of problems in 1st angle projections.)

6.5 A square prism, 30mm sides and 50mm high, with its axis vertical, is placed 10mm away from the planes of projections and one of its rectangular faces makes an angle of 30° with the vertical plane.

6.6 A triangular pyramid of height 40mm and 20mm sides of the base is placed co-axially with the axis vertical on the top of a hexagonal prism of height 20mm and 30mm sides. One of the sides of the base of pyramid is parallel to vertical plane and one of the rectangular faces of the prism makes 45° with the vertical plane.

6.7 A regular triangular pyramid 40mm high and 30mm base is placed with its axis parallel to vertical plane and one of the longer triangular faces parallel to horizontal plane. Apex of the pyramid is 20 mm from V and 35 mm from H. (Hint: Draw views with axis vertical and one side of the base perpendicular to V. Revolve this front view about the perpendicular side to get the required projection.)

6.8 A right hexagonal pyramid of 40 mm height and 25 mm base side is placed with its axis making 30° with H and parallel to V. (Hint: First draw the projection with the axis vertical, then rotate the axis to the desired angle for a hexagonal pyramid which is placed with its axis making an angle with V and parallel to H as shown in Fig. 6.8.)

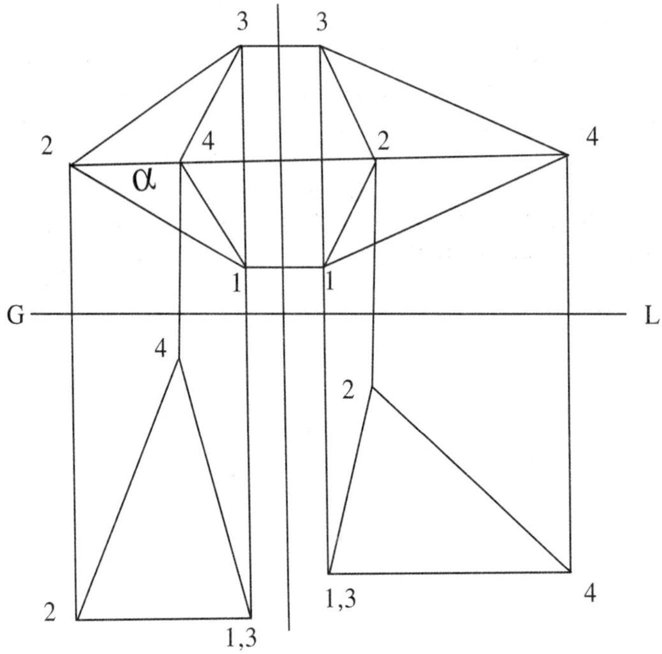

Hint for prob. 6.7

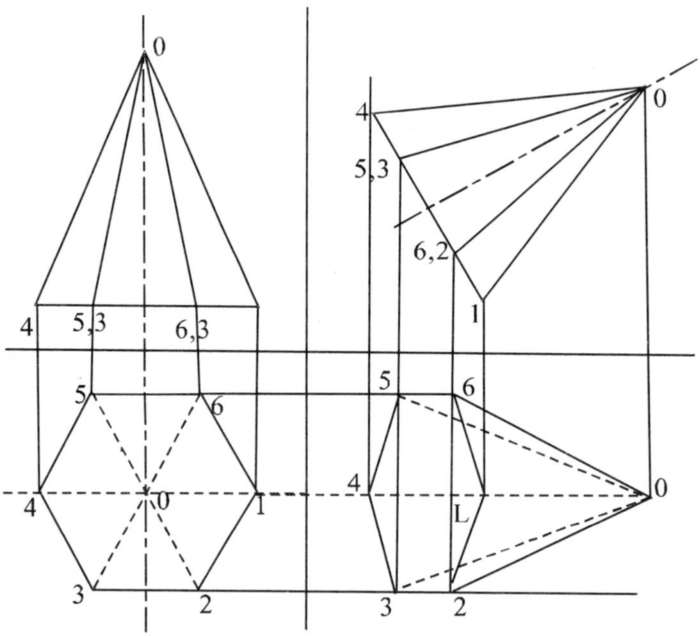

Hint for prob. 6.8

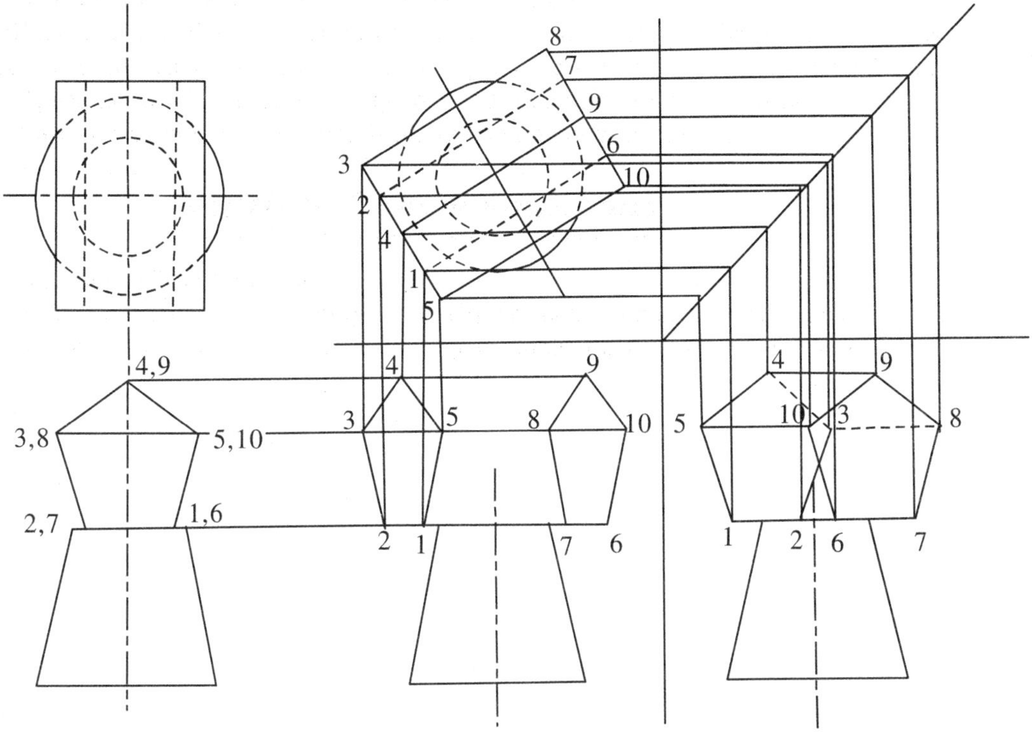

Hint for prob.6.9

6.9 A truncated hexagonal pyramid 36 mm high, 28 mm base side, 18 mm top sides is placed vertically with one side of the base parallel to V. A regular pentagonal prism of 20 mm sides, 40 mm long is centrally placed with one of its rectangular faces on the top of the truncated pyramid. The axis of the prism makes an angle of 30° with V. (Hint: Method is shown with a truncated cone and a pentagonal prism, Fig.6.9).

6.10 Two spheres of diameters 40 mm and 20 mm are placed touching each other such that the line joining the centers makes (i) an angle of 30° with H and parallel to V, (ii) an angle of 30° with H and 45° with V.

6.11 A 40 mm long 16 mm diameter cylinder, a 28 mm diameter sphere and a cone of base diameter 22 mm, 40 mm high are placed on the same horizontal plane touching each other. The cylinder axis is perpendicular to P-plane while the cone is vertical. One of the end faces of the cylinder and the axis of the cones are co-planer. The cylinder rests in front of the sphere and the cone.

Hint:

Draw the auxiliary view (top left) with the cone and the sphere both lying on the ground and touching each other. Measure the distance AB, which will remain constant in the top view when the sphere turns on the ground around the cone.

Draw the cone and the cylinder with its right hand end lying along the axis of the cone (apart of a front view).

Draw the side view of the cone and the cylinder such that the circle is tangent to the slant surface of the cone. Draw the projection of the sphere such that it touches the cylinder and the base of the cone.

The center lines on the top view are obtained as shown by rotation. Make A'B' = AB, then B' locates the center of the sphere on the top view.

6.12 A regular pentagon of 30 mm sides has one side on H. Its surface is at an inclination of 30° to H and the other side on which it rests on H is inclined at 30° angle to V.

6.13 A regular pentagonal plate ABCDE of 30 mm sides has its plane inclined at 45° to the ground. The line joining the vertex D to the mid-point of base AB is inclined at 30° to V.

6.14 A thin lamp shade in the form of a frustum of a cone with its larger end diameter at 200 mm, smaller end diameter at 100 mm and height 150 mm is lying on its slant side on the ground and the axis parallel to V.

(1)

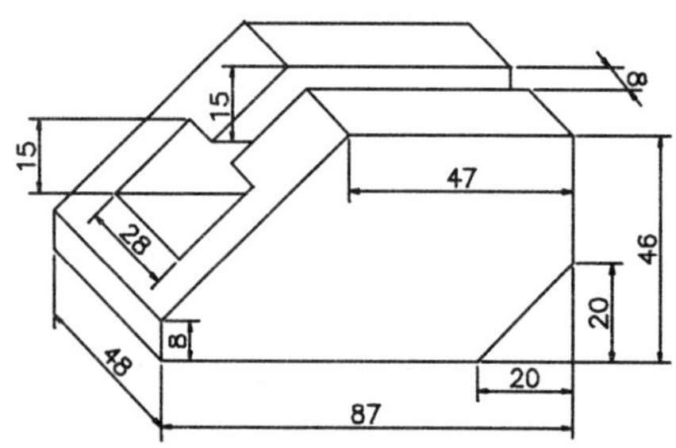

(2)

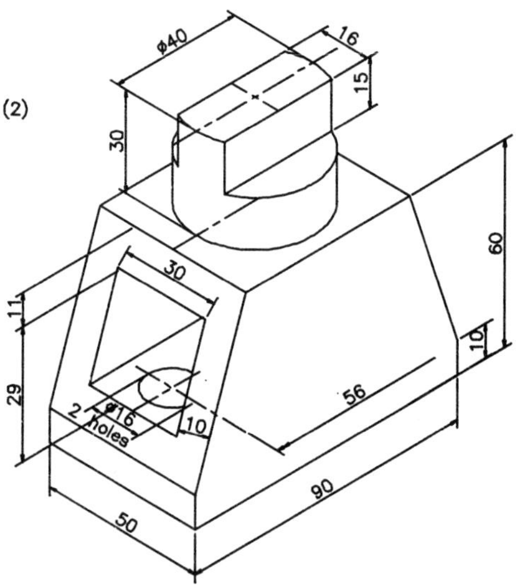

(3)

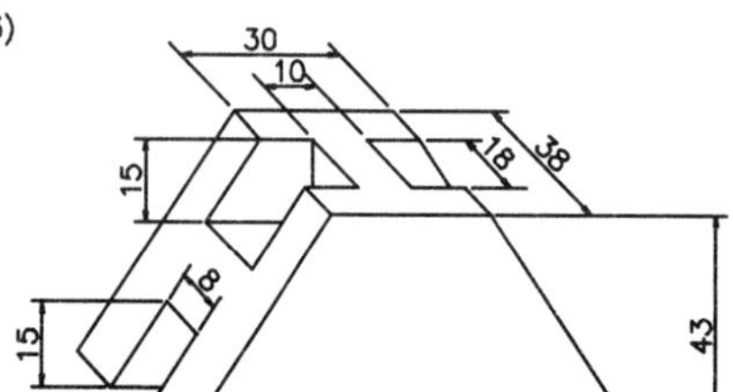

(4)

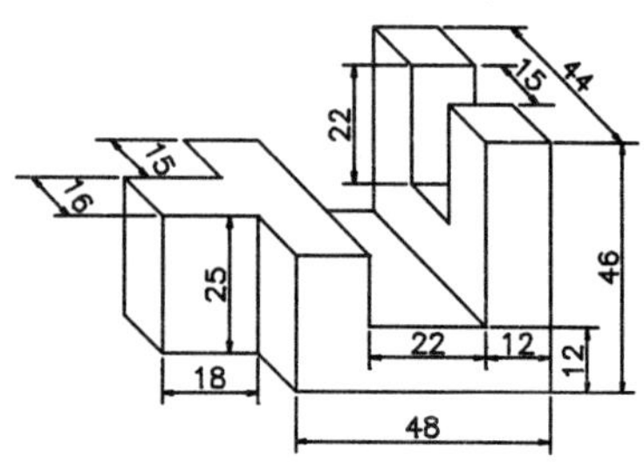

(5)

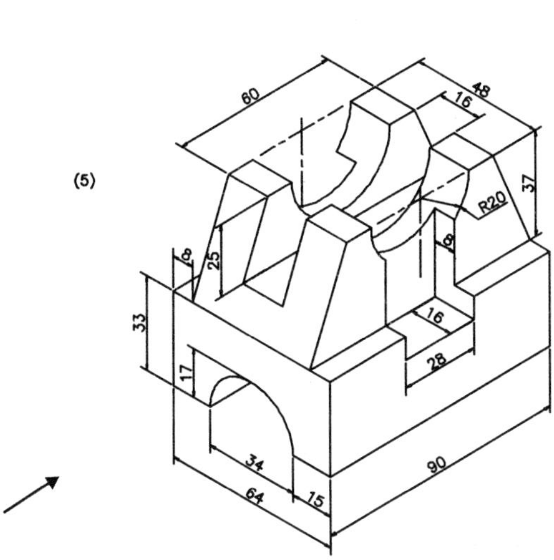

(6)

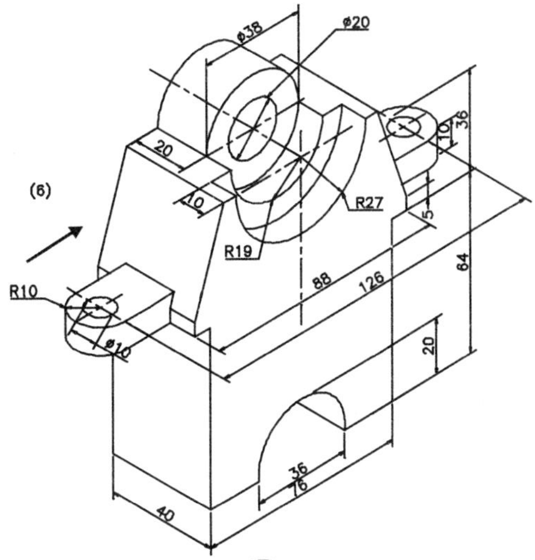

(7)

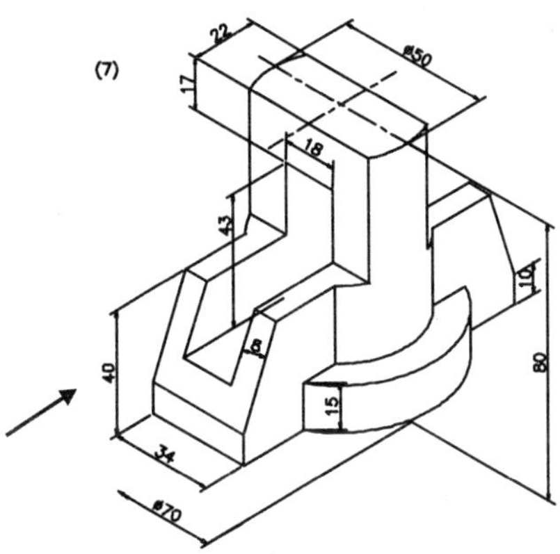

(8)

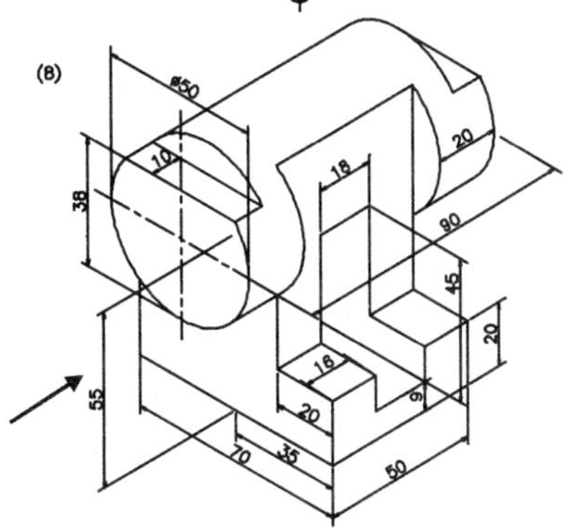

(9)

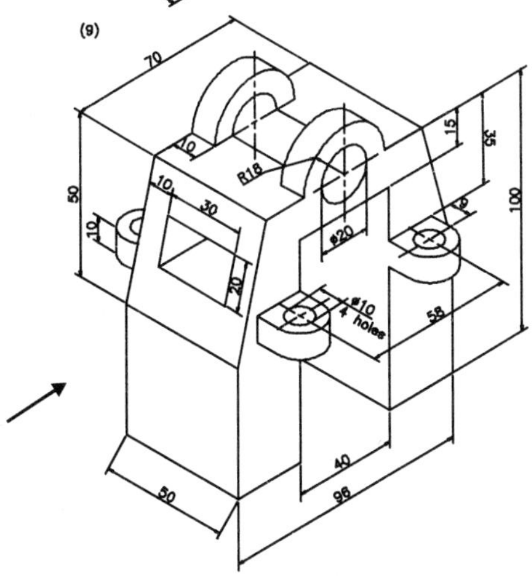

(10)

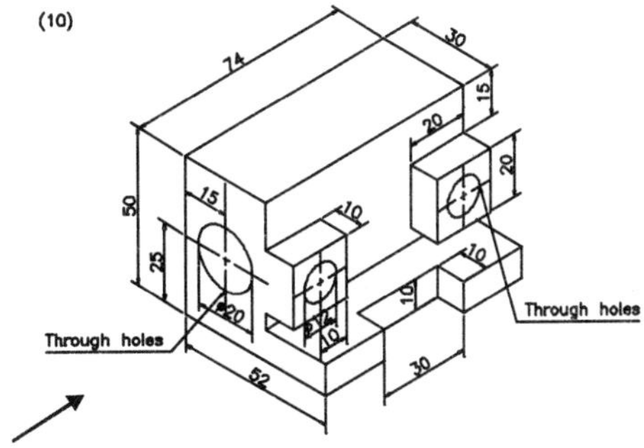

Probs. 6.18.1 to **6.18.10**

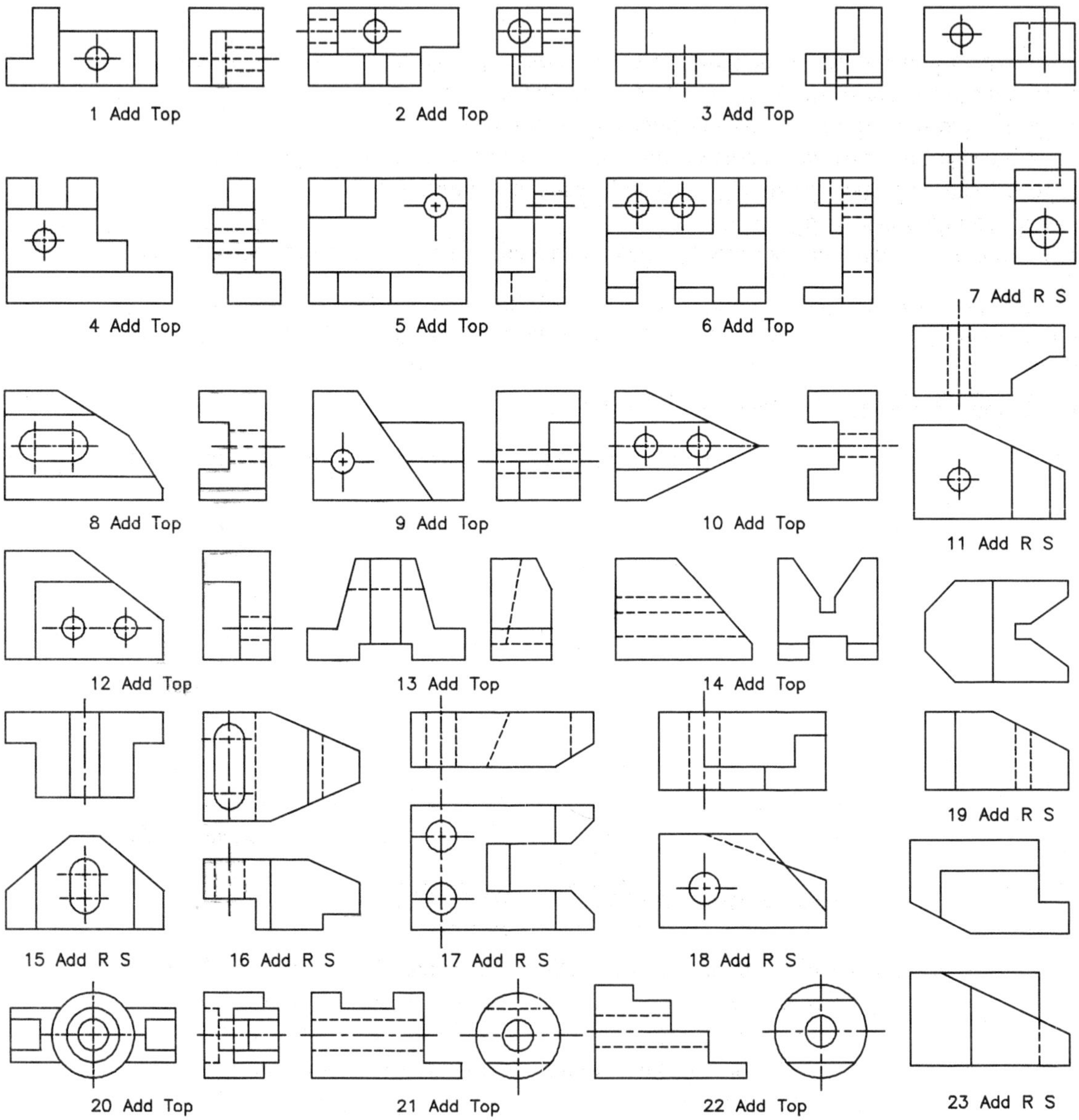

Probs. 6.19.1 to 6.19.23

6.15 The axis of a cylinder of 40 mm diameter and 200 mm length makes an angle of 30° with V and 45° with H.

6.16 A right circular cone of base 50 mm diameter and 60 mm height rests on the ground on a point on its base circle with the axis making an angle of 30° with both H and V.

6.17 Repeat problem no. 6.15 for a hexagonal prism of sides 20 mm and 200 mm length. One of the sides of the end face is kept parallel to H.

6.18. 1 to 6.18.10 Prepare complete orthographic projection drawings from pictorial drawing of the models shown. (Vide Probs. 6.18.1 to 6.18.10).

6.19. 1 to 6.19.23 Two complete views are given (Vide Prob. 6.19.1 to 6.19.23). Sketch the given views and add the views as indicated.

Scale 1:5. You may measure the drawing and draw in full scale.

CHAPTER – 7

ISOMETRIC DRAWING

In engineering drafting practice it is often necessary to prepare a drawing that can be read even by an individual who lacks the knowledge of orthographic representations. In those cases axonometric drawings *e.g.* isometric, dimetric or trimetric drawings are usually made according to a set of well defined principles that permit accurate size representation.

Axonometric projection*, although prepared in an entirely different manner, is basically an orthographic view of a titled part.

In isometric drawings the three principal faces and axes of an object are equally inclined to the plane of projection; in dimetric drawings two of the three principal faces and axes of the object are equally inclined to the plane of projection and in trimetric drawings all the three principal faces and axes of the object make different angles with the plane of projection (Fig. 7.1).

THEORY OF ISOMETRIC PROJECTION AND DRAWING

To explain orthographic isometric projection let us consider the simple case of a cube of which the orthographic projections are given, with its faces parallel and perpendicular to the planes of projection. First, turn the cube about the vertical axis through an angle of 45° (Top face kept parallel to H).

Then turn the cube for a second time, now about the horizontal axis (which is perpendicular to P-plane) through an angle of 35° (precisely, 35° 16'), then the projection obtained in the V-plane will be the isometric projection. It may be noted from Fig.7.2 (top) that the edges which are parallel to the axes OX, OY, OZ in the orthographic projection are now at an angle of 120° to each other, their lengths shortened to 0.82 of their true lengths.

---

* A projected view in which the lines of sight are perpendicular to the plane of projection, but in which the three faces of a rectangular object are all inclined to the plane of projection is called an axonometric projection.

Isometric Scale

Lengths on an isometric projection should be less than their true values. Fig.7.2 shows the isometric view of the cube, OX, OY and OZ being the isometric axes. The upper face of the cube slopes away from the viewer and so it is evident that its edges will appear foreshortened. The true size of ZP'X is given on the diagram as ZPX.

$$ZP' \cos 30° = ZP \cos 45°$$

$$\text{or} \quad ZP' = (\cos 45°/\cos 30°)\, ZP = \frac{\sqrt{2}}{3}\, ZP$$

$$= 0.8165$$

This conversion is obtainable geometrically as shown in Fig. 7.2. It should be clear that where a line does not lie parallel to any of the isometric axes *i.e.*, the line is non-isometric, its isometric length will not be equal to 0.8165 times its true length. In fact, the amount of reduction depends on the orientation of the line.

To make construction simpler, this shortening of length is not observed in engineering drawing, the true lengths of lines are drawn and marked off parallel to the isometric axes and the figure obtained is called the isometric drawing. This means that an isometric drawing will be larger than the isometric projection, but it will still be in the same proportion. Perspective drawing which is another form of pictorial representation, significantly differs form isometric drawing in that the receding lines coverage, while in isometric they remain parallel.

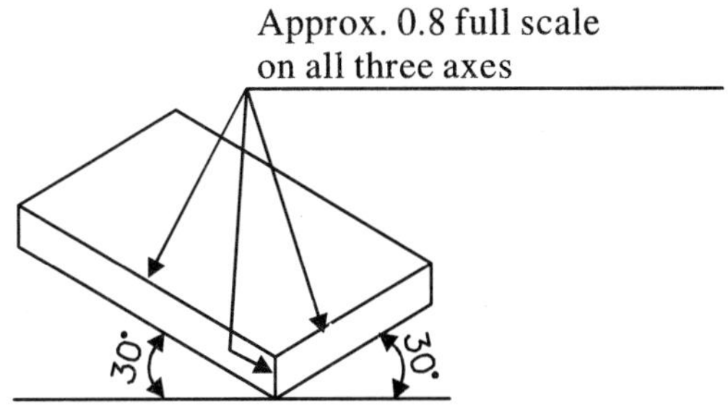

Isometric
Projection

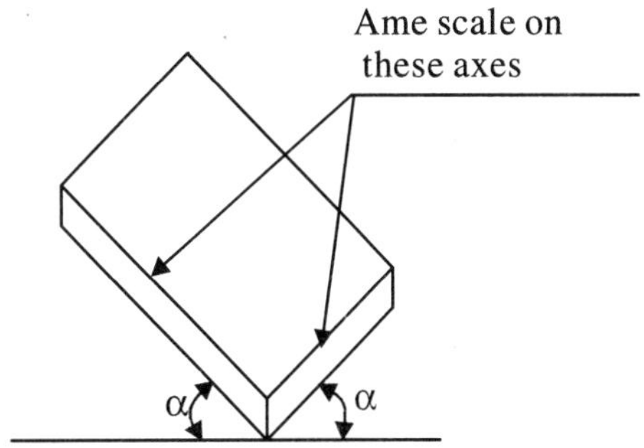

α Variable, upto 45° excepting 30°
dimetric projection

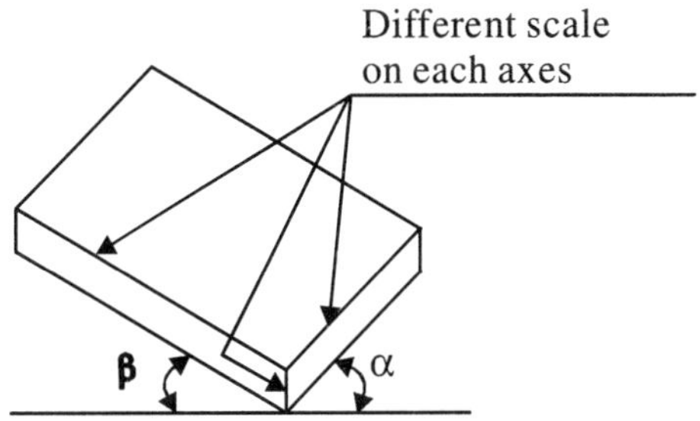

α, β Variable $\alpha \neq \beta$
$\alpha \neq 0, \beta \neq 0, (\alpha + \beta) < 90°$
Trimetric projection

Fig. 7.1

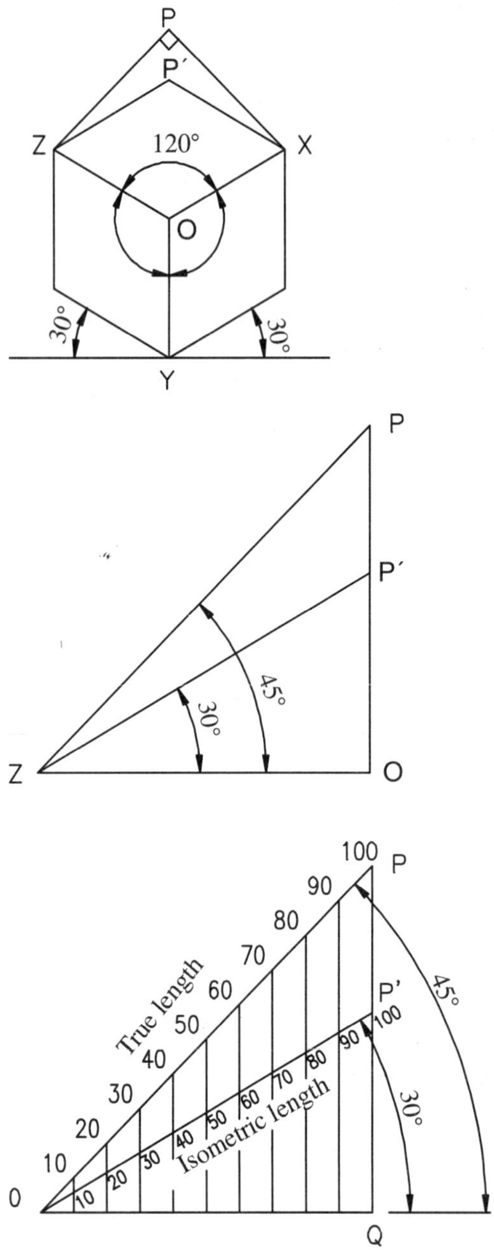

Fig. 7.2

Considering the obtained isometric projection (Vide Fig. 7.3) of the cube together with axes O´X´, O´Y´, O´Z´, the following conclusions are drawn:

(i) the axes O´X´, O´Y´, O´Z´ are at 120° with each other.

(ii) to prepare an isometric drawing of an object (say, a cube) the edges which were parallel to orthographic axes OX, OY, OZ must be laid of in directions parallel to the axes of isometry O´X´, O´Y´, O´Z´.

The isometric axes and line parallel to them are very easily drawn with the aid of 30° set-square.

PROCEDURE FOR ISOMETRIC DRAWING

When making an isometric drawing from orthographic views, follow the procedures given below:

1. Choose the direction from which the component is to be viewed so that the resulting drawing will show the detailed features of a component i.e., choose the appropriate pole.
2. Draw the outline of a box (construction lines) into which the component will just fit.
3. Use distances from the orthographic views to set out points within this box.
4. Make the lines prominent when the outline is completed.

NON-ISOMETRIC LINES

Edges whose projections or drawings are not parallel to one of the isometric axes are called *non-isometric lines*. It is most important to remember that measurements are to be made only on the drawings of isometric lines; *e.g.*, the diagonals of the face of a cube are non-isometric lines; although equal in length, their isometric drawings will be unequal. It is interesting to note that the length of the smaller diagonal in the isometric drawing is equal to the side of the cube. Sloping lines representing sloping surface appear as non-isometric lines. To draw them locate their end points found on the ends of isometric lines and join them.

DIMENSIONING ISOMETRIC DRAWINGS

At times, a simple object represented by an isometric drawing may serve as a working drawing. In such cases the necessary dimensions and specifications are placed on the drawing. Dimension lines, extension lines and the line being dimensioned should be in the same plane.

Since the isometric is a one-view drawing, it is not generally possible to avoid placing dimensions on the view or across dimension lines; however this practice should be avoided whenever possible.

Rectangular Solids

The step by step procedure in making an isometric drawing of a rectangular solid is shown in Fig. 7.4. Hidden lines are avoided in all forms of pictorial drawing, so also the center lines, unless the latter are required for proper dimensioning.

Equilateral Triangle

Consider the triangle ABC in Fig.7.5 contained in a plane ZOY and with base AC lying on the plane XOY. The X-coordinates of all its vertices are equal, Z-coordinates of A and C are zero. The equilateral triangle I whose plane is parallel to V-plane (XOZ) has equal Y coordinates for all its vertices and zero for Z-coordinates for the vertices lying in H-plane (XOY).

Regular Hexagon

Fig.7.5 shows the construction of a regular hexagon in isometric with the axes of coordinates passing through the axis of symmetry of the figure.

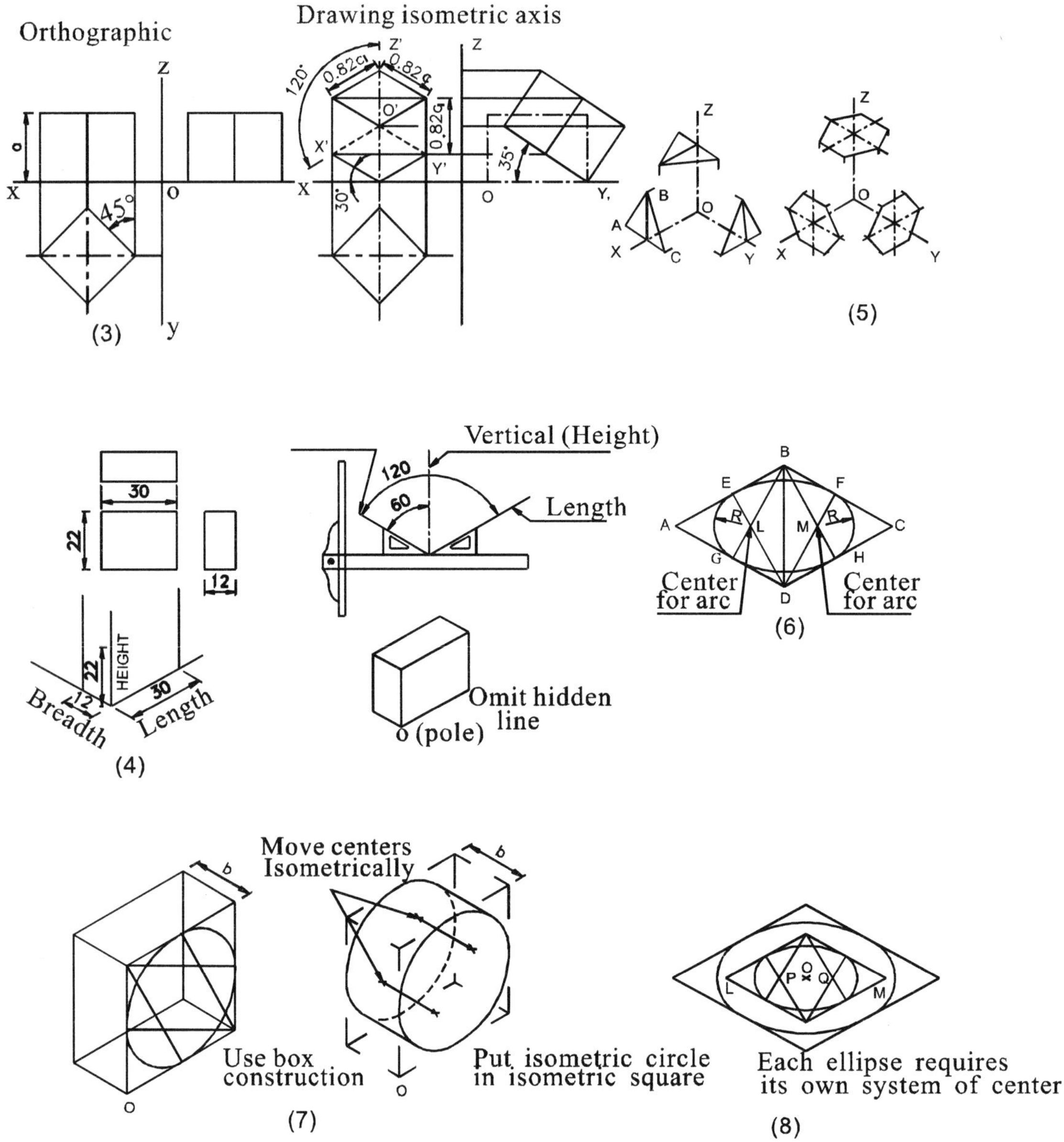

Fig. 7.3 to **7.8**

Circle

The circle takes an elliptical shape in isometric drawing when located in an isometric plane. Fig. 7.6 shows the method. First draw an isometric square ABCD in the required position, making the sides of ABCD equal in length to the diameter of the circle. Locate the points E, F, G and H at the mid points of the sides, then erect perpendiculars from these points. These perpendiculars intersect at M and L. Then draw circular arc as shown.

Cylinder

Figure 7.7 shows the box method of drawing isometric cylinders.

Figs 7.9 and 7.10 show the isometric drawings of a hexagonal prism and a pyramid respectively from which how isometric views have been developed from the two orthographic views, may be clearly seen.

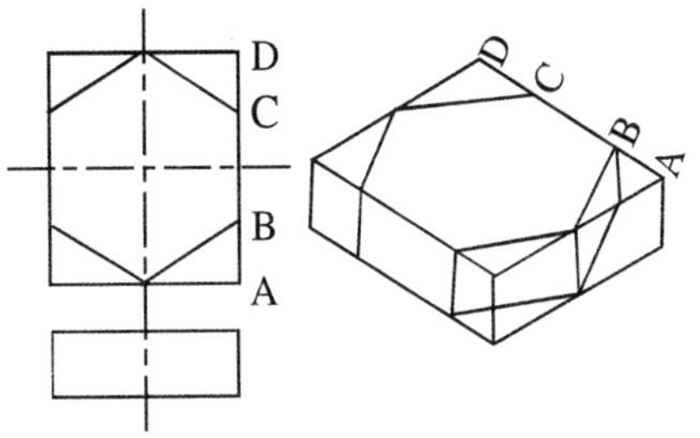

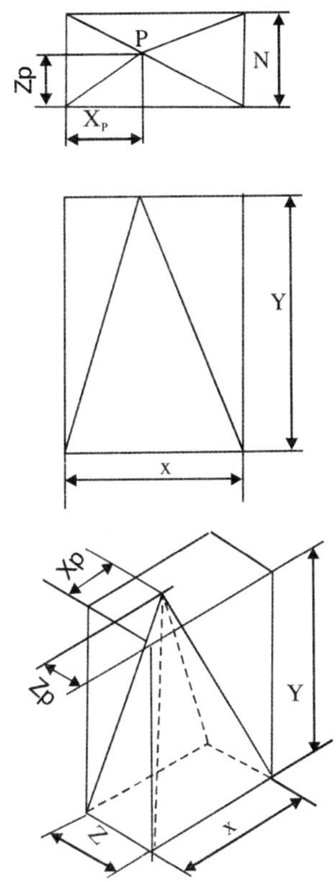

Fig. 7.9 Shows the deformation of 90° to some other angle in the isometric drawing

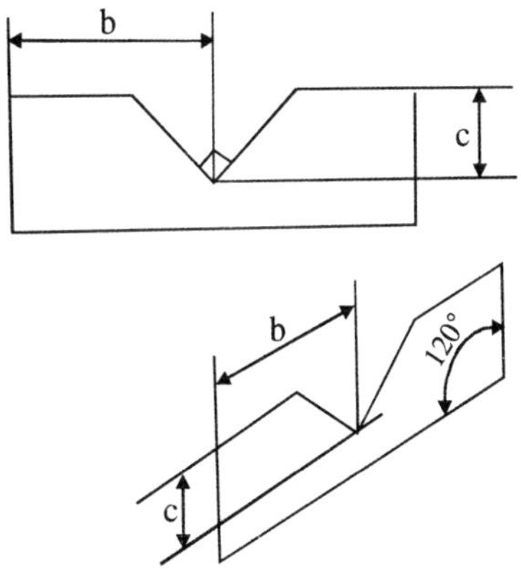

Fig. 7.10 Hidden lines show the edges; these hidden lines are not tolerated in the isometric drawing.

PROBLEMS

(Prob. 7.1 to 7.12 are drawn in scale 1:2. Scale the drawings to get the dimensions.)

7.1 – 7.4 The orthographic projections of the objects are shown in Prob. 7.1, 7.2, 7.3 and 7.4. Develop the isometric drawings of the objects on the indicated poles.

7.5 – 7.12 Construct the isometric drawings of the objects shown in Prob. 7.5 to 7.12. Choose appropriate poles.

7.13 – 7.14 Two Projections of a model are given. Construct a third one, and make an isometric sketch of the model

7.15 – 7.16 Two Projections of a model. Construct a third one, and make an isometric sketch of the model.

7.17.1 – 7.17.24 Three incomplete views to a scale of 1:2 are given below. Sketch the given views adding missing lines. Hence sketch isometric drawings of each.

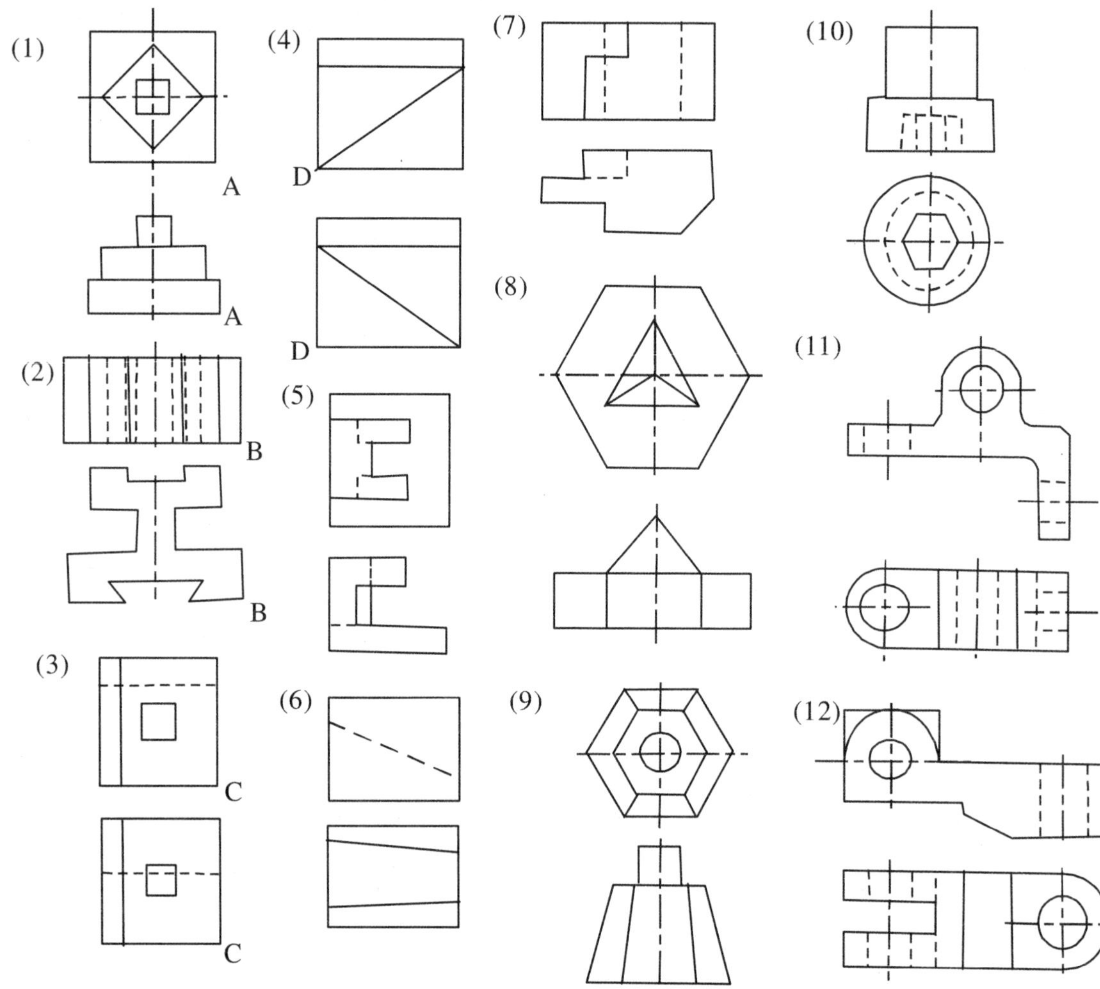

Prob. 7.1 - 7.12

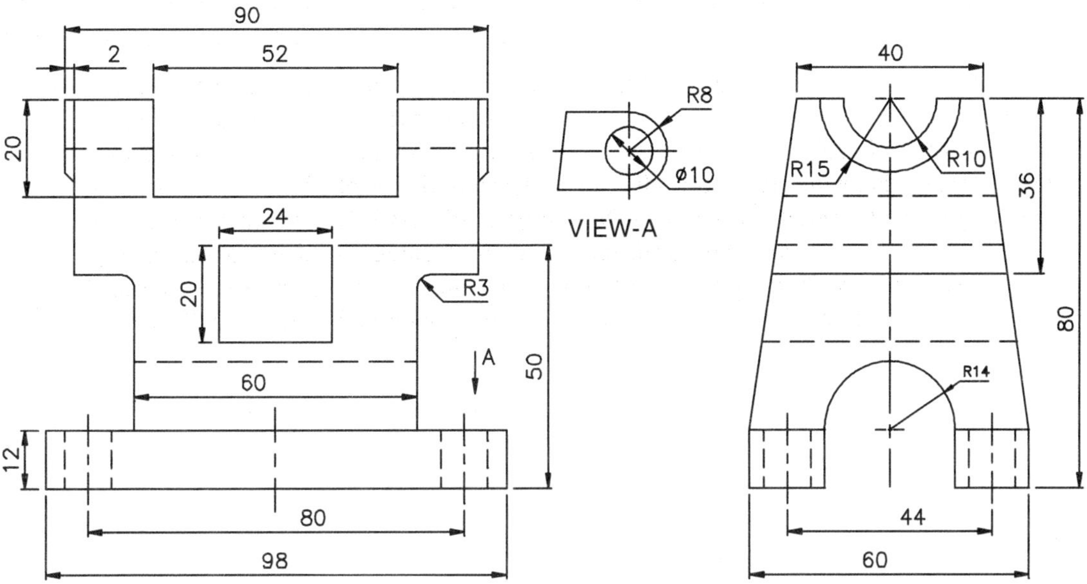

(13) Bed

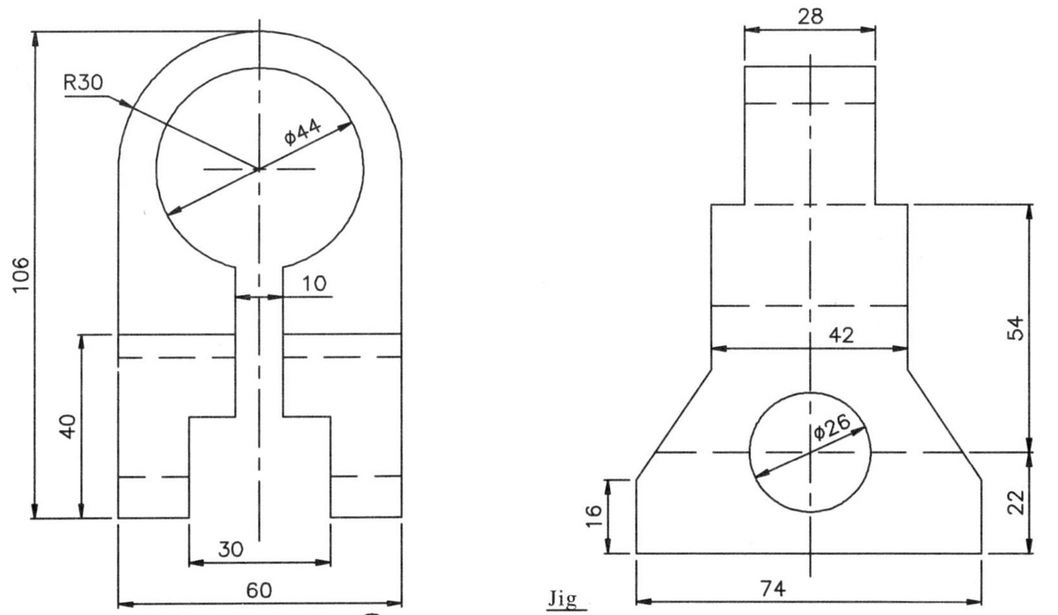

(14)

Prob. 7.13 - 7.14

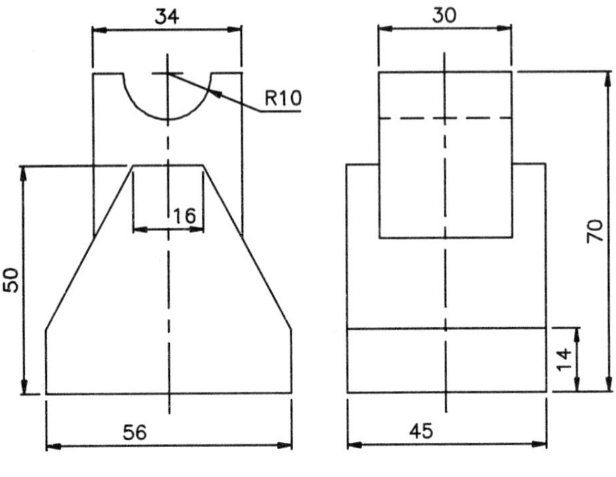

⑮

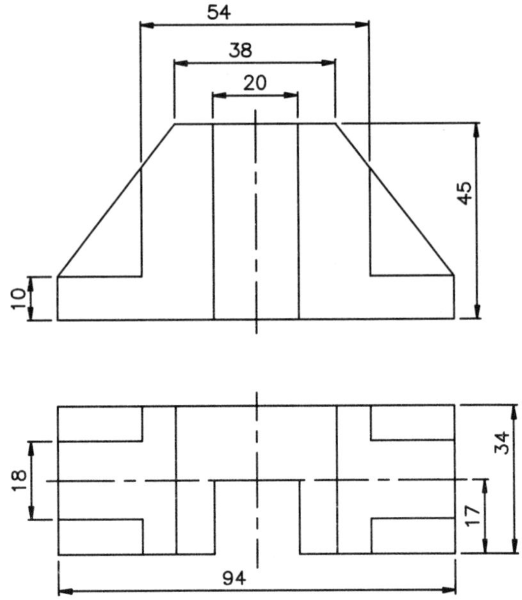

⑯

Probs. 7.15 - 7.16

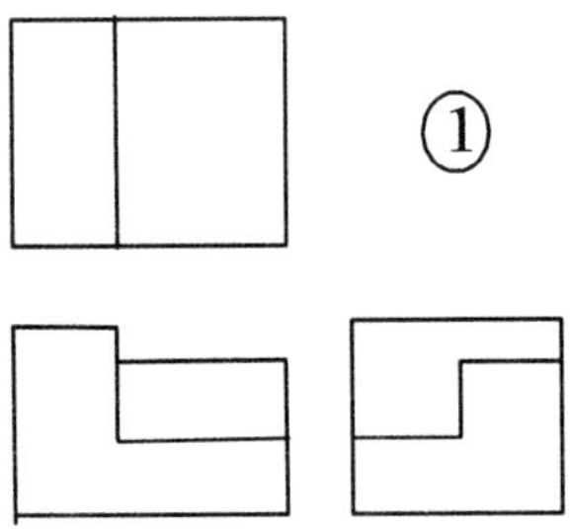

①

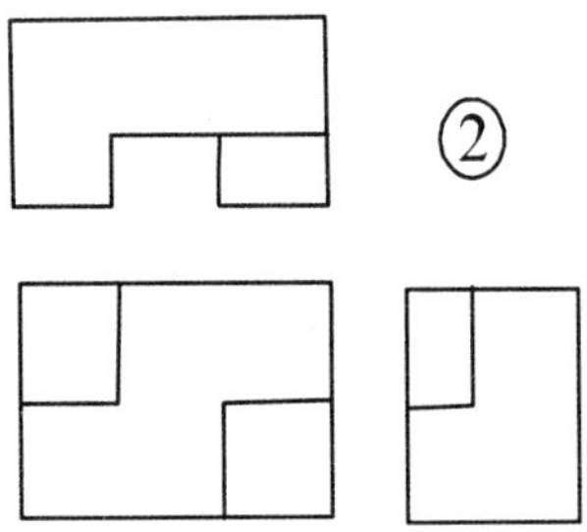

②

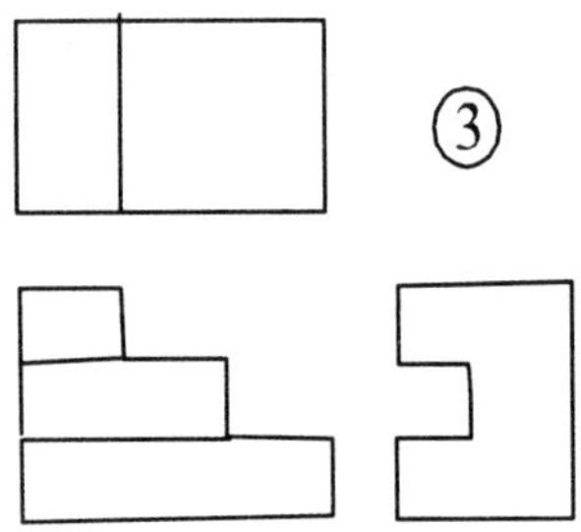

③

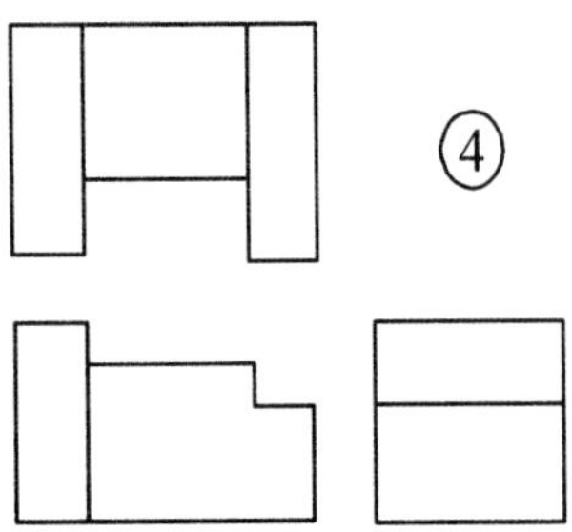

④

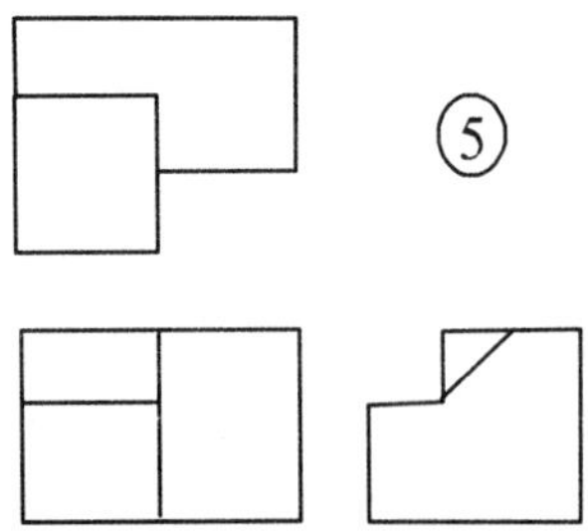

⑤

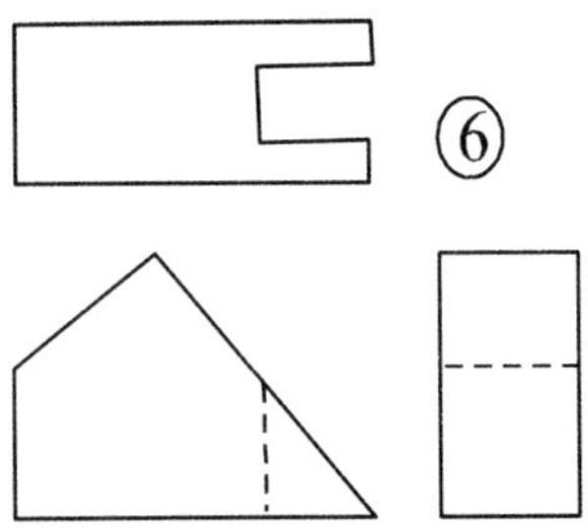

⑥

⑦

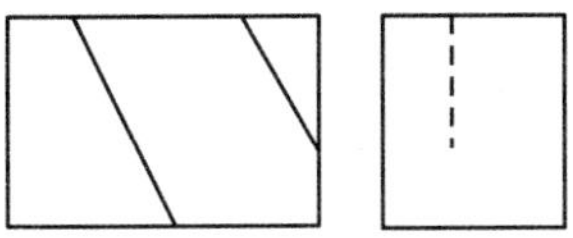

⑧

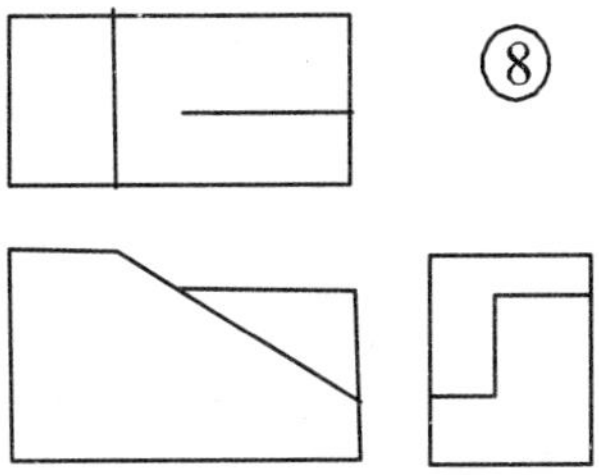

⑨

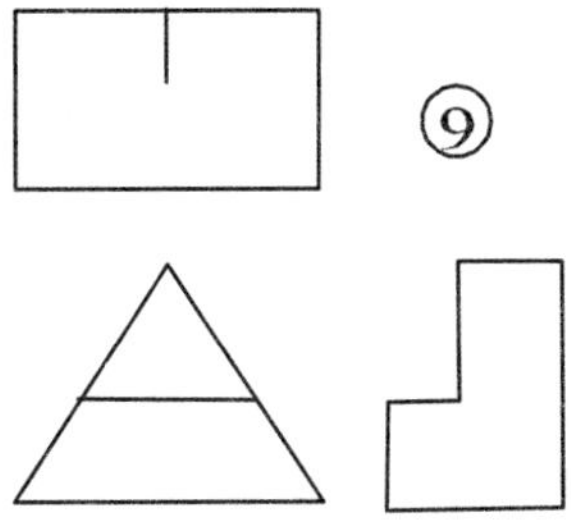

⑩

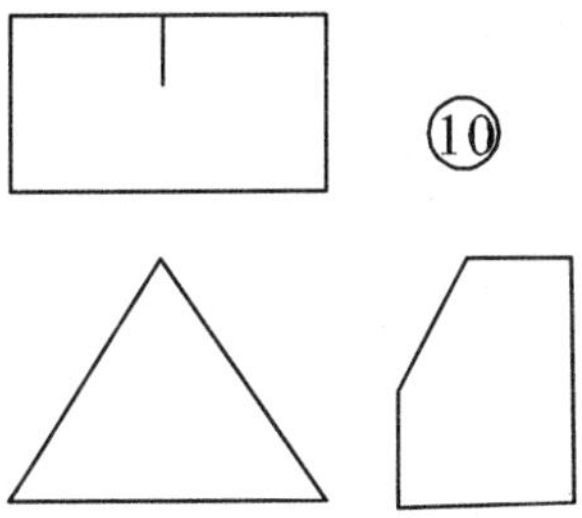

⑪

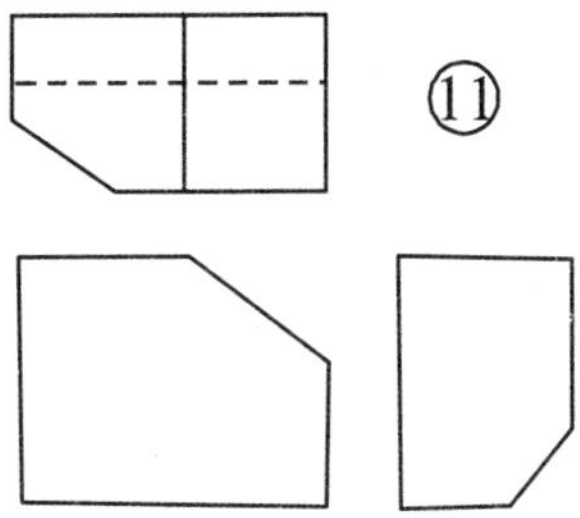

⑫

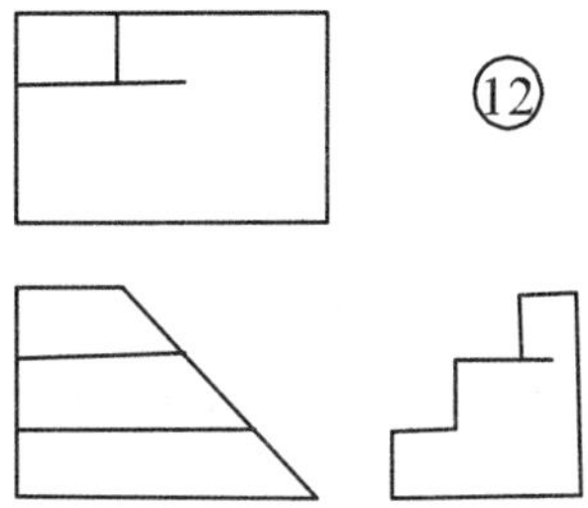

⑬

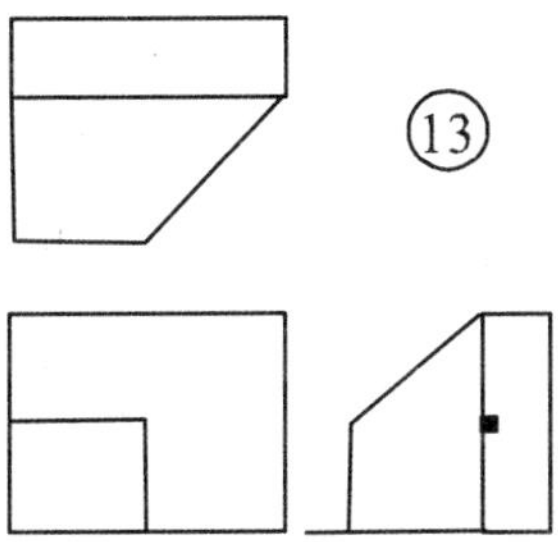

⑭

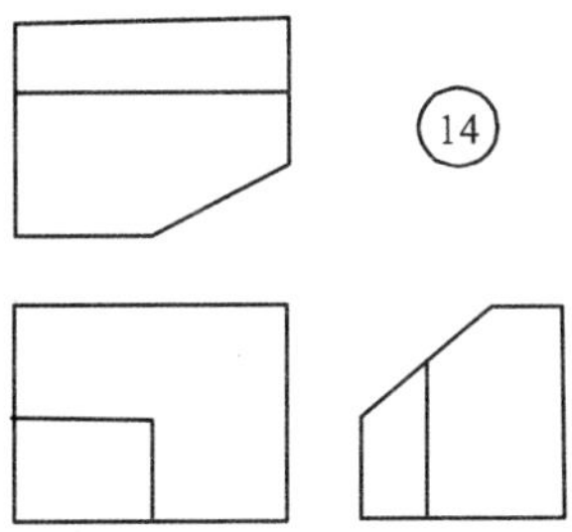

⑮

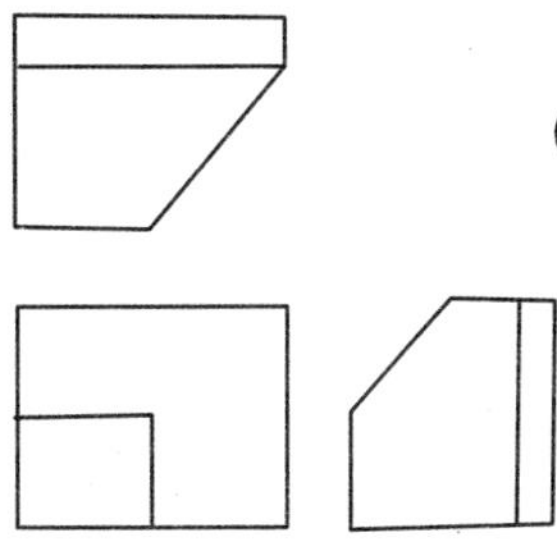

⑯

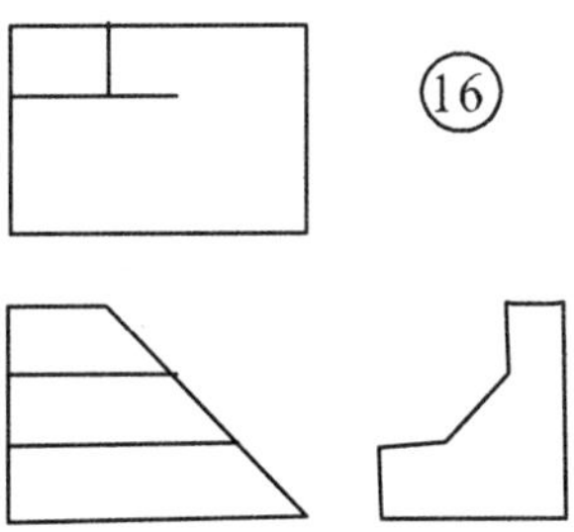

⑰

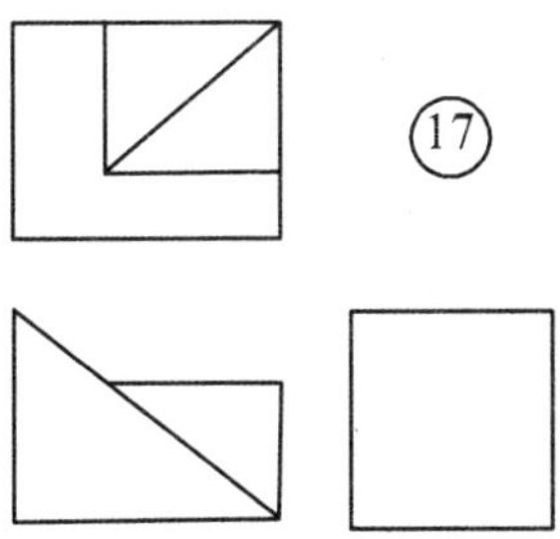

⑱

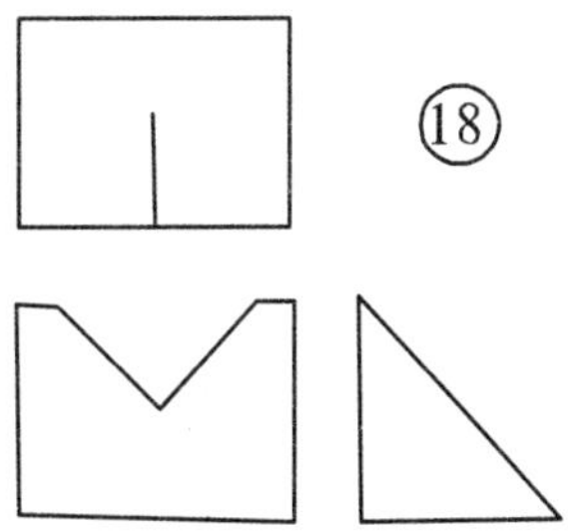

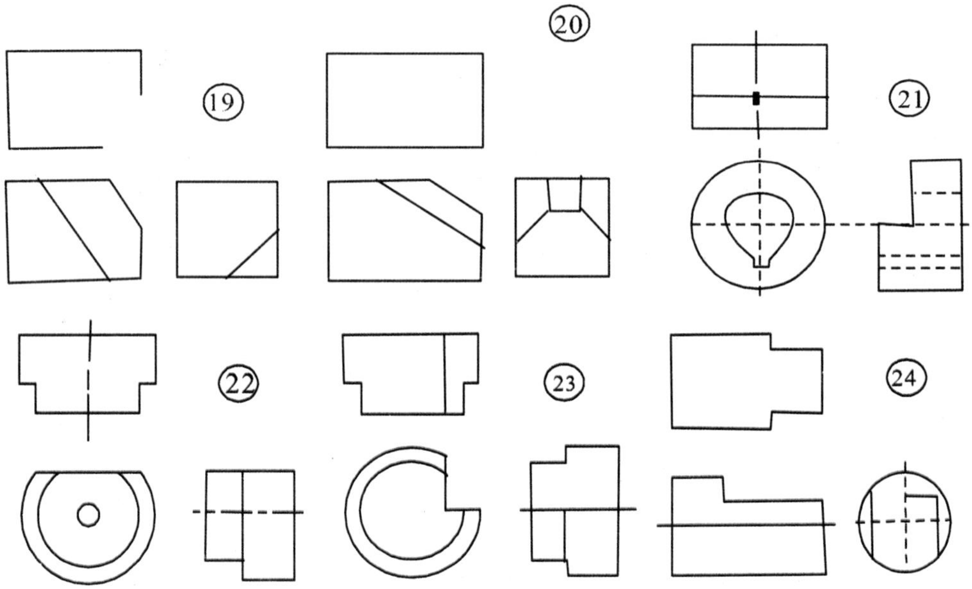

Probs. 7.17.1 – 7.17.24

CHAPTER – 8

SECTION AND SECTIONAL VIEWS

Sometimes an object is so complicated that its ordinary orthographic representation would result in a confusing bunch of dotted lines indicating hidden details. This situation can be avoided by making drawing "in section" which reveals the inner features while preserving the external identity of the component. To obtain a sectional view an imaginary cutting plane is passed through the object in a particular position and direction. The portion shown in a section drawing is the part of the object that would be seen if the segment nearest to the observer had been cut away and removed. The other views are drawn in the ordinary manner without any change except the inclusion of the symbolic cutting-plane line which indicates the location of the imaginary cutting plane.

FULL SECTION

When a longitudinal imaginary cutting plane extends through the entire object showing the whole object in section, the resultant view is known as a full section view (Fig.8.1).

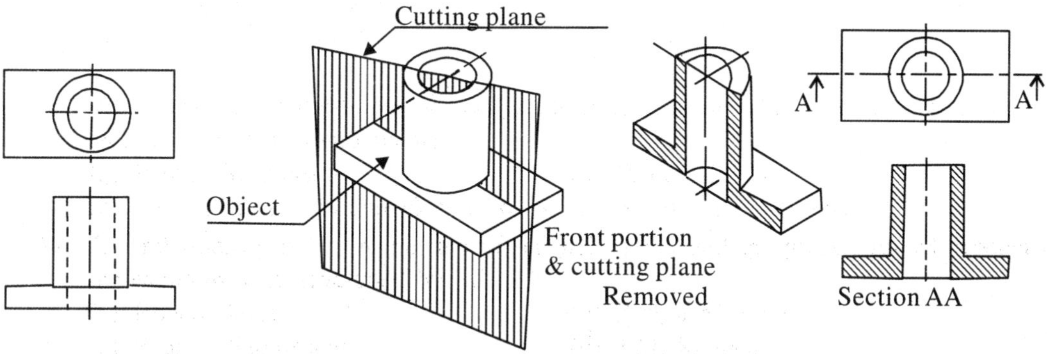

Fig. 8.1 Full section

If an object has an axis of symmetry then instead of a full section view half section view is admissible where the cutting plane extends upto the axis of symmetry (Fig.8.2).]

CUTTING PLANE LINE

Cutting planes are indicated by long chain lines, thickened at the ends, bends and change of directions and it should be thin elsewhere and are designated by capital letters, the direction of viewing being shown by arrows resting on the plane line. A simple way to avoid confusion is to remember that the arrows always point away from the discarded portion of the object.

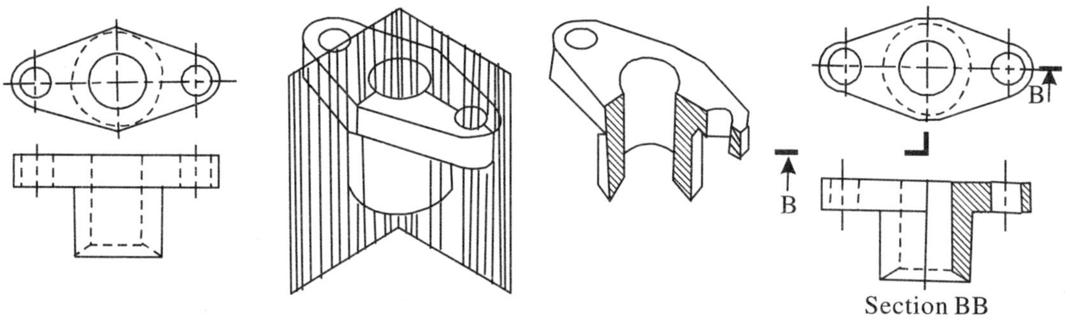

Fig. 8.2 Half section

OFFSET SECTION

A full section cutting plane need not always pass through the main axis or center line of an object. Its direction may be changed (*i.e.* offset) in order to show other details and features. Fig. 8.3 illustrates the use of an offset section.

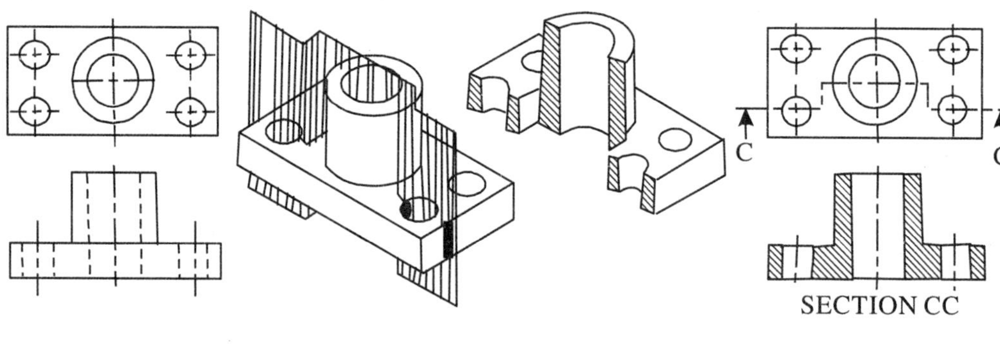

Fig. 8.3 Offset section

SPECIAL SECTION

Apart from the usual sections, some special sections *e.g.*, revolved section, removed section, local (partial or skin) section, successive section, *etc.* are also admissible. Fig. 8.4 indicates the above sections.

Sometimes sections in two intersecting planes are shown revolved into the plane of projection (Fig. 8.5). Section in two parallel in two parallel planes is shown in Fig. 8.6.

PARTIAL VIEW OF SYMMETRICAL OBJECTS

To save time and space symmetrical objects may be drawn in part, Fig.8.7. The line of symmetry is defined by two short parallel lines drawn at each end and perpendicular to the center line and these need not be shown when it is amply clear from the dimensioning.

SECTION LINES

The section line for each individual part should always be drawn at an angle of 45° to the axis or to the main outline of the section. A single piece is sectioned with sharp thin lines about 2.5 mm apart. In case of small pieces it may be 1.0 mm, while for large objects it may be 3 mm or even more. Hatching lines may be interrupted for dimensioning, if it is not possible to place dimensions out side.

OMISSION OF SECTION

Shafts, rods, nuts, bolts, rivets, keys, screws, studs, pins, gear teeth, webs, ribs, rolling elements (ball, roller, needle) of bearings, etc. which have no internal features, are in general not drawn in section if seen in profile (when their axes lie parallel to the plane of the paper). However, cross sections or broken out sections may be drawn if required.

THIN PARTS

The cross-section view of thin metal parts, gaskets or other material are shown in solid black.

SOME RULES

Since the prime reason for drawing section-view is to reveal the internal features, hidden lines should not be shown in section drawing unless they are absolutely necessary for further clarification or dimensioning a feature.

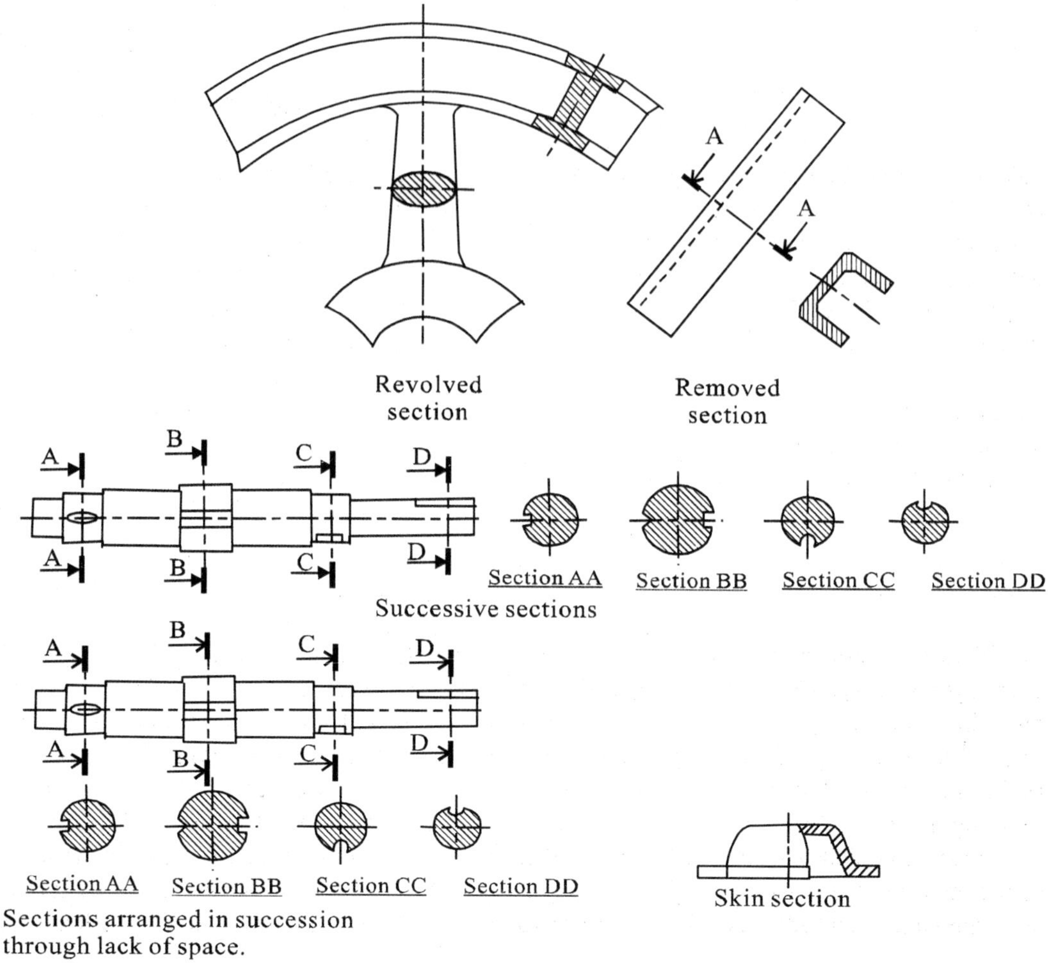

Fig. 8.4 Views in special section

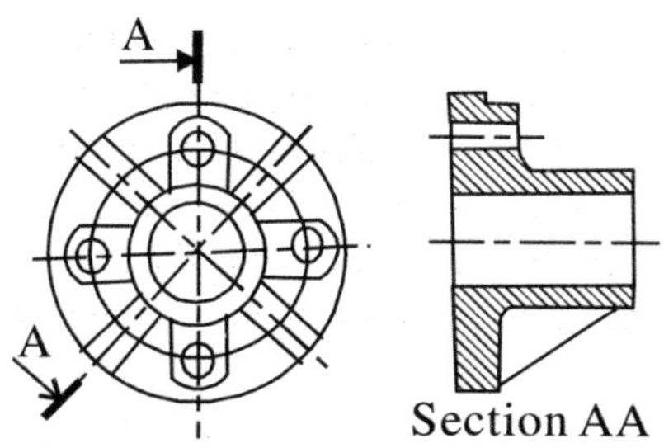

Fig. 8.5 Section in two intersecting planes

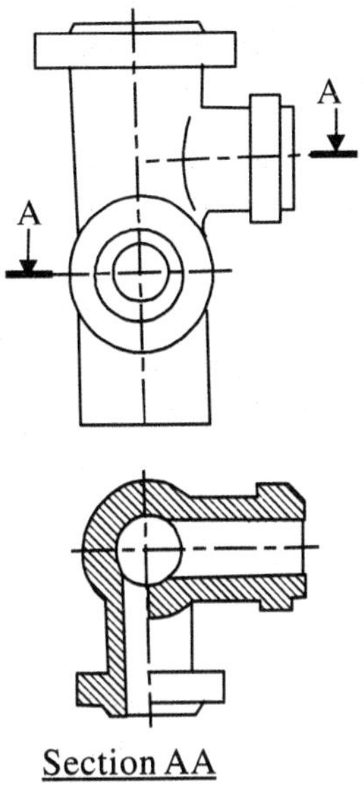

Fig. 8.6 Sections in two parallel planes

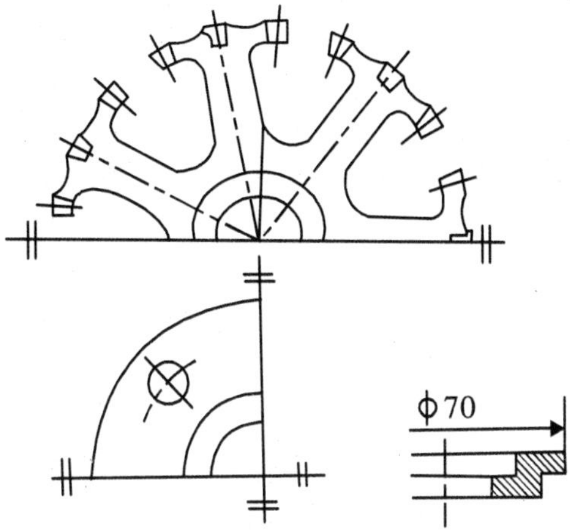

Fig. 8.7 Symmetrical objects

SECTION CODE

Different kinds of section lines are used to indicate various materials (*e.g.*, steel, lead, glass, asbestos *etc.*). Where several pieces are shown in section assembly it is useful to identify different parts and materials by using sectioning code suggested by BIS (Fig. 8.8).

In assembly drawing all portions of the section of a particular part are always shown with the section line running in the same direction. If two pieces are shown in assembly the section line of each are drawn at 45° running in opposite directions. When three or more pieces appear in assembly, angles 30°, 60° may be used or the pitch of section lines varied, as per U.S. practice; Indian standards have no specific recommendation.

CONVENTIONAL REPRESENTATION OF MATERIALS

TYPE	CONVENTION	DETAILS
Metals	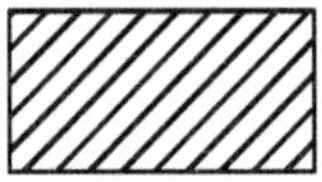	Steel, cast iron, coper and its alloys, aluminium and it alloys etc
	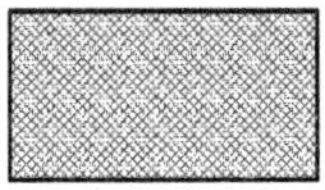	Lead, zinc, tin, white metal etc.
Glass and Ceramics	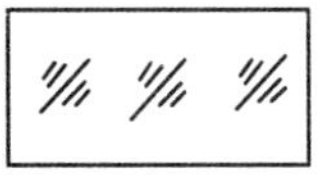	Glass
	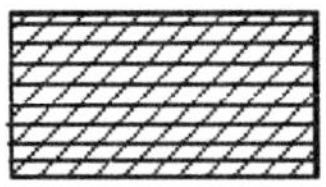	Porcelain, stoneware, marble, slate, etc.
Packing and insulating materials	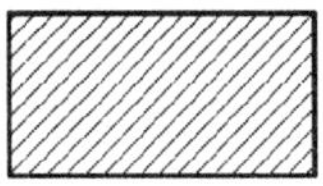	Aesbestos, fibre, felt, synthetic resin products, paper, cork, linoleum, rubber, leather, wax, inculating and filling materials.
Liquids	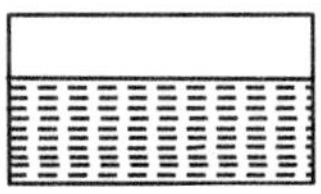	Water, oil, petrol, kerosene etc.
Wood	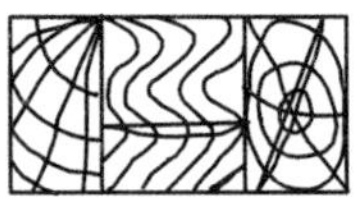	Wood, plywood, etc.
Concrete	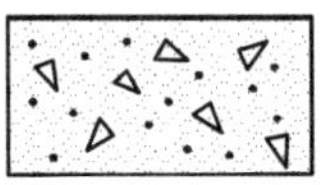	

Fig. 8.8

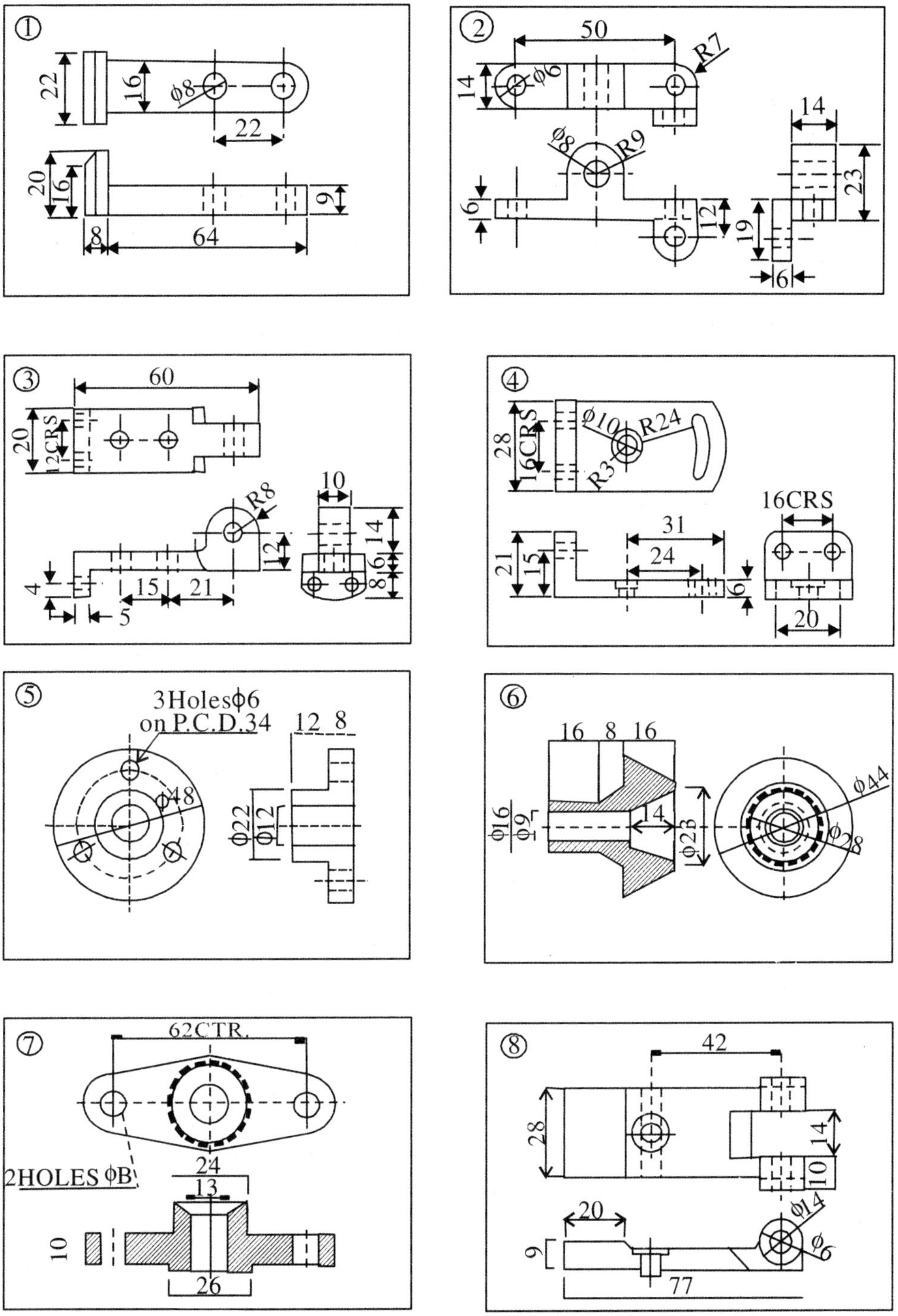

①
22
16
φ8
22
20
16
9
8
64
②
50
R7
14
φ6
14
φ8
R9
23
6
12
19
6
③
60
20
12CRS
10
R8
14
12
6
8
4
15
21
5
④
28
16CRS
φ10
R24
R3
16CRS
31
24
21
15
6
20
⑤
3Holesφ6
on P.C.D.34
12 8
φ48
φ22
φ12
⑥
16 8 16
φ16
φ9
14
φ23
φ44
φ28
⑦
62CTR.
2HOLES φB
24
13
10
26
⑧
42
28
14
10
φ14
20
9
φ6
77

Prob. 8.1 to **8.6**

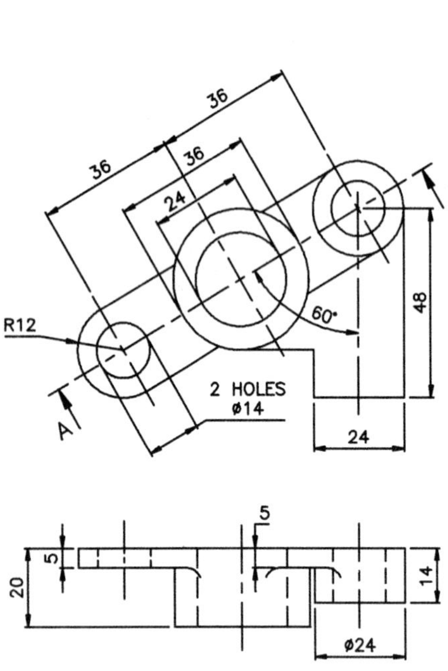

Bracket

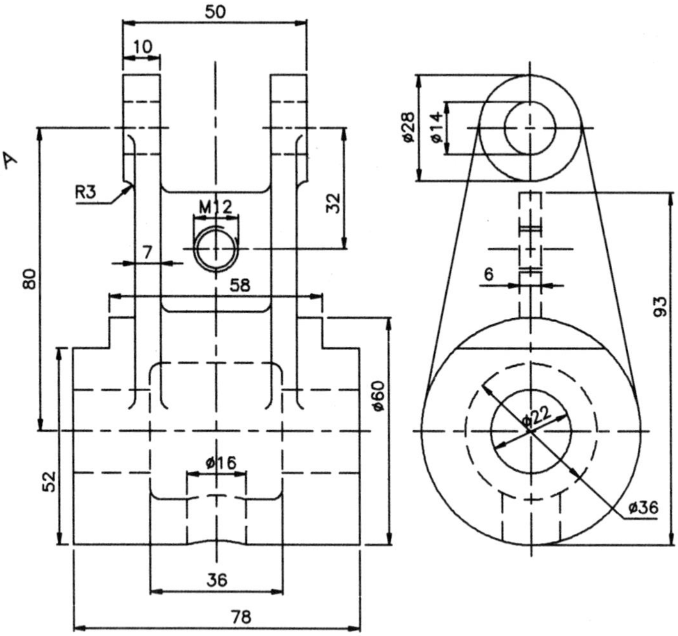

Fork

Prob. 8.7 to 8.8

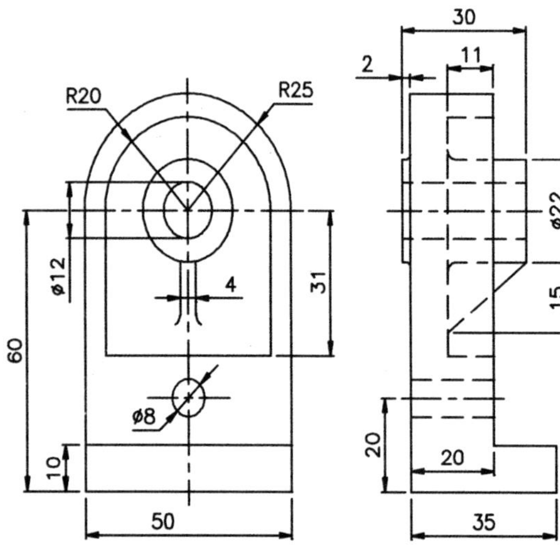

① Support

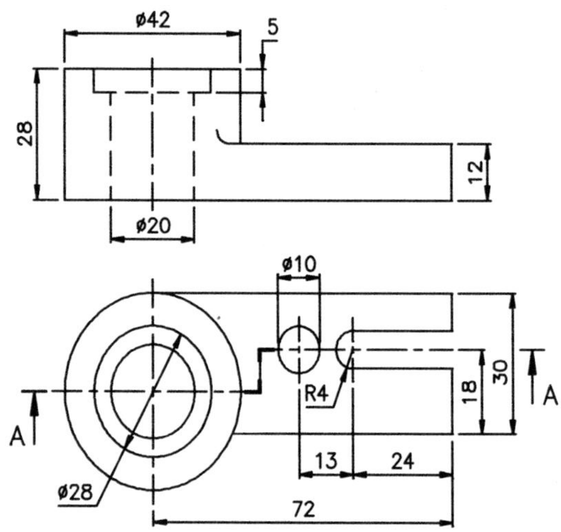

② Support

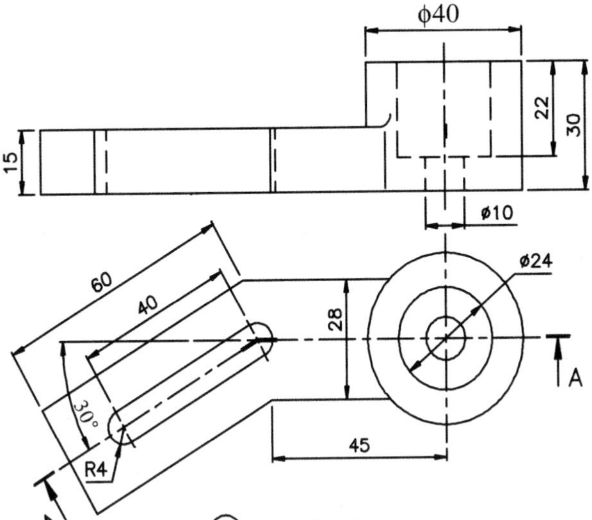

③ Crank

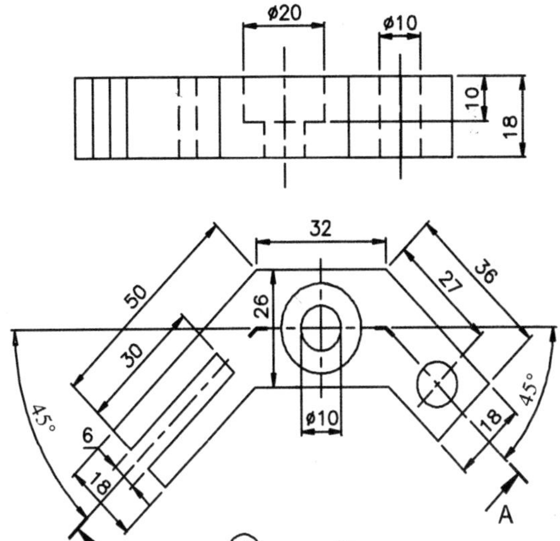

④ Clamp

Probs. 8.9 to **8.12**

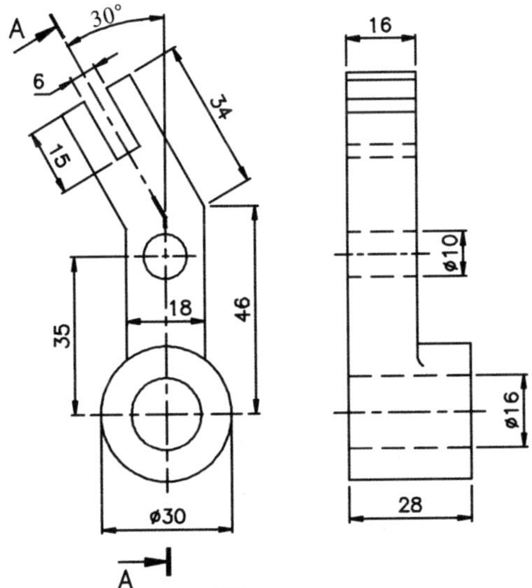

① Lever

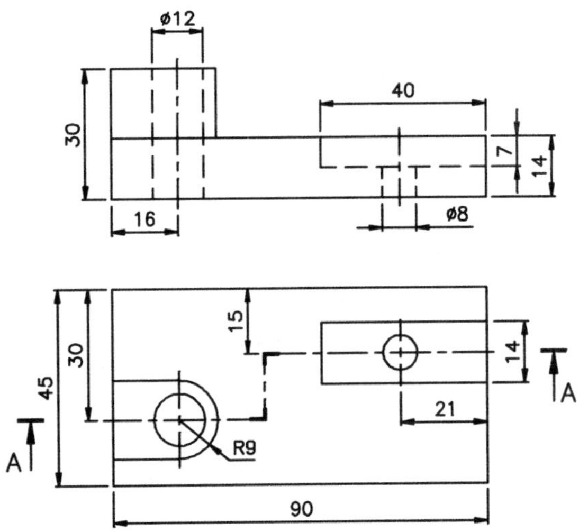

② Holder

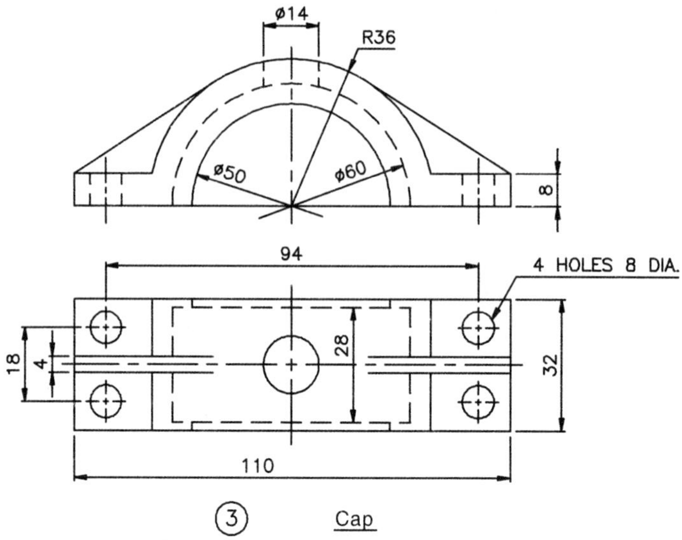

Probs. 8.13 to **8.15**

PROBLEMS

(Put necessary dimensions of the drawings of the Probs. 8.1 to 8.6. Assume the drawings are in scale 1:2.)

8.1 to 8.3 Two views of objects are shown in Probs 8.1, 8.2 and 8.3. Draw two principal views with one full section.

8.4 to 8.6 Draw top view and front view in half section of each of the objects shown in Probs. 8.4, 8.5 and 8.6, indicating the section planes. (Assume drawings are in scale 1:2.)

8.7 Two views of a component are shown. Draw the views in section as indicated on the drawing.

8.8 Given two views of a machine part construct the top view and make all the neccesary section.

8.9 Replace the left side view by a sectional view.

8.10 – 8.12 Replace the front view by sectional view.

8.13 Replace the left side view by a sectional view.

8.14 – 8.15 Replace the front view by full sectional view.

CHAPTER – 9

AUXILIARY VIEWS

A view not projected on one of the principal planes (e.g., Horizontal, Vertical and Profile) is an auxiliary view. A primary auxiliary view is a view projected on a plane perpendicular to one of the principal planes and which is inclined to the other two.

AUXILIARY VIEWS ARE GENERALLY USED TO SHOW

(a) the true length of an oblique line and

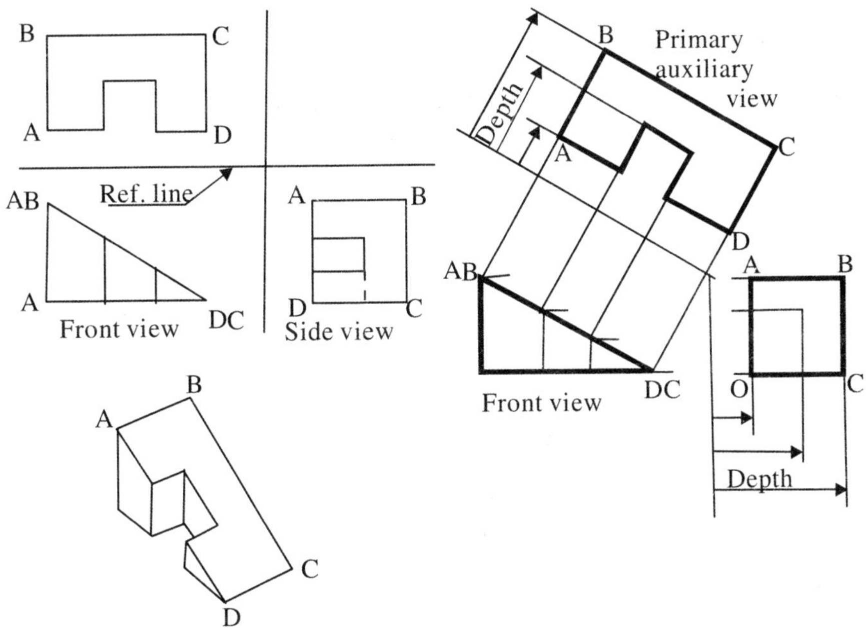

Fig. 9.1

A drawing should be dimensioned in the view where it represents its true shape and size. In many objects the bounded lines and planes may be inclined or oblique to the principal planes of projection and obviously those projected lines will be foreshortened in the principal views. To get their true shapes and complete description auxiliary views are drawn.

To find the true shape projection of an inclined surface of an object the auxiliary plane should be passed parallel to that surface.

TO CONSTRUCT AN AUXILIARY VIEW

The usual procedural steps in constructing an auxiliary view are shown in Fig. 9.1. The illustration should be carefully studied to understand the procedure.

AUXILIARY AND PARTIAL VIEW

Sometimes the use of an auxiliary view may allow elimination of one of the principal views or makes it possible to use a partial principal view. A break line is used at a convenient location to indicate as imaginary break for a partial view.

Fig. 9.2 shows a bracket. To dispense with the construction of several ellipses on the horizontal plane of projection, the orthographic drawing should be represented as in Fig. 9.2 (a) with the upper portion projected, on an auxiliary plane of projection.

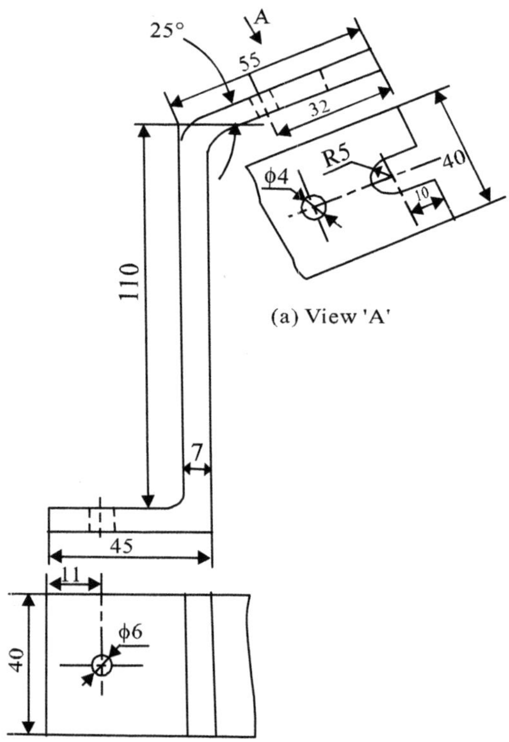

Fig. 9.2

Auxiliary views must be accompanied by the corresponding inscriptions, *e.g.*, view A, and the appropriate projection of the part must be referred to by an arrow indicating the direction of viewing.

To conveniently arrange all the necessary views on a standard sheet an auxiliary view may be turned through any desired angle Fig. 9.3.

Sometimes it is necessary to show certain spot on the surface of a machine part whose orthographic view gives no clear idea of this area. In such cases, the drawing is supplied with an additional view of the area, called a local view, Fig. 9.4.

A section view projected upon an auxiliary plane, is sometimes necessary to show the shape of an arm, rib and so forth, inclined to any two or all three of the principal planes of projection Fig. 9.5.

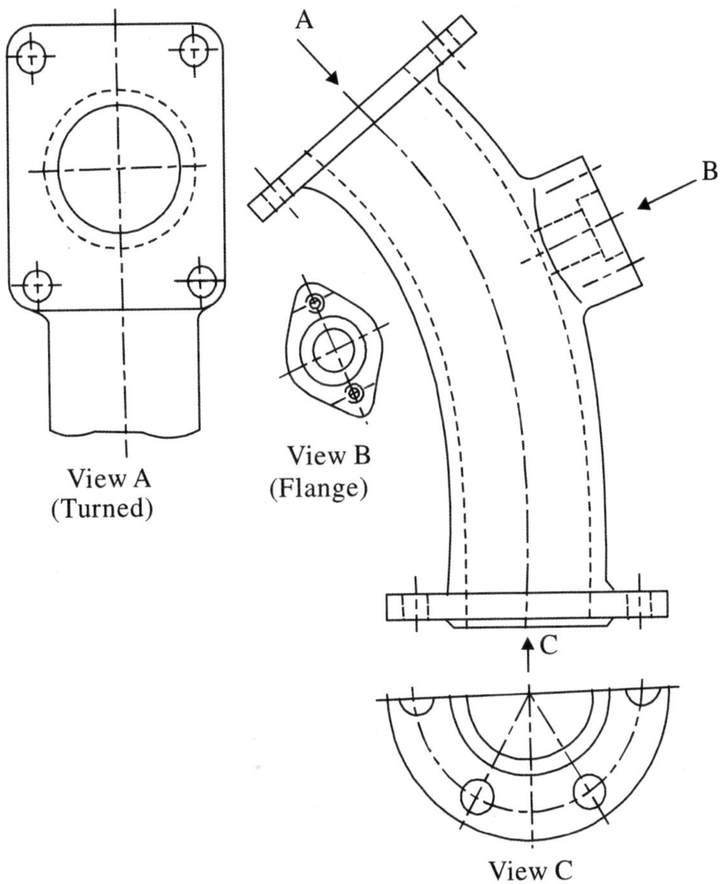

Fig. 9.3

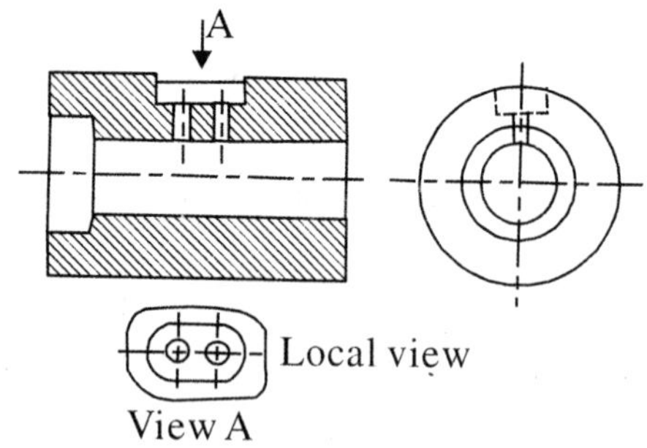

Fig. 9.4

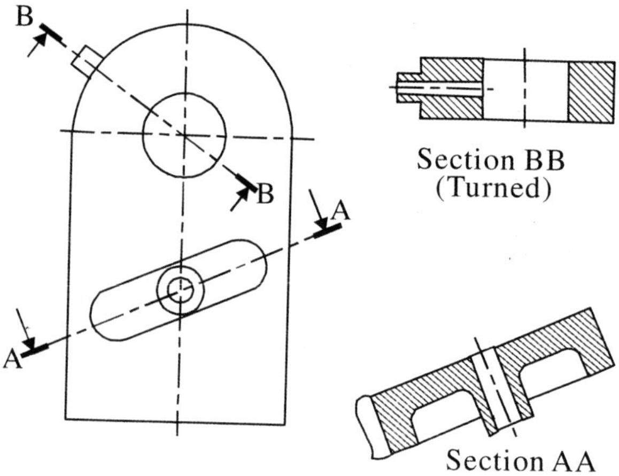

Fig. 9.5

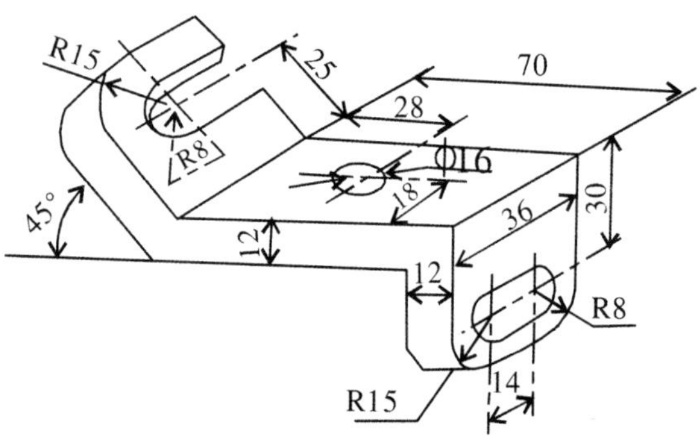

Fig. 9.6

Auxiliary view to complete a principal view: In many cases it is frequently necessary to project a foreshortened feature in one of the principal views from an auxiliary view. In the case of an object shown in Fig. 9.6, the foreshortened projection of the inclined face in the top view can be projected from the auxiliary view. The elliptical curves are obtained by projecting points from the auxiliary view to the front view and from there to the top view. It is obvious that the location of these points in the top view with respect to the center line will be same as that in the auxiliary view with respect to the auxiliary center line.

PROBLEMS

9.1 to 9.3 Reproduce the given front view and the true shape auxiliary view. Develop the top view.

9.4 to 9.5 Redraw the given top and front views of the truncated cylinder and hexagonal prism. Show the true shapes of the inclined surfaces in the auxiliary views.

9.6 Front view and incomplete right hand side view of a flag staff are shown in Prob. 9.6. Complete the front view as a sectional view and side view as an ordinary orthographic projection.

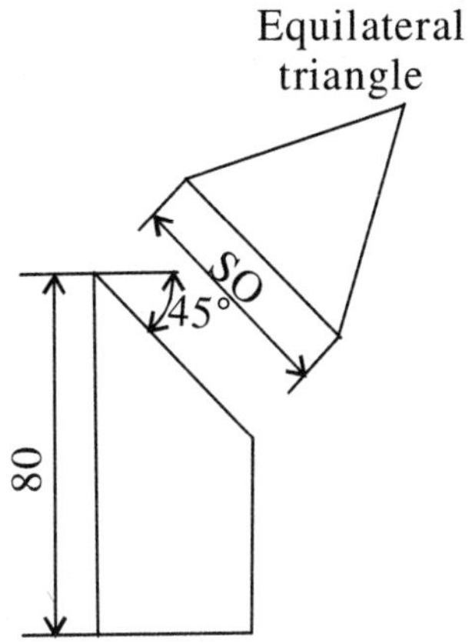

(1)

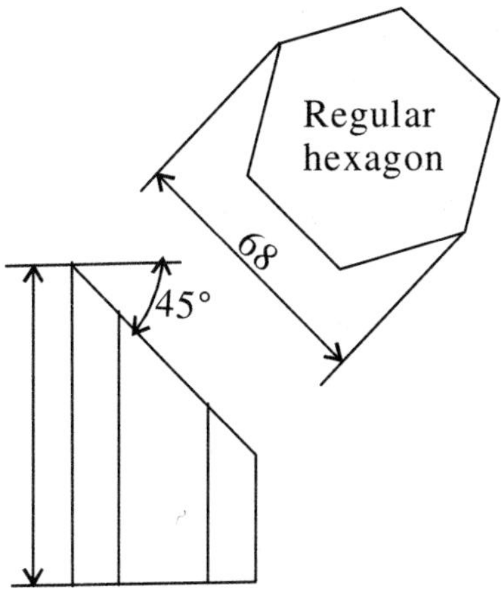

(2)

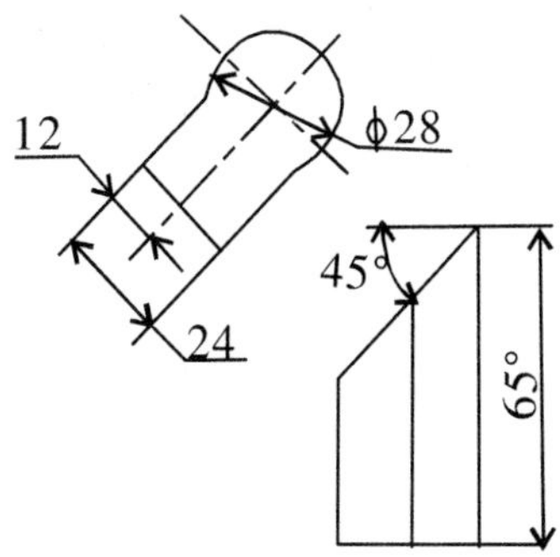

(3)

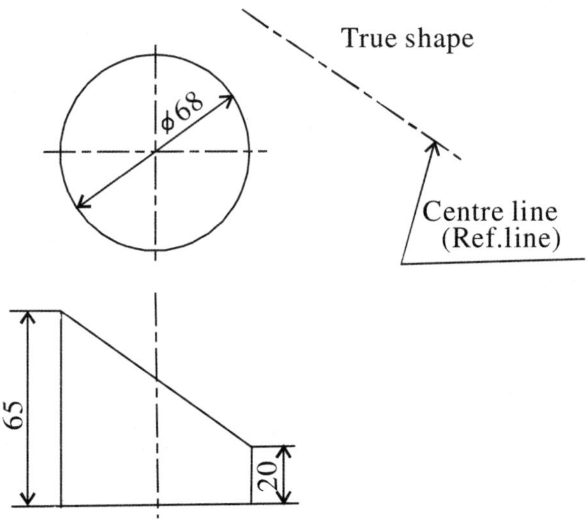

(4)

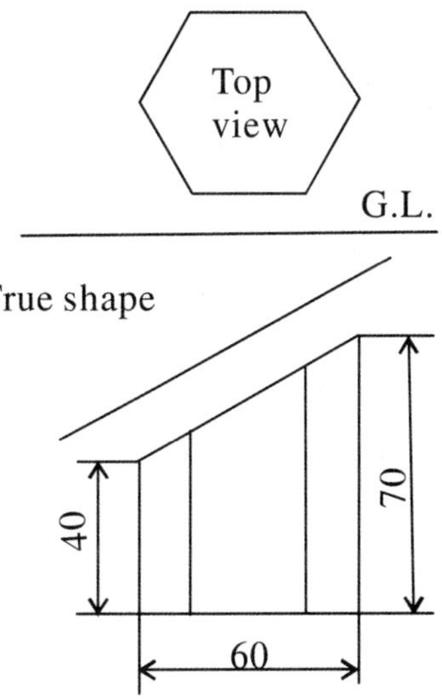

(5)

Probs. 9.1 – 9.5

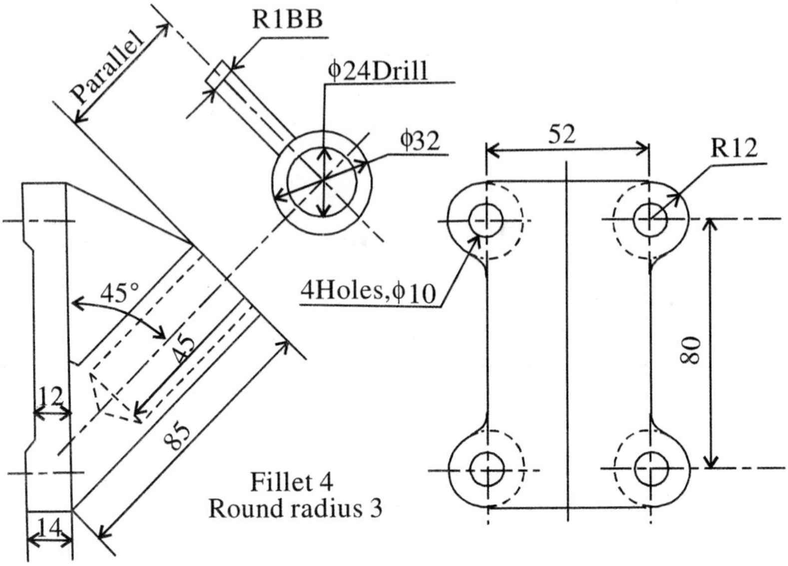
R1BB
Parallel
ϕ24Drill
ϕ32
52
R12
45°
4Holes,ϕ10
45
80
12
85
Fillet 4
Round radius 3
14

Prob. 9.6

CHAPTER – 10

PERMANENT JOINTS

Amongst the common permanent fastening methods riveting and welding are widely used in engineering practice. Before going into details, a brief description of materials and structural shapes used for joints is given below.

MATERIAL

Materials mainly used in engineering construction are Cast Iron, Wrought Iron and Steel. Besides these, other materials like Copper, Brass, Bronze, Cast Steel are also frequently used. In recent times various types of plastics and composite materials are also becoming popular.

Cast Iron is hard but brittle, very strong in resisting compressive forces, but in tension or pulling apart it is rather weak and not reliable. It is easily melted and in the fluid state can be cast into the mould easily in any desired shape. Due to this reason Cast Iron which is comparatively cheap with good strength, is widely used for the frame work of all general machinery and stationery engines.

Wrought Iron and "mild'" or "soft" steel are malleable, ductile and non-brittle materials. They are a bit difficult to melt but can easily be made quite soft by heating and in this condition they can be "rolled" into plates, sheets, bars and rods, "drawn out" into tubes and wires, "forged" into parts like crank, tyres, levers etc. These materials can be easily welded, gas cut and machined to various shapes and sizes. Mild steels have good tensile and shear strengths and various structures, girders are usually made of the same.

Available forms of wrought iron and mild steel consist principally of plates of different thickness, sheets and strips and long rolled bars of various cross sections; some of the common sections are shown in Fig. 10.1.

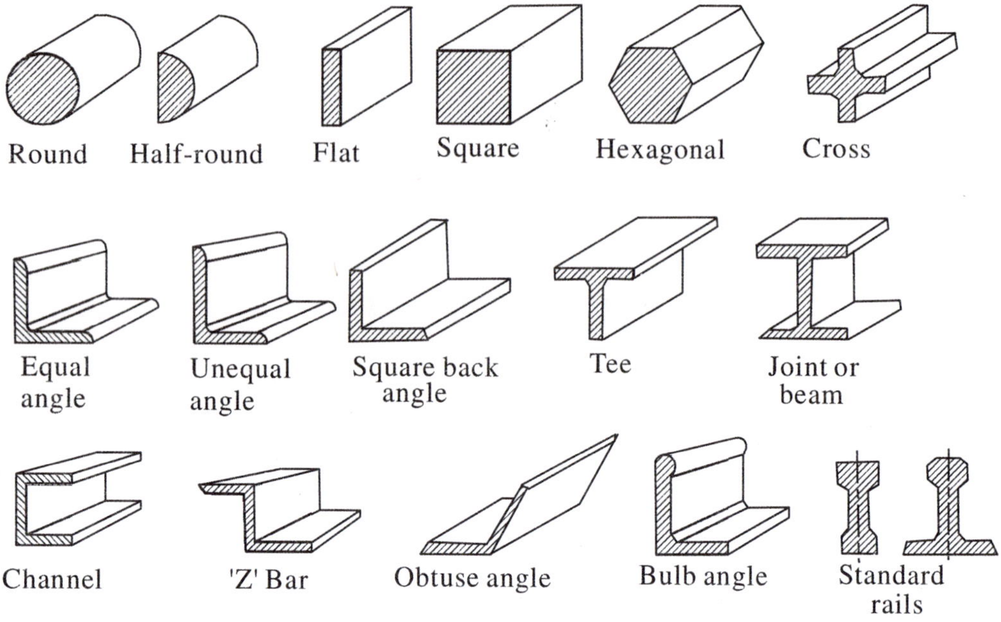

Fig. 10.1

RIVETS

A rivet is a cylindrical shank of circular section with a set head at one end. The rivet shank is inserted into a drilled or punched hole in structural elements to be joined and the set head of the rivet is pressed firmly against the metal element by a bucker. While riveting, the protruding end of the shank is upset by repeated hammer or sledge blows or by pressure of special riveting machine to obtain the desired shape, thus forming another head called the closing head.

RIVET HEADS

There are various types of rivets. They are distinguished by the shape of their heads. The most common type is the snap (button) head rivet. If a flush surface is required, countersunk rivets are used; if, however a slight protrusion is allowable countersunk oval head may be used. Various types of rivet heads depending on sizes and their use in structure or boiler are shown in Figs. 10.2a, 10.2b & 10.2c.

RIVETED JOINT

Depending on the arrangement of rivets, the joints are classified as single riveted joint or multiple riveted joint (double-riveted, triple-riveted, *etc.*)

Rivets may be arranged in a staggered manner or in parallel rows. When metal plates are connected with rivets then it is called a riveted joint. Depending on the position of the jointed plates the joints are termed *lap joints or butt joints* with one or two cover straps. Various types of joints are shown in Fig. 10.3.

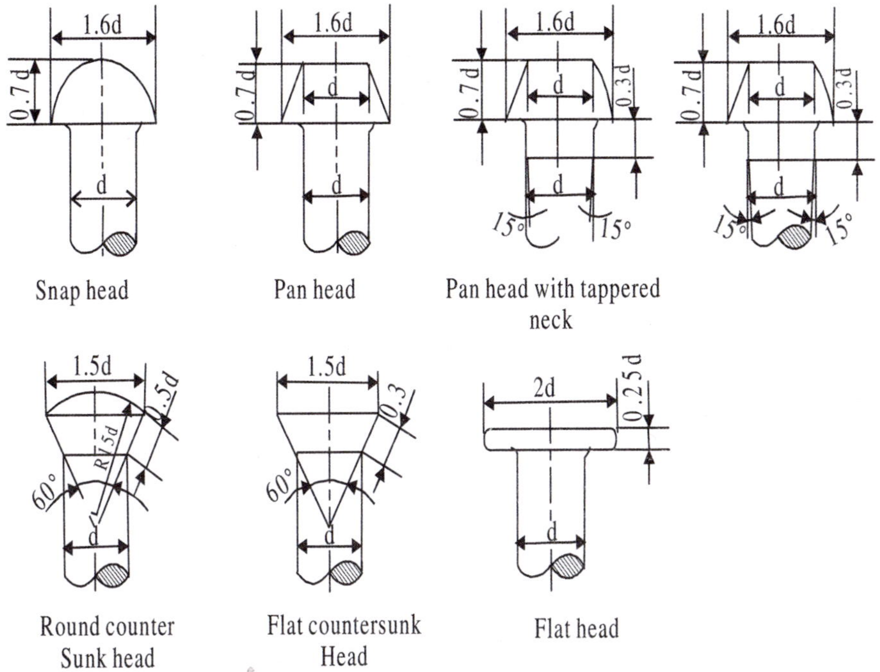

Fig. 10.2 (a) Rivets for general purpose (12 to 48 mm diameter)

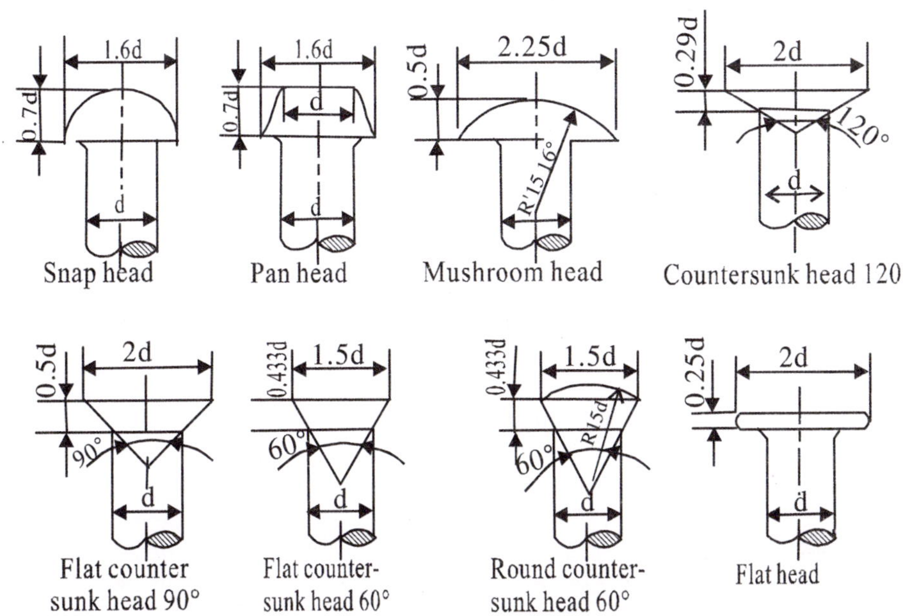

Fig. 10.2 (b) Rivets for general purpose (below 12mm diameter)

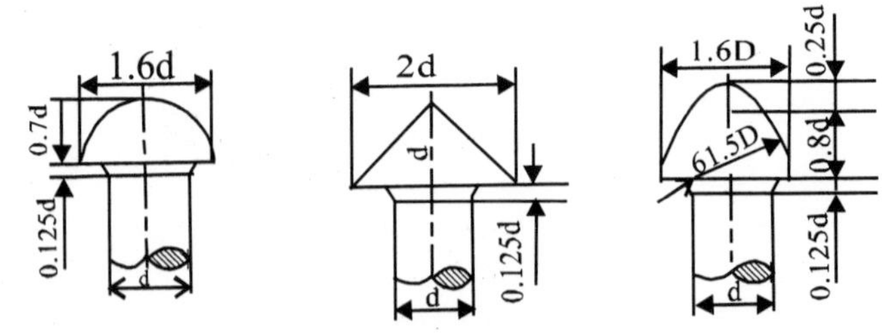

Snap head Steeple head Conical head

Fig. 10.2 (c) Boiler rivets (12 to 48mm diameter)

Joints may be

1. Strong joints where strength is the main consideration (*e.g.*, beams, trusses, bridges *etc.*)
2. Tight joints, which must ensure pressure tightness (*e.g.*, reservoirs, tank *etc.*)
3. Strong tight Joints, which must be both strong and pressure tight (*e.g.*, steam boilers, gas tanks *etc.*)

Tightness is achieved by providing sufficient initial tension to the rivet. Caulking and/or fullering is resorted to for making rivet joints leak proof (Fig. 10.3).

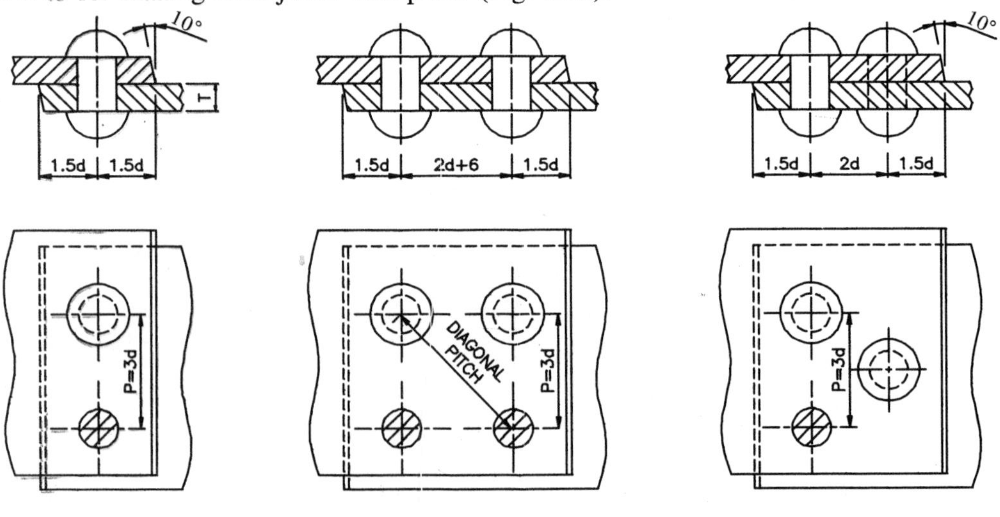

Single riveted Double riveted (Chain) Double riveted (Zig-Zag)

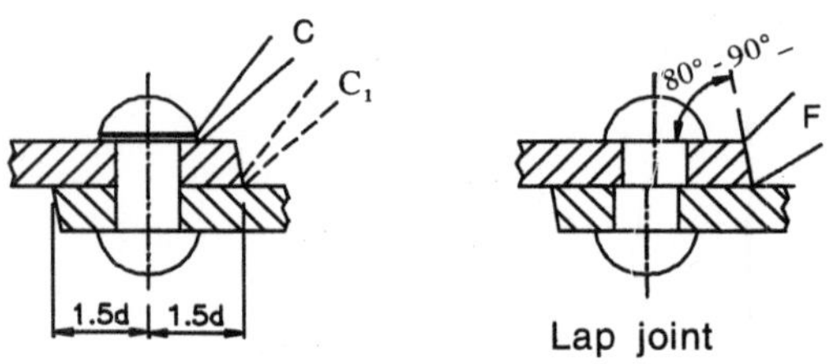

Lap joint

'C' and 'C1' are caulking tool. 'F' is a fullering tool. (The Rivet is shown in single shear.

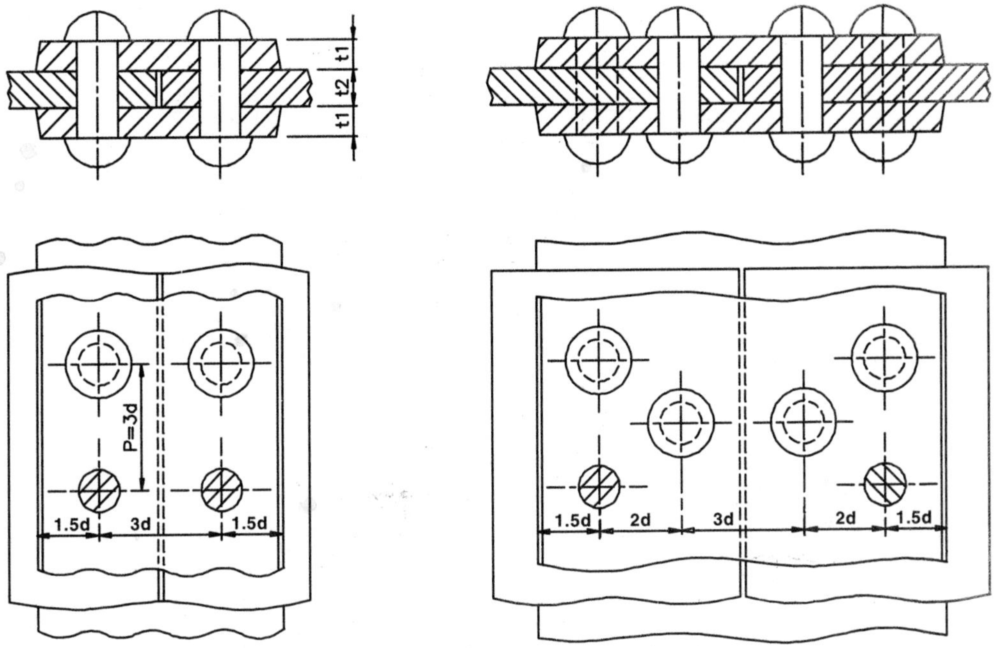

Single riveted

Double riveted

Butt joint

Fig. 10.3

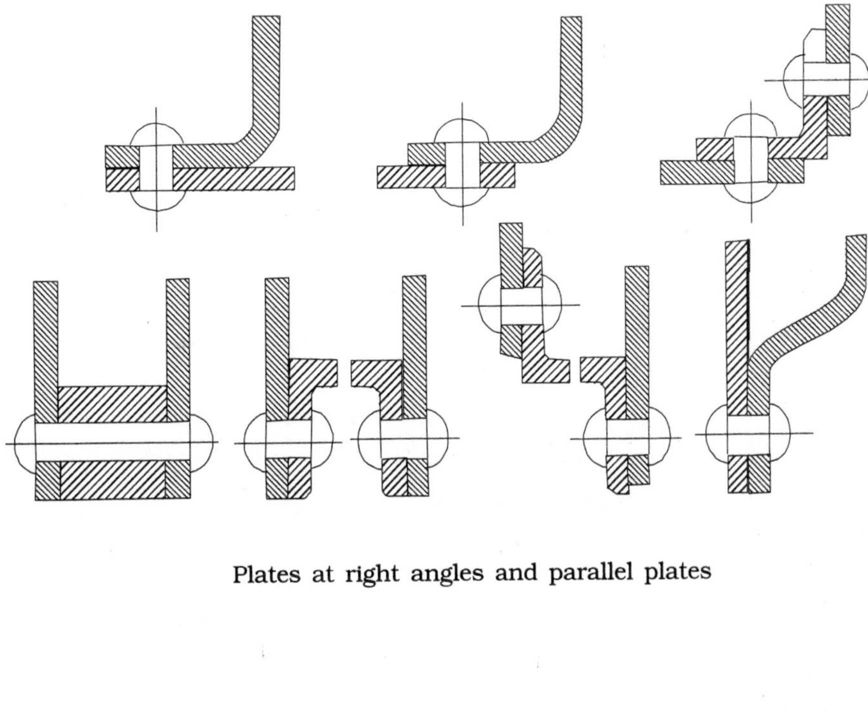

Plates at right angles and parallel plates

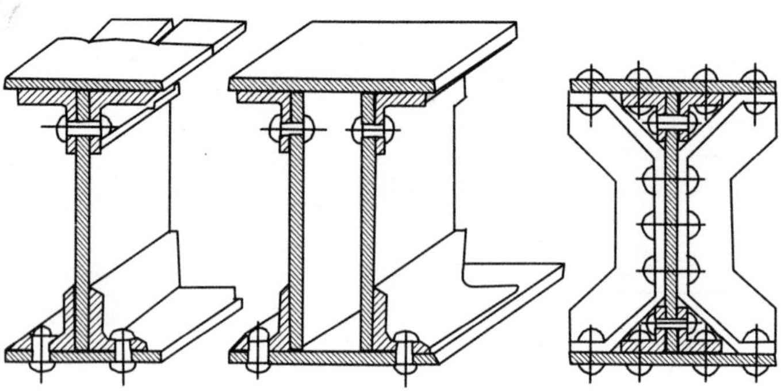

Fig. 10.4. Built up girders

Table 10.2: Standard thickness of plates, (mm) — 3, 5, 8, 10, 12, 16, 20, 25, 32, 40

Standard thickness of sheets, (mm)

0.40, 0.50, 0.63, 0.90, 1.00, 1.10, 1.25, 1.40, 1.60, 1.80, 2.00, 2.24, 2.50, 2.80, 3.15, 3.55, 4.00

RIVETED WORK

Various types of riveted work consisting of standard rolled sections or plates are shown in Fig. 10.4 which are varieties of common 'built up' girders, joints, boiler connection etc. in use.

WELDED JOINT

Permanent Joints obtained by welding have won universal recognition in modern engineering fabrication. *Welding* is the process of joining a number of metal pieces by raising the temperature at the joint (which may be achieved by a gas-flame, an electrical energy, an energy beam striking the work piece, friction etc.) so that the metal become plastic / molten followed by application of pressure / the use of a filler metal to join the parts. Sometimes no filler metal is needed. Electric arc welding is most widely used.

Aluminium can be cold welded by applying pressure only.

Welds can be classified as butt and fillet type. Butt welds are joints connecting the butts of parts located in one plane. When the plate-thickness exceeds a certain value (which, in turn, depends on the welding process) the edges of the butts are to be prepared specially to facilitate access of the filter metal.

WELDED JOINTS

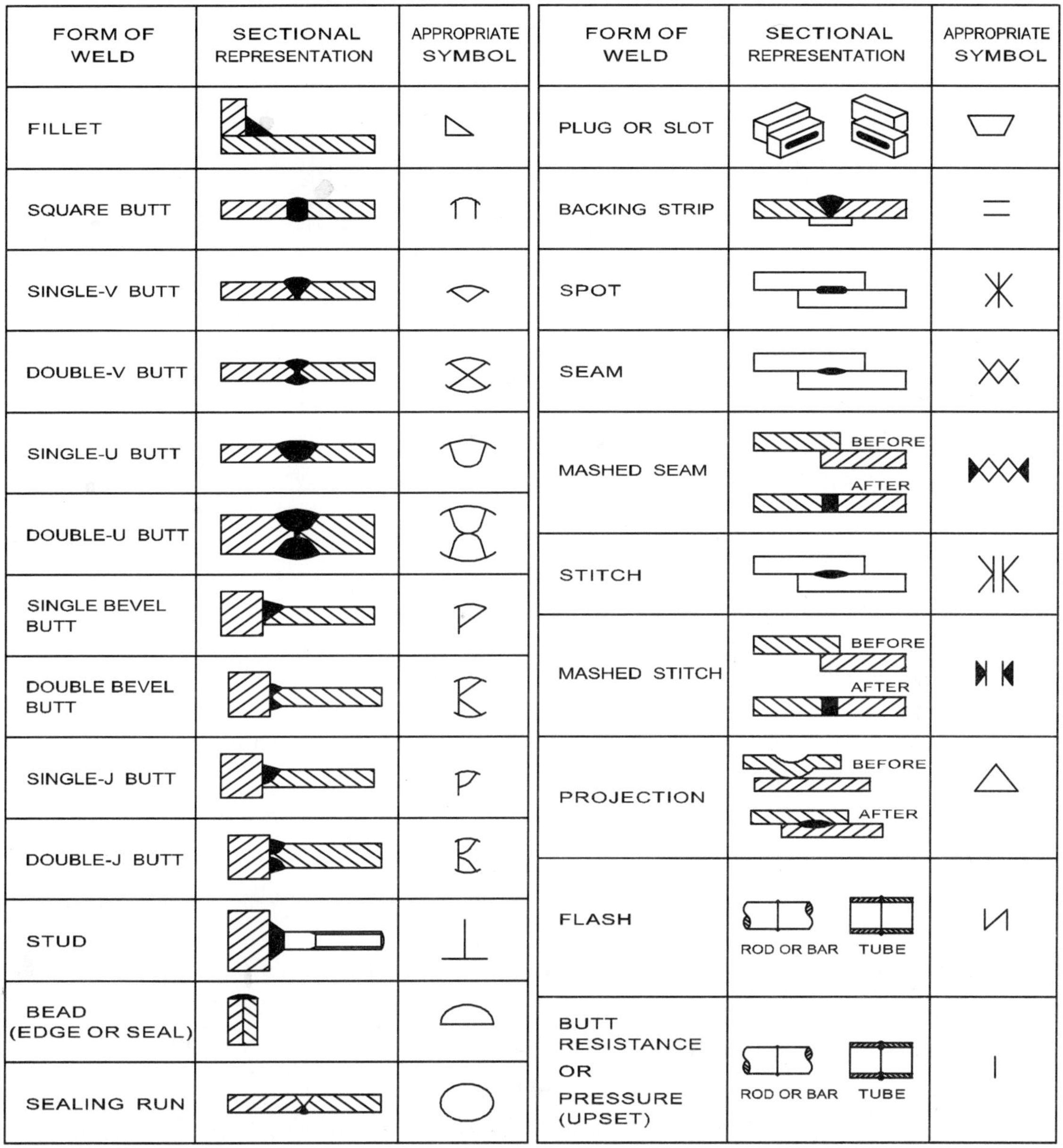

Fig. 10.5 Types of welds and their symbols

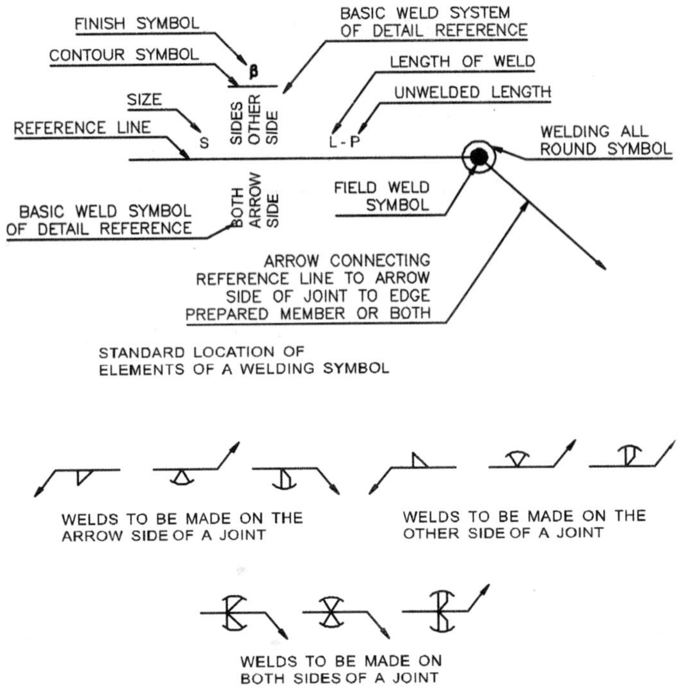

SPECIAL INSTRUCTION	DRAWING REPRESENTATION	SYMBOL
WELD ALL ROUND		○
SITE WELD (FIELD WELD)		●
FLUSH CONTOUR		—
CONVEX CONTOUR		⌒
CONCAVE CONTOUR		◡
GRINDING FINISH	G	G
MACHINING FINISH	M	M
CHIPPING FINISH	C	C

Fig. 10.6

Lap joints are made with fillet welds and are termed transverse or longitudinal according to the direction of the load-application.

Various types of welded joints with their standard symbols are shown in Fig. 10.5. Welds are shown on drawings as in Fig. 10.6.

The above symbol hitherto used has been modified a bit to dispense with the confusion as to whether the weld (weld face) is on the arrow side of the joint or on the other side, Fig. 10.6.

A dashed line (as thick as the Reference line) is drawn parallel to the Reference line either above or beneath the same.

The symbol is placed on the continuous Reference line if the weld is on the arrow side of the joint and the weld is on the other side if the symbol is placed on the dashed line.

For symmetrical welds, the dashed line is unnecessary and should be omitted.

Location of welds having their linear dimensions transverse to the lines of stress are approximately 30% stronger per average unit length than welds having linear dimensions parallel to the lines of stress. Symmetrical joints are preferable. The shape of the sections joined determines the assembly keeping in view that the weld called for are accessible for welding and for non-destructive testing and repairing whenever necessary.

GENERAL RECOMMENDATIONS FOR WELDED COMPONENTS

Reduce the number of welded members, welds and ribs. Apply double welds only for severest acting loads. Never locate seams in the flange parting planes. Make inner welds only in heavy housings. Make the rim from flat steel and butt-weld so as to locate the weld in the tooth gap for fabricated gears. Do not machine the hub and rim before welding. Do not cut out the ribs but form them from flat steel. The rim should overhang the ribs. Cut off the rib to eliminate the weld intersection. Locate the sealing welds inside. Do not locate the weld base in the zone of tension. Size of welding should be minimum consistent with strength. Over welding enhances the cost and is as bad as under welding.

Table 10.3: Recommended minimum size of fillet welds

Plate thickness (mm)	Minimum size of weld (mm)
3 to 5	3
6 to 8	5
10 to 16	6
18 to 20	10
21 to 35	15
over 36	20

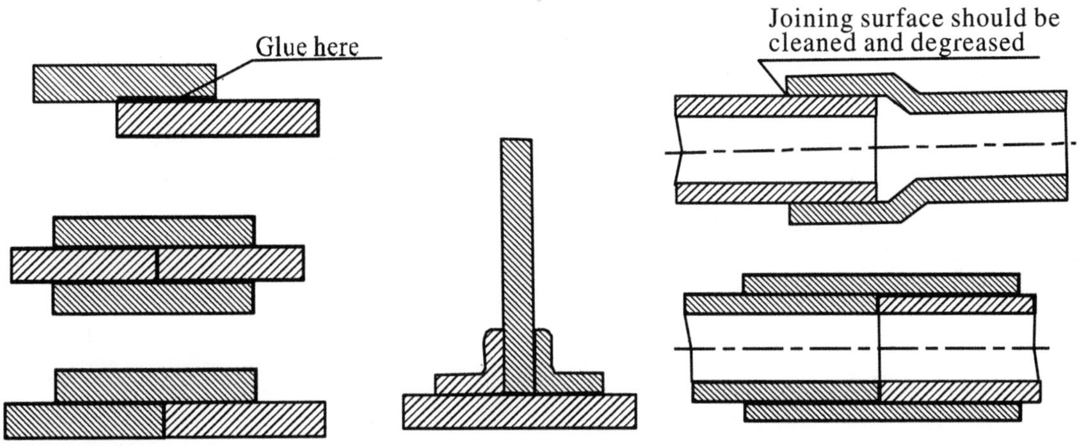

Fig. 10.7 Glued joints

GLUED JOINTS

In recent times various types of adhesives have been developed which are used to join metallic elements and nonmetallic elements (textolite, foam plastics, wood *etc.*) and two non-metallic elements. In practice a thin layer of glue is applied between the parts to be joined. Some commercial glues available in the market are Fevicol, Dendrite, Quickfix, Araldite *etc.* Glued joints for built-up sections and tubes are shown in Fig. 10.7.

PROBLEMS

Rivet & Riveted Joint

10.1 Draw two views of a single-riveted lap joint taking d = 24 mm and standard plate thickness.

10.2 Draw two views of a double zig-zag riveted lap joint using 24 mm rivet.

10.3 Draw two views of a double-riveted butt joint with two cover plates as shown in Fig. 10.3. Take d = 20 mm, t = 10 mm, t_1 = 8mm.

10.4 The flat D-shaped end cover of a pressure vessel is connected to the cylindrical shell by means of riveting. The shell and end plates are both 10 mm thick. The inner diameter of the shell is 1200 mm. A flush outer surface is desired. Select the appropriate type and size of rivets and draw two views of the joint.

10.5 Draw two views of the following:
- (i) Built-up plate girder.
- (ii) Built-up box girder.

The following specifications are given in mm:

Web Plate	800 × 12.0 (width × thickness)
Flange Angles	150 × 150 × 18.0 (ISA 150150)
Flange Plate	400 × 20

10.6 Show by freehand how the following built-up beams or columns can be formed:
- (i) Box beam using channels.
- (ii) Columns using double joints (H-Section).

Welding

10.7 An I-beam is to be fabricated by welding. Represent the object on drawing using IS: 813-1961 symbols.

10.8 Solve problems 10.5 and 10.6 using welding in place of riveting.

10.9–10.10 The frontal diametric view of the yoke as shown in Prob. 10.9 and the isometric view of the angle bracket (Prob. 10.10) are to be fabricated by welding. Draw sketches of the components you propose to use to build up the yoke/bracket. Show welding symbolically on the assembled drawing of the yoke/bracket.

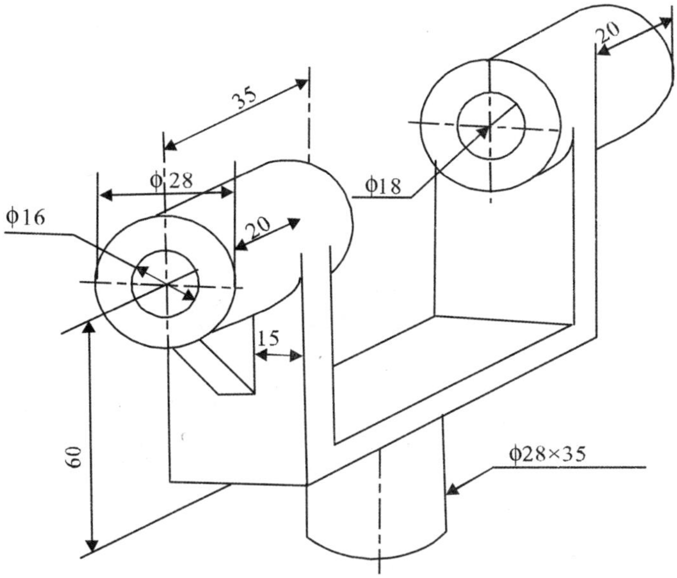

YOKE

Prob. 10.9

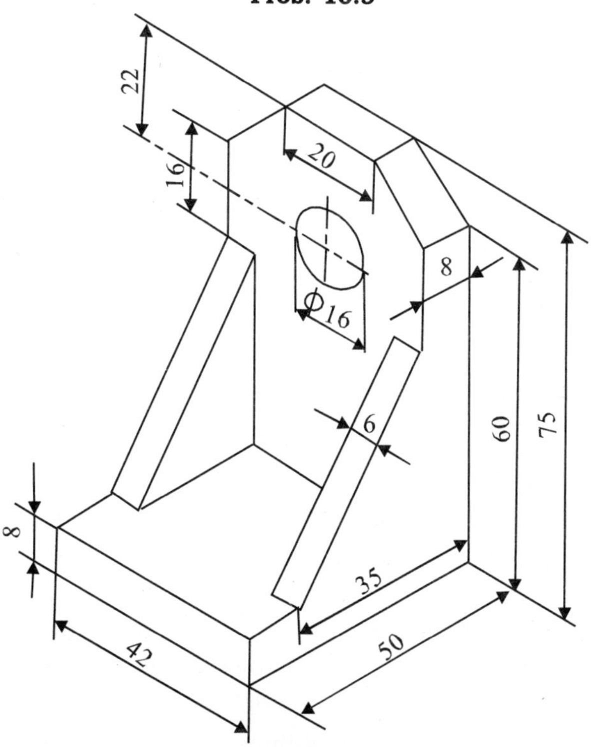

BRACKET

Prob. 10.10

CHAPTER - 11

THREADED JOINTS

SCREW THREAD

When a helical groove is cut or generated over a cylindrical or conical piece, screw threads are formed. The screw thread may be single start or multiple start.

HELIX

When a point moves on the surface and parallel to the axis of a rotating cylinder or cone held between centers a helix is generated.

PITCH

Pitch of a thread is the axial distance from one point on the thread to the similarly situated corresponding point on the next tread.

LEAD

It is the axial advance per rotation of the moving member of the screwed pair (i.e. screw and nut). In case of single threaded screw, lead and pitch are equal. The lead, l and pitch, p are related as l = np, where n is the number of starts of thread.

Thread forms are in general of two types

(1) Power threads *e.g.* square thread, acme thread, knuckle thread.

(2) Fastening threads *e.g.* V-thread like B.S.W. (British Standard Whitworth) and I.S.O. (International Standard Organization) metric thread profile.

Square threads are the most efficient for power screws. Acme threads are used in power screws where a nut split lengthwise is desirable for wear adjustment. The buttress thread is also a power thread used in vises where power is transmitted in one direction only. Efficiency of a V-thread is lower than that of a power thread. V-threads are suitable for fastening so that nuts will not shake loose so easily. Thread profiles are shown in Fig. 11.1.

Metric V-threads may be coarse or fine, depending on the purpose. A fine metric thread differs from a coarse metric thread of same major diameter. In that its pitch and depth are smaller than those of the latter.

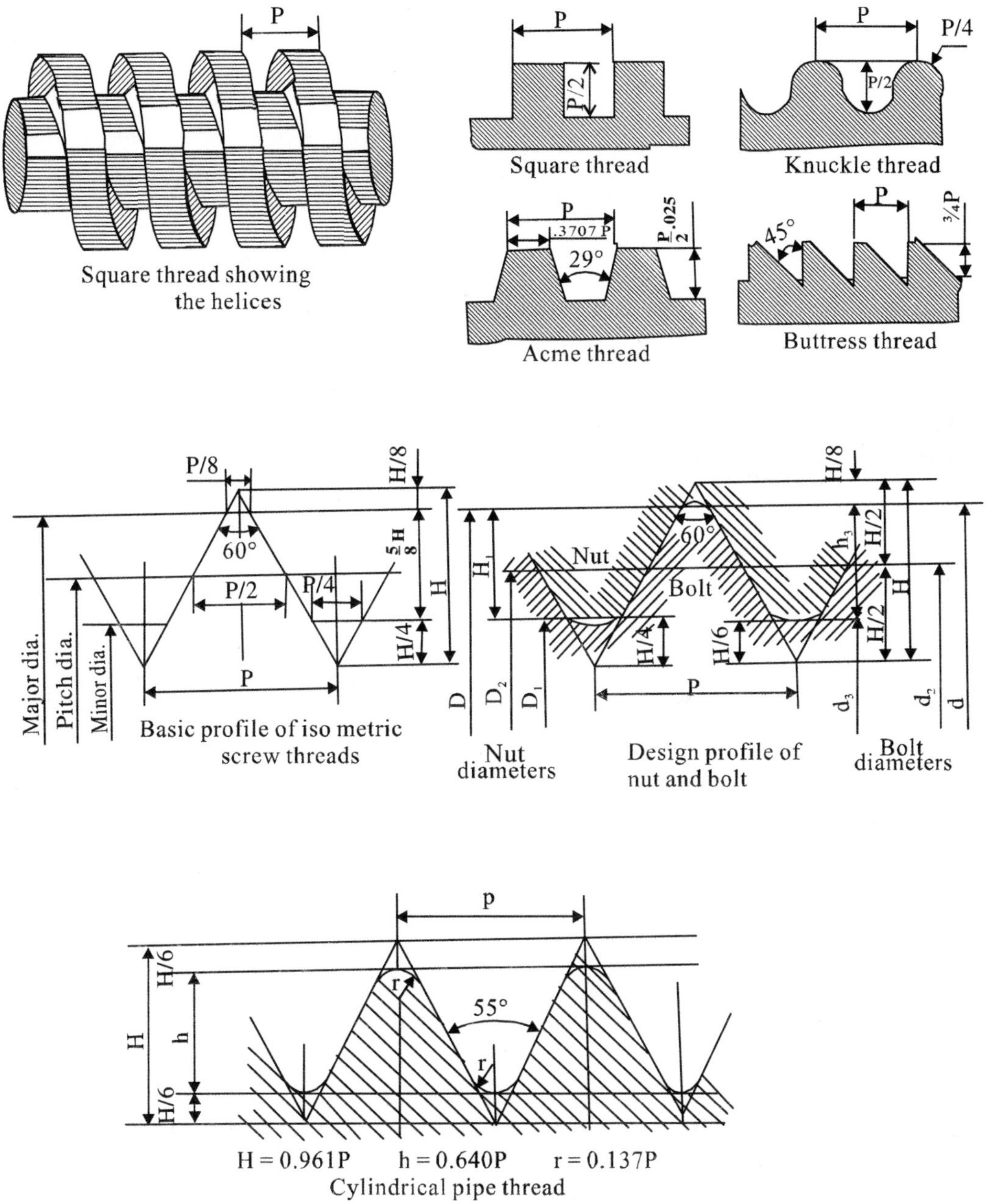

Fig. 11.1

CYLINDRICAL PIPE THREAD

This type of threads is used in pipe lines where leakage of fluid is a problem. The threads are cut at 55° with a pitch considerably smaller than that of the coarse thread. The dimensions are given in standard pipe thread, Table No. 11.4. Note that the Size Designation is the 'Nominal Bore' of the pipe in inches whose outside surface is threaded.

Table 11.1: Metric screw threads (with I.S.O. metric profile)

Fine series		Coarse series	p, mm
Designation	pitch, p, mm	M1, (M 1.1), M1.2	0.25
M 8 × 1	1	M (1.4)	0.3
		M. 1.6, (M 1.8)	0.35
M 10 × 1.25	1.25	M 2	0.4
		(M 2.2), M 2.5	0.45
M 12 × 1.25	1.25	M 3	0.5
		(M 3.5)	0.6
(M 14 × 1.5)	1.5	M 4	0.7
		(M 4.5)	0.75
M 16 × 1.5	1.5	M 5	0.8
		M6,(M 7)	1.0
M 18 × 1.5	1.5	M 8, M 9	1.25
		M 10, M 11	1.5
M 20 × 1.5	1.5	M 12	1.75
		(M 14), M 16	2.0
M 22 × 1.5	1.5	M 18, M 20 (M 22)	2.5
		M 24 (M 27)	3.0
M 24 × 2	2	M 30, (M 33)	3.5
		M 36, (M 39)	4.0
(M 27 × 2)	2	M 42, M 45	4.5
		M 48, M 52	5.0
M 30 × 2	2	M 56, M 60	5.5
M 36 × 3	3	M 64, M 68	6.0
(M 39 × 3)	3		

(Second preference sizes are shown within brackets.)

Table 11.2
Square threads
(Normal Series)

Nominal dia mm	Pitch mm
22 24 26 28	5.0
30 32 (34) 36	6.0
(38) 40 (42) 44	7.0
(46) 48 50 52	8.0
55 (58) (60) (62)	9.0
65 (68) 70 (72)	
	10
75	10
(78) 80 (82)	

Table 11.3
Clearance holes for metric bolts

Bolt thread diameter d, mm	Clearance hole diameter, A, mm Fine	Medium	Coarse
1.6	1.7	1.8	2
2	2.2	2.4	2.6
2.5	2.7	2.9	3.1
3	3.2	3.4	3.6
4	4.3	4.5	4.8
5	5.3	5.5	5.8
6	6.4	6.6	7
7	7.4	7.6	8
8	8.4	9	10
10	10.5	11	12
12	13	14	15
14	15	16	17
16	17	18	19
18	19	20	21
20	21	22	24
22	23	24	26
24	25	26	28
27	28	30	32
30	31	33	35
33	34	36	38
36	37	39	42
38	40	42	45

(Second preference sizes are shown within brackets.)

Table 11.4: Standard pipe threads
(All dimensions in millimeters)
(Refer to cylindrical pipe thread in Fig. 11.1)

Size designation	Pitch p, mm	Basic Diameter, mm Internal & External Threads		
		Major	Pitch	Minor
FP 1/8	0.907	9.728	9.147	8.566
FP 1/4	1.337	13.157	12.301	11.445
FP 3/8	1.337	16.662	15.806	14.940
FP 1/2	1.814	20.995	19.793	18.631
FP 3/4	1.814	26.441	25.279	24.117
FP 1	2.309	33.249	31.770	30.291
FP 1 1/4	2.309	41.910	40.431	38.952
FP 1 1/2	2.309	47.803	46.324	44.842
FP 2	2.309	59.614	58.135	56.656
FP 2 1/4	2.309	65.710	64.231	62.752
FP 2 1/2	2.309	75.184	73.705	72.226
FP 3	2.309	87.884	86.405	84.926
FP 3 1/2	2.309	100.330	96.851	94.372
FP 4	2.309	113.030	111.551	110.72
FP 5	2.309	138.40	136.951	135.472
FP	2.309	163.830	162.351	160.872

Designation: An internal pipe thread for fastening purposese of size 2 shall be designated as Int-FP2 and an external pipe thread as Ext. FP 2.

IS: 2543 - 1984

THREADED ELEMENTS

Bolts, studs, nuts and screws are the common threaded elements used for detachable fastening of machine parts.

Fig. 11.2 illustrates various types of threaded elements. A bolt is characterized by the shape of its head *e.g.* hexagonal, square or hexagonal socket *etc.*, which facilitates tightening by an appropriate wrench or spanner. Similarly nuts may have different shapes.

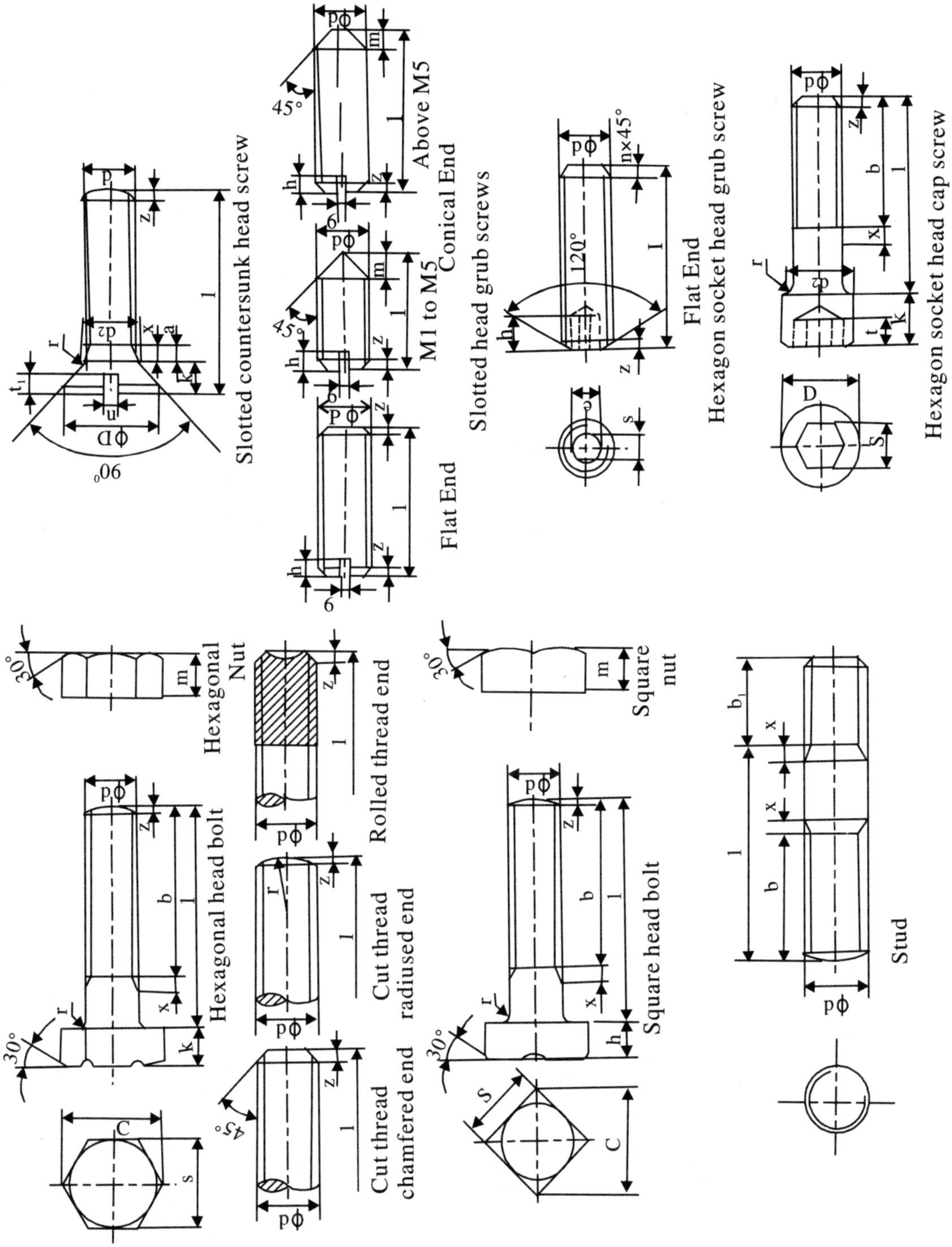
Slotted countersunk head screw
Slotted head grub screws
Conical End
Above M5
M1 to M5
Flat End
Flat End
Hexagon socket head grub screw
Hexagon socket head cap screw
Hexagonal head bolt
Hexagonal Nut
Rolled thread end
Cut thread radiused end
Cut thread chamfered end
Square head bolt
Square nut
Stud

Fig. 11.2

A stud is a cylindrical rod with threads on both ends while screws are comparatively smaller size threaded elements having different shapes of head. Screws and studs are driven into threaded holes in one of the two pieces to be joined.

Dimensions of the threaded length, ends, heads *e.g.*, b, z, k *etc.* are standardized. IS: 2389 specifies dimensions for Bolts, Screws, Nuts and Locknuts; IS:1365 for slotted countersunk Head and Slotted Raised countersunk Head Screws (Diameter Range 1.6 to 20 mm); IS:2585 for Black Square Head Bolts, Screws and Nuts (Diameter Range M6 to M39); IS:2388 for Slotted Head Grub Screws *etc.*.

BOLTED JOINT

A bolted joint, in general, consists of three parts besides components to be joined; a bolt, a nut and a washer which may be dispensed with if the mating surface is properly machined. A bolt consists of a cylindrical shank V-threaded partly or fully and a chamfered square or hexagonal head.

A representative bolted joint with standard conventional proportions is shown in Fig. 11.3

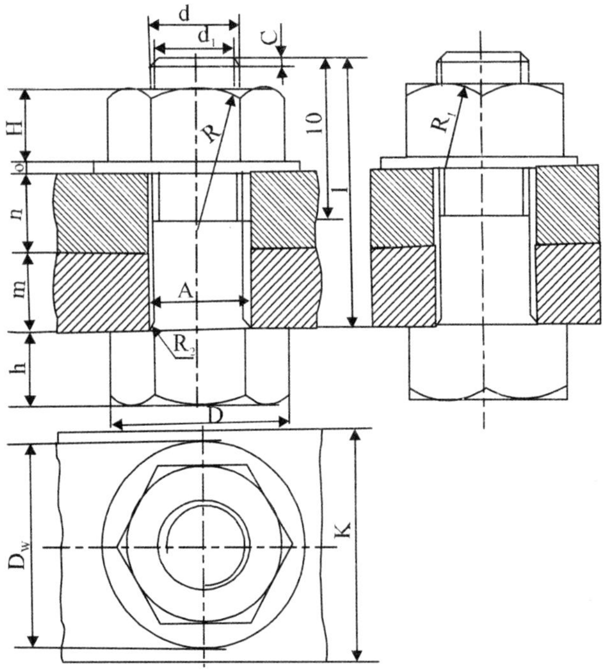

d_1 = 0.85d

D = 2d

H = 0.8d

h = 0.7d

D_w = 2.2d

s = 0.15d

A = 1.1d

l_0 = 2d to 2.5d

R = 1.5d

R_1 = d

R_2 = 0.5c

c = S to 2S

K = 3d

l = as required

chamfering circle dia = 1.75d

Fig. 11.3 Bolted joint with conventional ratio

STUD JOINT

Dimensions of studs are given in the appropriate IS. One end of the stud is screwed into a threaded hole in a part to be joined. Other part is provided with a hole drilled to a diameter somewhat larger than the diameter of the stud. A nut is screwed on to the other end of the stud. Usually a washer is placed under the nut. A stud will be used when one of the pieces to be joined is too thick to use ordinary headed bolts or where there is no room for bolt-head.

Steps followed in making a threaded hole for a stud and studded joint is known in Fig. 11.4.

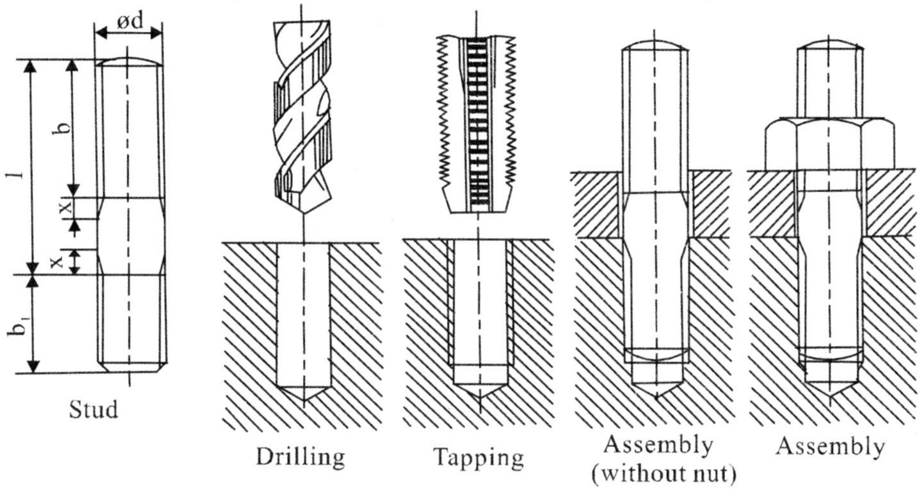

Fig. 11.4

A representative stud joint with conventional ratios is shown in Fig. 11.5. It may be noted that values of l_2 vary with material of the internally threaded part; the weaker the material the greater the length. Normally studs are made of mild steels. In representing bolted and studded joints on assembly drawing, the dimensions may be given following conventional ratios.

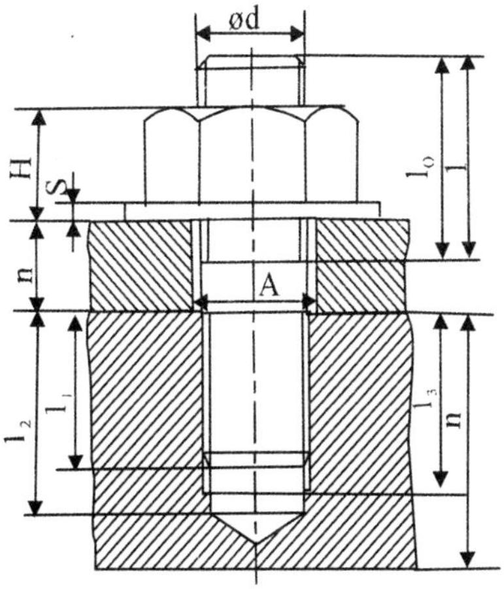

l & l_0 (= b) = As tabulated in Table 11.5

l_1 (= b_1) = 1.25d $\quad$ $l_2 = l_1 + 0.5d$

$l_3 = l_1\ 0.4d$ $\quad$ Height of Nut = (H – S) = 0.8d

s = 0.15d $\quad$ A = 1.1d

Fig. 11.5 Stud joint with conventional ratio

Table 11.5: Dimensions for stud (Diameter range 3 to 39 mm)
(All dimensions in millimeters)
Refer to Stud Fig. 11.4

Nominal Size, d	M3																		
	*12	13	14	15	16	18	22	26	30	34	38	42	46	50	54	60	66	72	78
b(=10)	+-	-	-	-	-	-	-	32	36	40	44	48	52	56	60	66	72	78	78
	**-	-	-	-	-	-	-	-	-	57	61	65	69	73	79	85	91	97	103
b1	Max.4.00	4.70	5.20	5.70	6.20	7.20	9.50	11.50	13.80	15.80	17.80	19.80	22.10	24.10	26.10	29.10	32.10	35.50	38.50
Type A	Min.3.00	3.50	4.00	4.50	5.00	6.00	8.00	10.00	12.00	14.00	16.00	18.00	20.00	22.00	24.00	27.00	30.00	33.00	36.00
b1	Max.5.70	6.20	7.20	8.50	9.00	10.50	13.80	16.80	19.80	23.10	26.10	29.10	32.10	35.50	38.50	43.00	47.50	52.00	67.00
Type B	Min.4.50	5.00	6.00	7.00	7.50	9.00	12.00	15.00	18.00	21.00	24.00	27.00	30.00	33.00	36.00	40.00	45.00	49.00	54.00

(Type B has a longer mental end than Type A)

*For lengths upto 130 mm

+For lengths above 130 mm and upto 200 mm

**For lengths above 200 mm

Note – Non-perferred diameters of studs are given within brackets.

For x refer to IS: 1369-1969: 'X' can also be calculated from the relation. x=0.075d/tanα

Throat angle α being _> 15° (large) > 221/2" (medium) and > 30° (small)

IS: 1862-1967

IS: 1862-1967

Table 11.6: Dimensions for Hexagonal Socket Head Cap Screws (Diameter Range 3 to 24 mm)

(All dimensions in millimeters)

Refer to Fig. 11.2

Size		M3	M4	M5	M6	M8	M10	M12	M14	M16	M18	M20	M22	M24
b	*	12	14	16	18	22	26	30	34	38	42	46	50	54
	†	-	-	-	-	28	32	36	40	44	48	52	56	60
d (h 13)	Nom	3	4	5	6	8	10	12	14	16	18	20	22	24
	Max	3.00	4.00	5.00	6.00	8.00	10.00	12.00	14.00	16.00	18.00	20.00	24.00	
	Min.	2.86	3.82	4.82	5.82	7.78	9.78	11.73	13.73	15.73	17.73	19.67	21.67	23.67
D (h 13)	Nom	5.5	7	8.5	10	13	16	18	21	24	27	30	33	36
	Max	5.50	7.00	8.50	10.00	13.00	16.00	18.00	21.00	24.00	27.00	30.00	33.00	36.00
	Min.	5.32	6.78	8.28	9.78	12.73	15.73	17.73	20.67	23.67	26.67	29.67	32.67	35.60
K (h 13)	Nom	3	4	5	6	8	10	12	14	16	18	20	22	24
	Max	3.00	4.00	5.00	6.00	8.00	10.00	12.00	14.00	16.00	18.00	20.00	22.00	24.00
	Min.	2.86	3.82	4.82	5.82	7.78	9.78	11.73	13.73	15.73	17.73	19.67	21.67	23.67
S (D 12)	Nom	2.5	3	4	5	6	8	10	12	14	14	17	17	19
	Max	2.61	3.11	4.15	5.15	6.15	8.19	10.19	12.23	14.23	14.23	17.73	17.73	19.28
	Min.	2.52	3.02	4.03	5.03	6.03	8.04	10.04	12.05	14.05	14.05	17.05	17.05	19.06
t	Max	1.70	2.40	3.10	3.78	4.78	6.25	7.50	8.70	9.70	10.70	11.80	12.40	14.00
	Min.	1.30	2.00	2.70	3.30	4.30	5.50	6.60	7.80	8.80	9.80	10.70	11.30	12.30
r	Min.	0.1	0.2	0.2	0.25	0.4	0.4	0.6	0.6	0.6	0.6	0.8	0.8	0.8
da	Max	3.6	4.7	5.7	6.8	9.2	11.2	14.2	16.2	18.2	20.2	22.4	24.4	26.4
z	Max	0.50	0.70	0.80	1.00	1.25	1.50	1.76	2.00	2.00	2.50	2.50	2.50	3.00

$x = 0.075/\tan \alpha$

* For lengths upto 130 mm. † For lengths over 130 mm and upto 220 mm.

SCREW FASTENING

A screw is driven into a threaded hole in one of the pieces to be joined (a nut is absent in this case). Fig. 11.2 shows four types of screw in use. Dimensions are available in IS. A thread in a screw hole is represented in the same manner as in a stud-joint. Depending on their use screws are manufactured having various styles of heads and points. Principal of the majority screws is in fixing the position of one part with respect to another.

LOCK NUTS

Nuts tend to get loosened under vibration or variable forces. It is therefore necessary in such cases to provide for the securing of nuts in position by some kind of locking device. Fig.11.6 shows various types of locking arrangements.

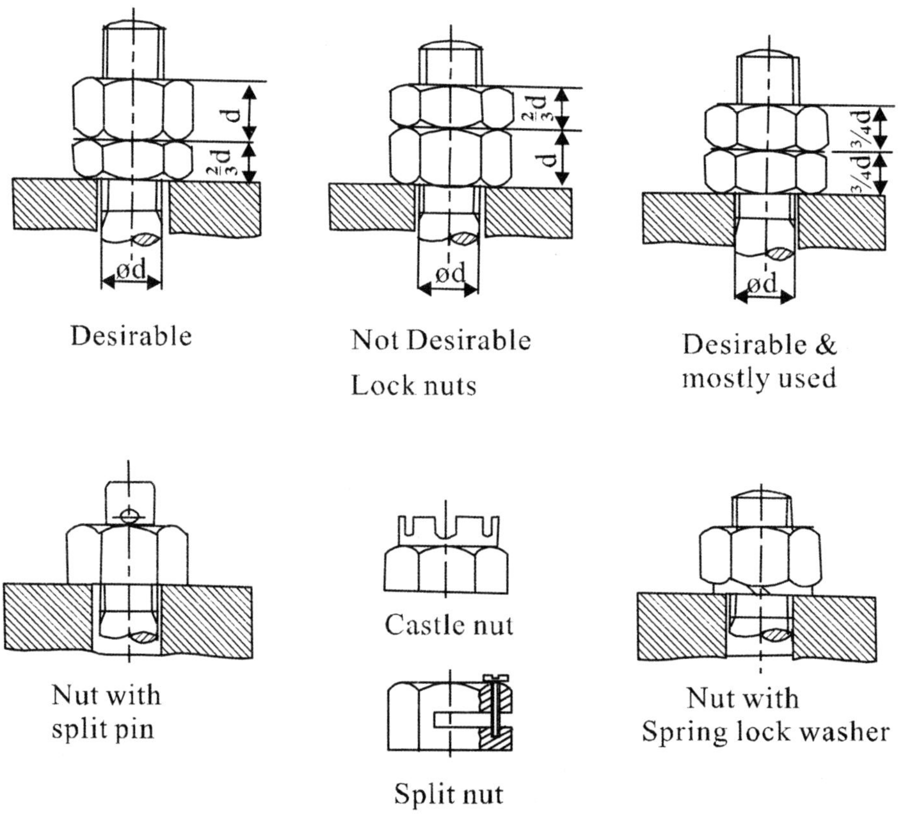

Fig. 11.6 Locking devices

DESIGNATION OF BOLTS, SCREWS & NUTS

A typical example of bolt designation is shown below:

Hexagonal bolt M4 × 12-IS: 2389-P-4D

This means a bolt of nominal diameter 4 mm having a length of 12 mm and made of steel having mechanical properties conforming to symbol 4D. If a bolt is made of a material other than steel, the symbol '4D' shall be replaced by the words 'Brass', 'Aluminium', *etc.*, as applicable.

Symbol 4D indicates that the mechanical (tensile) strength is equal to 37 kgf/mm^2 (minimum).

Yield point = 21 kgf/mm^2 (minimum)

Elongation % = 25 (minimum)

P-refers to precision grades.

Other grades are: T – Turned, B – Black

BLACK SQUARE HEAD BOLT

Square bolt with a square nut – Square bolt M 10 × 30 N-IS: 2585
(thread size M 10, length 30 mm)

Square Bolt only	Square Bolt M 10 × 30 –IS:2585
Square Screw	Square Screw M 10 × 30 –IS:2585
Square Nut	Square Nut M 10 × 30 –IS:2585
Square Bolt with Hex. Nut	Square Bolt M 10 × 30 –HN –IS:2585

PROBLEMS

Relevant tables given earlier in this chapter should be consulted whenever necessary.

11.1 Draw the following thread profiles taking a pitch of 20 mm.

(i) Square thread (ii) Knuckle thread

(iii) Acme thread (iv) Buttress thread

(v) Design profile of a I.S.O. metric screw thread.

(vi) Cylindrical pipe thread.

11.2 Draw three views of a hexagonal bolt M 20 × 75 – IS: 2389-B-4-D fitted with an appropriate nut. (*Hint:* see the conventional ratios of metric bolt - vide Fig. 11.3).

11.3 Fit a square bolt with Hex. Nut, M 28 × 88 HN to complete the pin joint shown in Prob. 11.3 and complete the views.

11.4 Fit tap bolts in the flange-joint as shown in Prob. 11.4. Draw the two views.

11.5 Two FP 1 pipes are to be joined to the elbow as shown in Prob.11.5. Draw the assembled view.

①

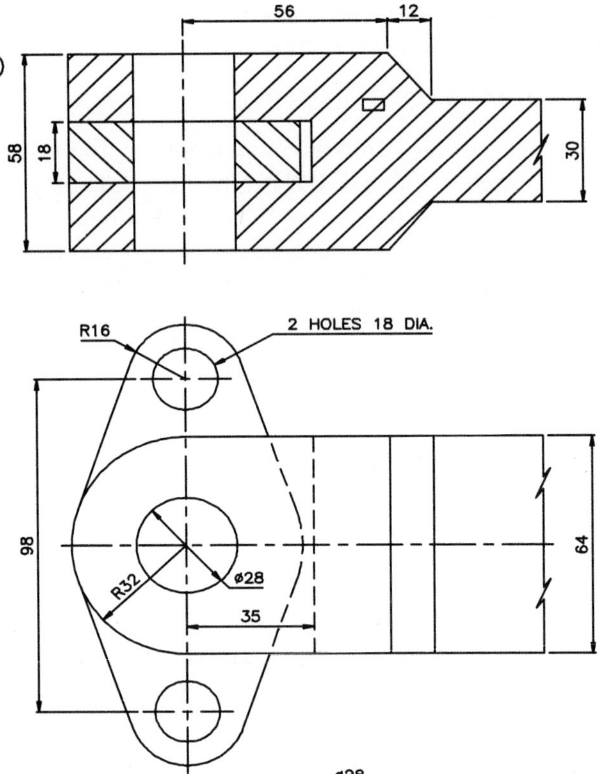

②

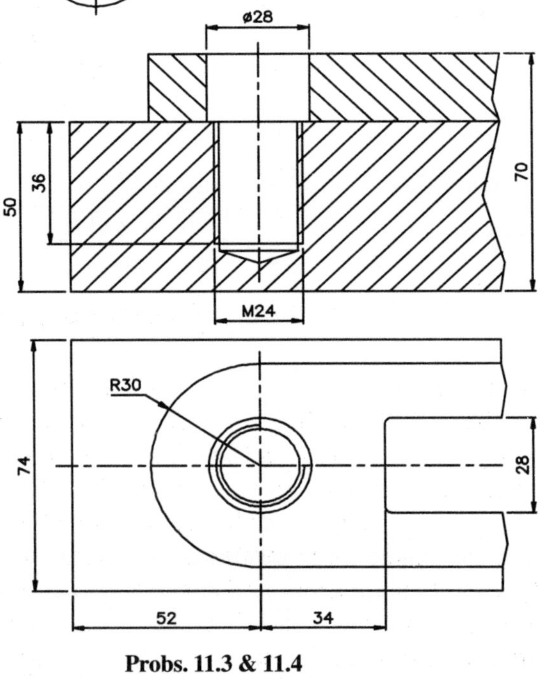

Probs. 11.3 & 11.4

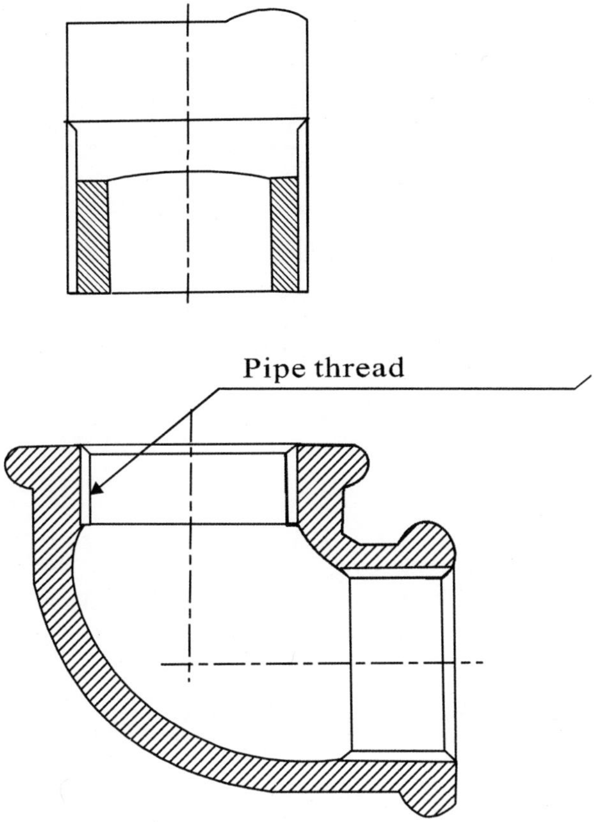

Prob. 11.5

CHAPTER – 12

MISCELLANEOUS MACHINE ELEMENTS

KEYED JOINT

The major function of keys is to prevent relative rotation between the members connected by keys and keyways e.g. shaft and pulley, gears, couplings, chain sprockets, crank, crank-shaft etc. Keys, in addition, prevent relative axial movement. Keyed joints are simple in design and easy to assemble or dismantle.

Key: A key is usually a rectangular or slightly tapered piece – square, rectangular or round in cross section – made of steel (C 40, C 50) in general. Rotary motion is transmitted by means of the working faces of the key which mate with the keyway on the components. A small radial clearance between the key and keyway in the hubs is provided. Fig. 12.1 shows different forms of keys in position.

(a) **A Saddle key** is used for light loads. When this is fitted on a flat surface made on the shaft, it is called a flat key.

(b) **A Sunk key** is used for large power transmission. A sunk key may be square (d/4×d/4) or rectangular in cross section, where d is the diameter of the shaft.

(c) **A Woodruff key** is used for transmitting low power, e.g., automobile, tractors, machine tools.

(d) **A Tangential key or Kennedy** key is used for high power transmission as in rolling mills. To lock shafts subjected to reversal of direction two keys are used in two locations.

(e) **A Gib-head key** is an ordinary regular key with gib at one end. It can be easily withdrawn by putting a wedge between the gib and the hub. This feature is very important if the small end of the key is not accessible.

(f) **A Round key or pin** (not shown) is usually tapered, but may be cylindrical. This is used in low power drives and also for locking of drilling bushes from rotation.

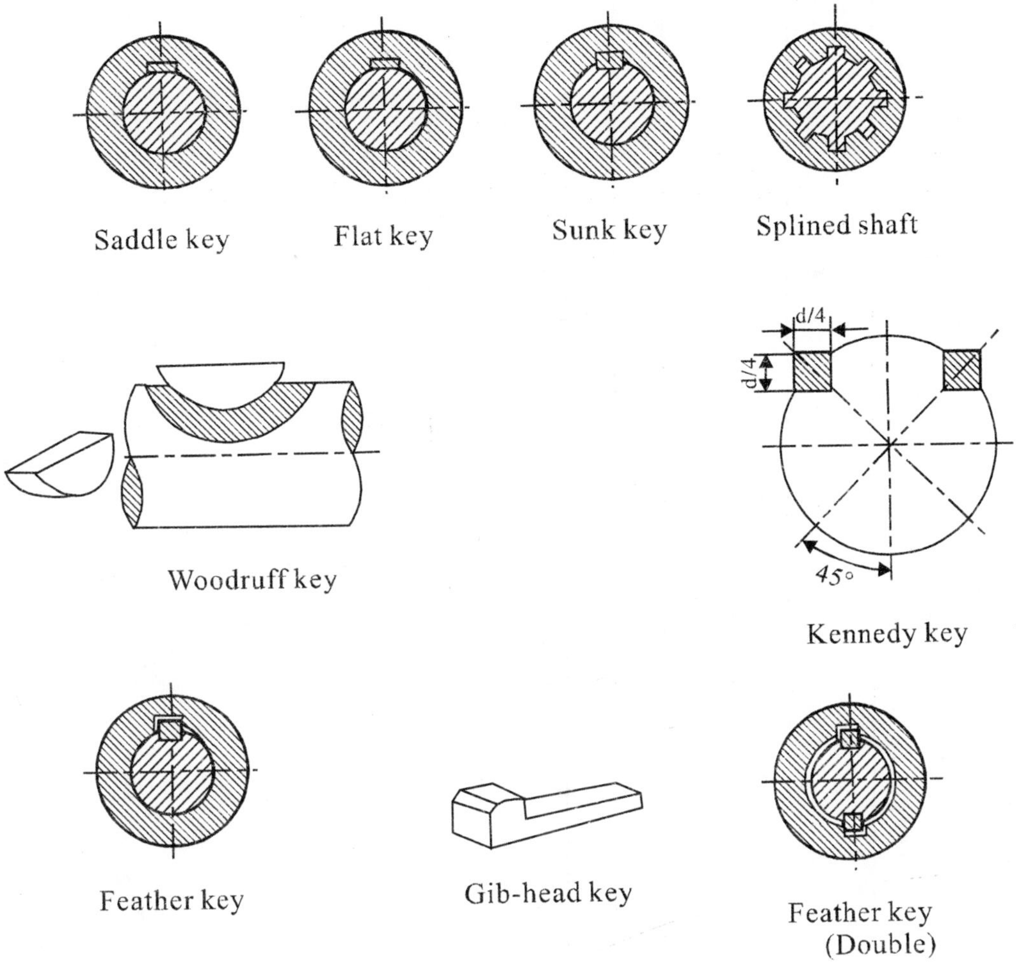

Fig. 12.1

(g) **A Feather key** permits the hub to move axially on the shaft while still rotating with it, (Fig. 12.1). Small radial and tangential clearances are provided in the hub. Two keys may also be used,

Splines are multiple feather keys with the keys machined integral with the shaft. They are particularly useful where radial space has to be conserved. They are used in machine tool gear boxes, automobile speed gear box etc. to permit hubs to move axially on the shaft while still rotating with it.

Various types of keys and keyways have Indian standards. **Representative Tables for "Dimensions of parallel keys and keyways"** and **"Gib head keys and keyways"** are given for ready reference.

Table 12.1: Dimensions of parallel keys and keyways
(All dimensions in mm)

For shaft	above Upto	6	8	10	12	17	22	30	38	44	50	58	65	75	85	95
diameters		8	10	12	17	22	30	38	44	50	58	65	75	85	95	110
Key cross	Width	2	3	4	5	6	8	10	12	14	16	18	20	22	25	28
Section	Height	2	3	4	5	6	7	8	8	9	10	11	12	14	14	16
Keyway	in Shaft t_1	1.2	1.8	2.5	3	3.5	4	5	5	5.5	6	7	7.5	8.5	9	10
Depth	in Hub t_2	1	1.4	1.8	2.3	2.8	3.3	3.3	3.3	3.8	4.4	4.9	5.9	5.4	6.4	7.4
(nominal)																
Tolerance on	f_1			+ 0.1						+ 0.2						
Keyway	depth t_2						+ 0.1					+ 0.2				
Chamfer or	Max			0.25			0.35		0.55	0.80			1.30		2.00	2.95
radius of key	Min			0.16			2.25		0.40	0.60			1.00		1.60	2.50
Keyway radius	r_2 Max			0.16			0.25		0.40	0.60			1.00		1.60	2.50
Length of	L Min	6	6	8	10	14	18	22	28	36	45	50	56	63	70	80
key	Max	20	35	45	56	70	90	110	140	160	180	200	220	250	280	320

Designation: A parallel key of width 10 mm. height 8 mm and shall be designated as:

Parallel key 10 × 8 × 50 IS : 2084-1962

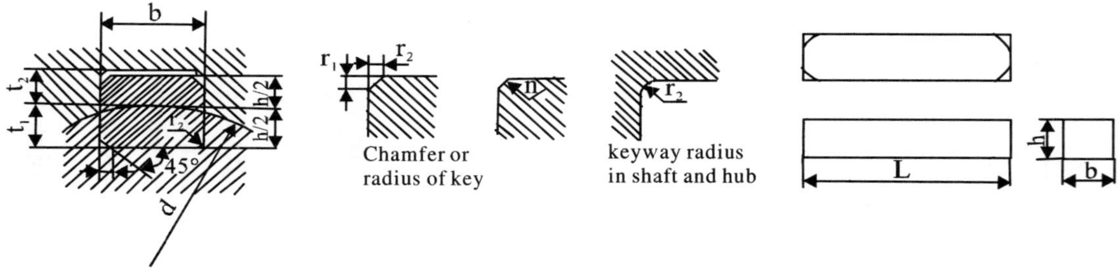

Table 12.2: Shaft diameter keys and keyways

SHAFT DIAMETER, d		KEY					KEYWAY IN SHAFT AND HUB					
Above	Upto & including	Width b (h9)	Height h (Normal)	Tol. on h	Height of Gibhead h	Chamfer or Radius r (Min)	Width of in keyway t_1 (D10)	Depth in shaft	Tol. on t	Depth in Hub t_2	Tol. on t_1	Radius at bottom of Keyway t_2 (max.)
10	12	4	4		7	0.16	4	2.5		1.2		0.16
12	17	5	5		8	—	5	3		1.7		
17	22	6	6	+0.1	10	0.25	6	3.5		2.1		0.25
22	30	8	7		11		8	4	+0	2.5	+0.1	
30	38	10	8	—	12	—	10	5		2.5		—
38	44	12	8	—	12	—	10	5		2.5		—
44	50	14	9		14	0.40	14	5.5		2.9		0.40
50	58	16	10		16		16	6	—	3.4	—	
58	65	18	11		18		18	7		3.6		
65	75	20	12	+0.2	20	—	20	7.5		3.8		—
75	85	22	14		22		22	8.5		3.8		
85	95	25	14		22	0.60	25	9		4.3		0.60
95	110	28	16		25		28	10		5.3	+0.15	
110	130	32	18		28		32	11		6.2		—
130	150	36	20	—	32	—	36	12		7.2		
150	170	40	22		36		40	13	+0.15	8.2		1.00
170	200	45	25		40	1.00	45	15		9.2		
200	230	50	28		45		50	17		10.1	-	
230	260	56	32		50	—	56	19		11.1		
260	290	63	32	+0.3	56	1.60	63	20		12.1		1.60
290	330	70	36		63		70	22		13.1	+0.3	
330	380	80	40		70	—	80	25		14.1		—
380	440	90	45		75	2.50	90	28		16.1		2.50
440	500	100	50		80		100	31		18.1		

PIPE JOINTS

Pipe and/or tubes are generally made of cast iron, steel, brass, copper, reinforced concrete, Plastic, fibre reinforced plastics for conveying fluid and/or fluid-solid mix in case of pneumatic and hydraulic conveying.

Pipe joints may be of various types as discussed below:

(a) *Threaded Joints with or without Socket /Ferrule* (Fig. 12.2)

These are most widely used forms of connection. It is important that the threads are thoroughly cleaned and that both the pipes and the fitting are lubricated before assembling. Properly made, these joints are economical and satisfactory for water as well as for steam and air.

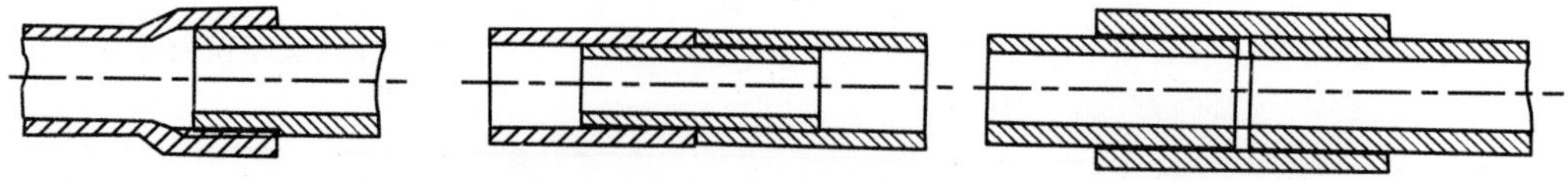

Fig. 12.2

(b) *Flanged Joint* (Fig. 12.3)

This is used for connection of two co-axial pipes. The face of each flange is machined square to the axis of the pipe. Packing in the form of thin disc is placed between the flanges to make the joint leak-proof.

(c) *Spigot and Socket Joint* (Fig 12.3)

This joint is used where slight flexibility is desired as in the case of underground gas and water mains and sewer pipes. The annular space between the spigot and the socket is wound with yarn chord and a ring of lead is cast in the mount of the socket, which is retained in position by the groove.

(d) *Expansion Joints*

These joints are incorporated in pipelines to allow for a considerable elongation or contraction of the rigidly fastened pipes due to temperature variation.

(i) U-bend type

(ii) Circle bend type

(iii) Gland type (Fig.12.4)

In the simple type of expansion joint suitable for steam pressure joint, the sleeve slides in a socket and the leakage is prevented by an asbestos filled stuffing box.

(e) *Brazed Joints*

(f) *Welded Joints*

CAST IRON PIPES AND JOINTS

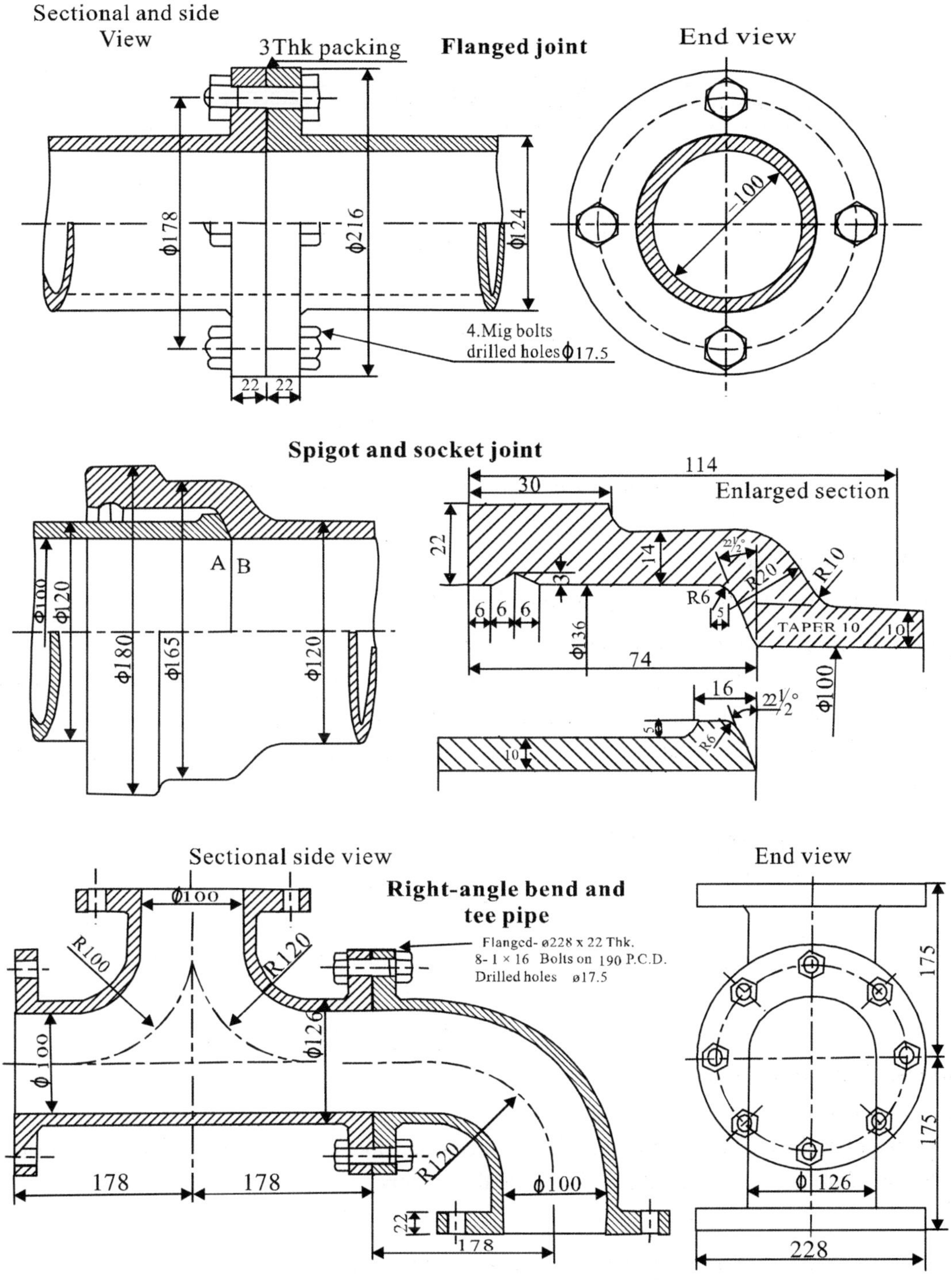

Fig. 12.3

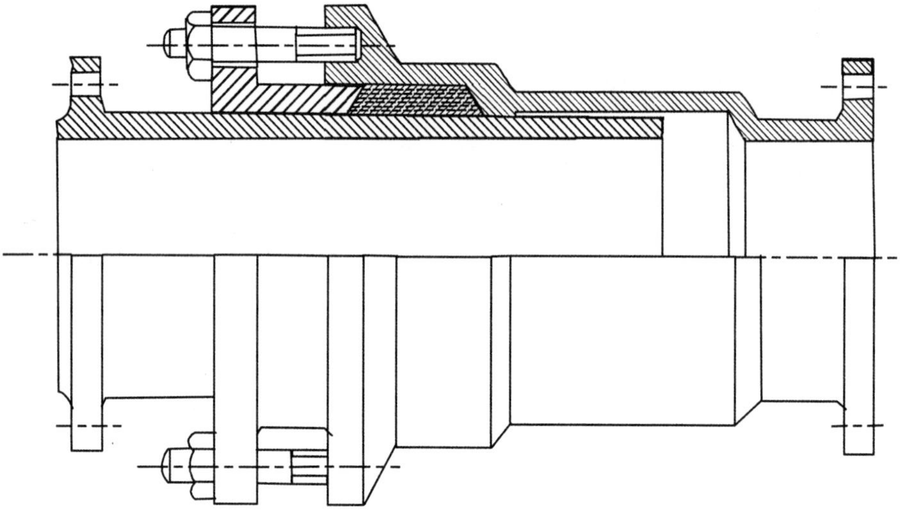

Fig. 12.4 Gland type expansion joint

JOINTS FOR RODS

Various joints are shown in Figs. 12.5 and 12.6.

Adjustable Joint with Turn Buckle

The facing ends of rods are threaded, one left handed, the other right handed. The turn buckle with matching internal threads at the ends is rotated to adjust the gap between the rods. If the rods to be joined are other than circular in cross-section, the facing ends are forged to round shape (Fig. 12.5).

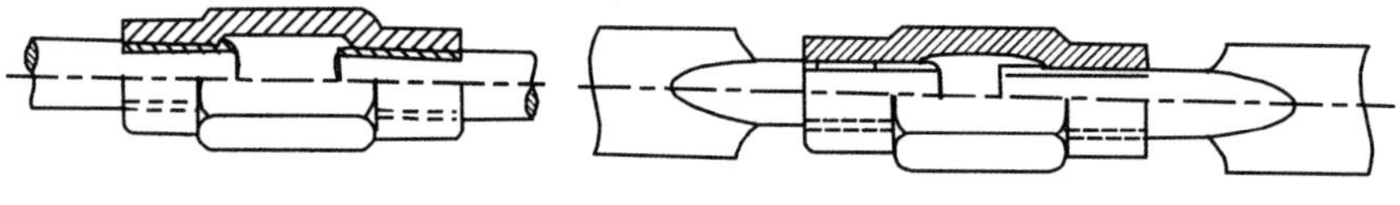

Adjustable joint for rods　　　　Adjustable joint for flat rods

Fig. 12.5

Cottered Joint (for round rods)

The end of two rods are pushed inside steel sleeve through which the cotters are inserted through the rectangular slots. The cotters compress the rods and the sleeve on opposite sides. Clearance as shown must be left to facilitate 'draw' of the cotters.

Use:

Piston rod and cross head assembly, valve rods, fixtures *etc.*

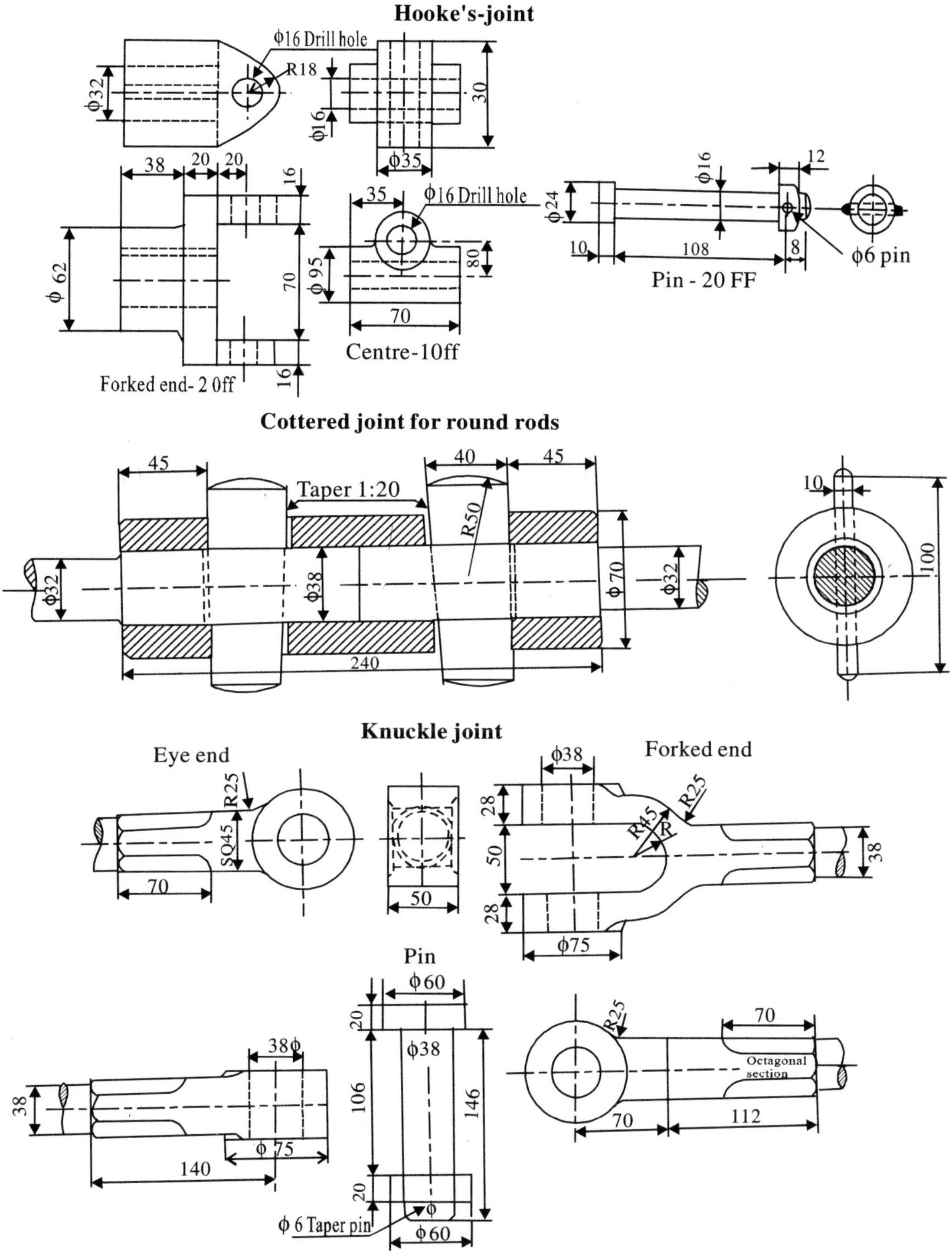
Hooke's-joint
ϕ16 Drill hole
R18
ϕ32
ϕ16
30
ϕ35
38
20
20
16
ϕ 62
70
16
Forked end- 2 0ff
35
ϕ16 Drill hole
ϕ95
80
70
Centre-10ff
ϕ24
ϕ16
12
10
108
8
ϕ6 pin
Pin - 20 FF
Cottered joint for round rods
45
40
45
Taper 1:20
R50
ϕ32
ϕ38
ϕ 70
ϕ32
240
10
100
Knuckle joint
Eye end
R25
SQ45
70
50
ϕ38
Forked end
R25
R45
R
28
50
28
38
ϕ75
Pin
ϕ60
20
ϕ38
106
146
20
ϕ 6 Taper pin
ϕ60
38ϕ
38
ϕ 75
140
R25
70
Octagonal section
70
112

Fig. 12.6

Knuckle Joint

These joints are used to connect two or more rods together, where relative angular movement is desired. It consists of one eye end and a forked end through which a pin is passed and held in position using split or taper pins as shown in figure 12.6.

Use:

Connecting rod joined with cross head, valve rods, tie rods of roof truss, minor joints in chains, window stays and in numerous other applications even in spectacles.

Hooke's Joint

Hooke's Joint or Universal Coupling is used to couple shafts whose axes are at an axial misalignment of upto 30°, but it is not suitable for transmission of heavy torque. In very common form of Hooke's joint forks are made at the ends of the shafts at right angles to each other.

PULLEYS

Pulleys are normally required for transmission of power. Most common material used is cast iron. For high speeds, fabricated mild steel pulleys are used.

A cast iron pulley as per BIS:1691-1968 is shown in Fig. 12.7. The width of the pulley will depend on the width of the belt. According to BIS, the pulley widths are given below in Table 12.3.

Table No. 12.3. Pulley width [BIS 2122 (Pt I)-1973]

Belt width, mm	Pulley to be wider than belt width by mm
Upto and including 125	13
Above 125 to 250	25
Above 250 to 375	38
Above 375 to 500	50

The face of a pulley rim is generally crowned to keep the belt on the pulley, since inaccurate alignment of the pulleys causes the belt to run off side ways. One of the pulleys in a drive should be crowned.

Crown height, h = about 1.5 cm per meter of face width. (Otherwise to obtain crown height 'h' draw a curve with the center on the vertical axis of pulley and radius 3 to 5 times the face width 'a').

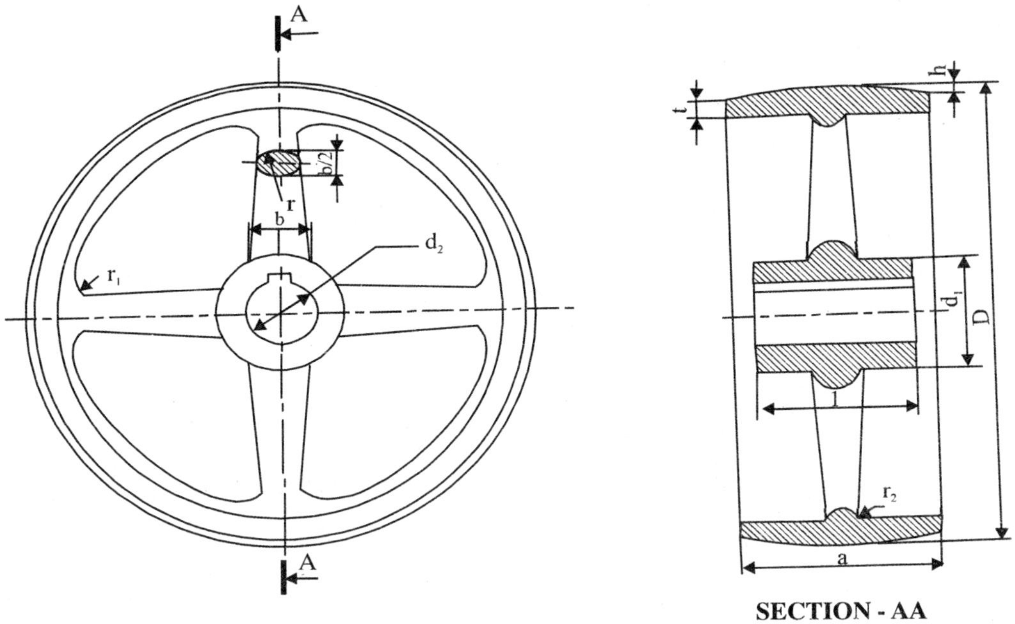

Fig. 12.7

(i) Rim thickness of CI pulley:

$$t = (D/200) + 3 \text{ mm for single belt.}$$
$$t = (D/200) + 6 \text{ mm for double belt.}$$

(ii) $(d_1 - d_2)/2 = 0.412\ (aD)^{1/3} + 6\text{mm}$ for single belt and
$= 0.529\ (a)^{1/3} + 6\text{mm}$ for double belt.

(iii) Thickness of arm, b, near boss $= 2.94\ (aD/4n)^{1/3}$ for single belt, and
$= 2.94\ (aD/2n)^{1/3}$ for double belt.

(iv) Thickness of arm, b, near rim $= (2/3)b$.

(v) No of arms, n = 4 for D varying from 200 mm to 450 mm,
and = 6 for D above 450 mm.
(For D upto 200 mm, use web)

(vi) Minimum length of bore, l = (2/3) a (it may be more for loose pulley but in no case it exceeds a).

(vii) Radius of arm section, r = (3/4)b.

(viii) Radius near rim, r_1 and r_2 = b/2.

Step Pulley

Figure 12.8 shows a cast iron step pulley which is used for speed variation in various machine tools or in countershafts or other machineries.

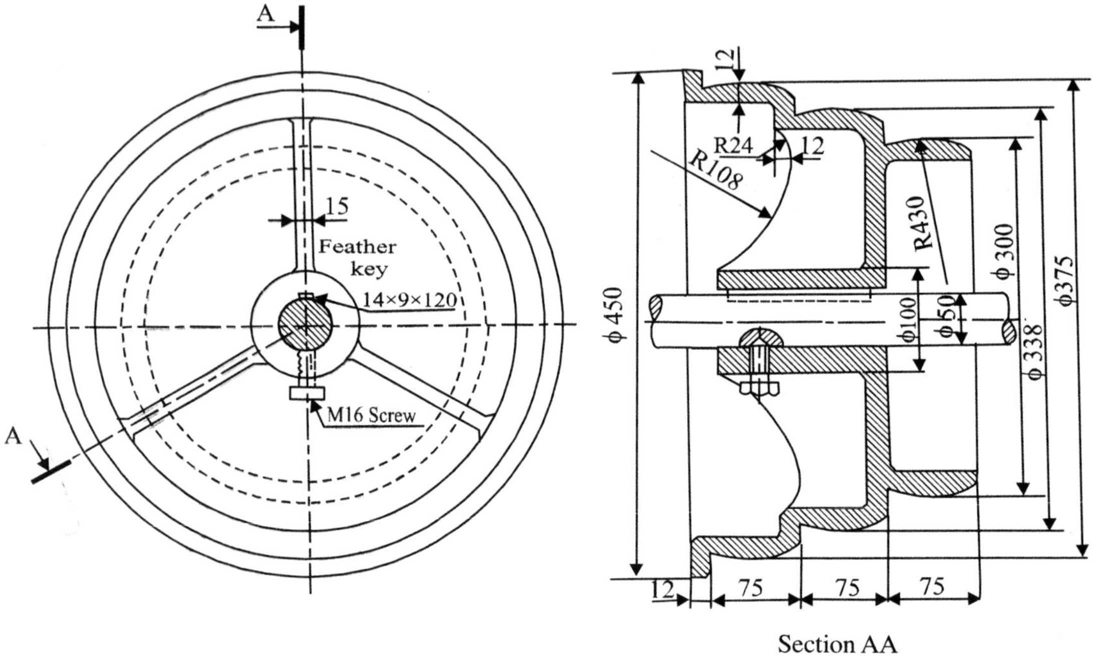

Fig. 12.8 Step pulley

SHAFT COUPLING

Shaft couplings are used to connect one shaft to another and may be of various types e.g., Rigid-Coupling (Sleeve Coupling, Clamp Coupling, Flange Coupling), Flexible Coupling (Rubber Bush Type, Oldham's Coupling) and Slip Coupling (this is used to permit relative rotation or slip between the connected shafts).

Figure 12.9 shows a Rigid Flange Coupling. Standard proportions are given below:

Shaft diameter = d

Boss diameter, D_1 = 1.8d to 2d.

Bolt circle diameter, D_2 = 2.6d to 3d.

Bolt diameter = $[0.423d / \sqrt{n} + 8]$mm

Total length of coupling, L = 2l = 2.5d to 5d

Overall diameter, D = 3d to 5.8d

Fitting boss or spigot or register,

Diameter, D_f = d + d/4

Fitting boss height, h = 4 to 10 mm

Clearance, t_c = 2 mm

Flange thickness, $(I - t_1)$ = 0.5 d

Thickness of protective circumferential flange, t_2 = 0.25 d

No. of bolts, n = 3 for shaft diameters upto 40 mm.

= 4 for shaft diameters above 40 mm upto 100 mm.

= 6 for shaft diameters above 100 mm upto 180 mm.

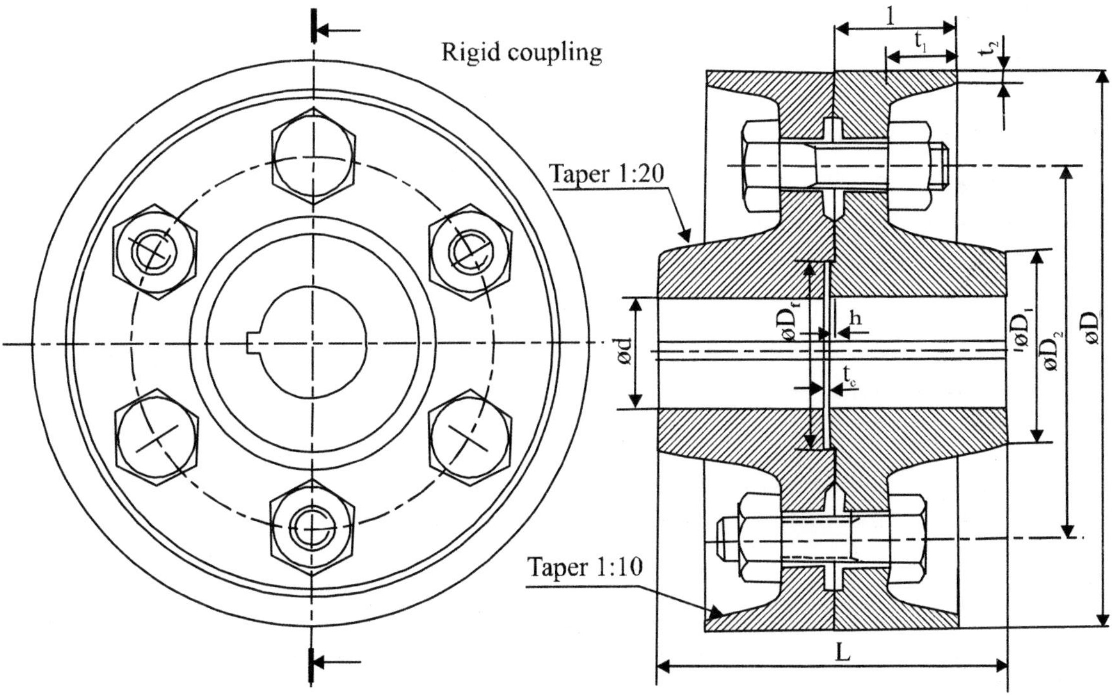

Fig. 12.9 Shaft coupling

PROBLEMS

12.1 Draw different forms of keys in position as shown in Fig. 12.1. (It is desirable to take some standard dimensions from IS.)

12.2 Draw a threaded joint (a) with a socket, (b) without a socket as per Fig. 12.2 for FP 2.

12.3 Draw the details of the flanged pipe joint shown in Fig. 12.3.

12.4 Draw the assembled view in half section showing the leak-proofing arrangement of the spigot joint shown in Fig. 12.3. Also draw the details of the joint.

12.5 Draw full sectional front view, top view and left hand side view of the right angle bend and tee pipe shown in Fig. 12.3.

12.6 Hand out the sketch of the details of Cotter Joint shown in Fig. 12.6.

12.7 Draw the three principal views (with local section, if necessary) of the Knuckle Joint Assembly, the details being given in Fig. 12.6.

12.8 Draw the front view and top view of the assembled Hooke's Joint (Fig. 12.6) showing the forked ends with shafts. (Square key, having a dimension of $d/4 \times d/4$, where d = shaft diameter, may be used.)

12.9 Draw two views of the (i) Adjustable joint for Round Rods and (ii) Adjustable Joint for Flat Rods shown in Fig. 12.5. The threaded portion of the rod is 24 mm in diameter. All other dimensions are to be taken proportionately from the figure.

12.10 Draw two principal views of a 450 mm diameter flat pulley which accommodates a belt 200 mm wide. The pulley will be fitted on a shaft 30 mm in diameter (vide Fig. 12.7).

12.11 Fig. 12.8 shows a cast iron stepped pulley fitted to a diameter 50 mm shaft with a key and set screw. Hand out two views of (a) the pulley, (b) the key and (c) the set screw.

12.12 Fig. 12.9 shows a rigid shaft coupling. Draw appropriate views of the flanges and the bolt with nut for a 75 mm diameter shaft coupling.

12.13 Prob. 12.13 shows a ribbed flange on a ϕ54 mm splined spindle having 8 equispaced splines of 9 mm width and 4 mm depth. Sketch out the views of the parts separately indicating details of the spline.

12.14 Hand out a proportionate sketch of the expansion joint as shown in Fig. 12.4 considering 55 mm dia. pipe. Take other dimensions proportionately from the figure.

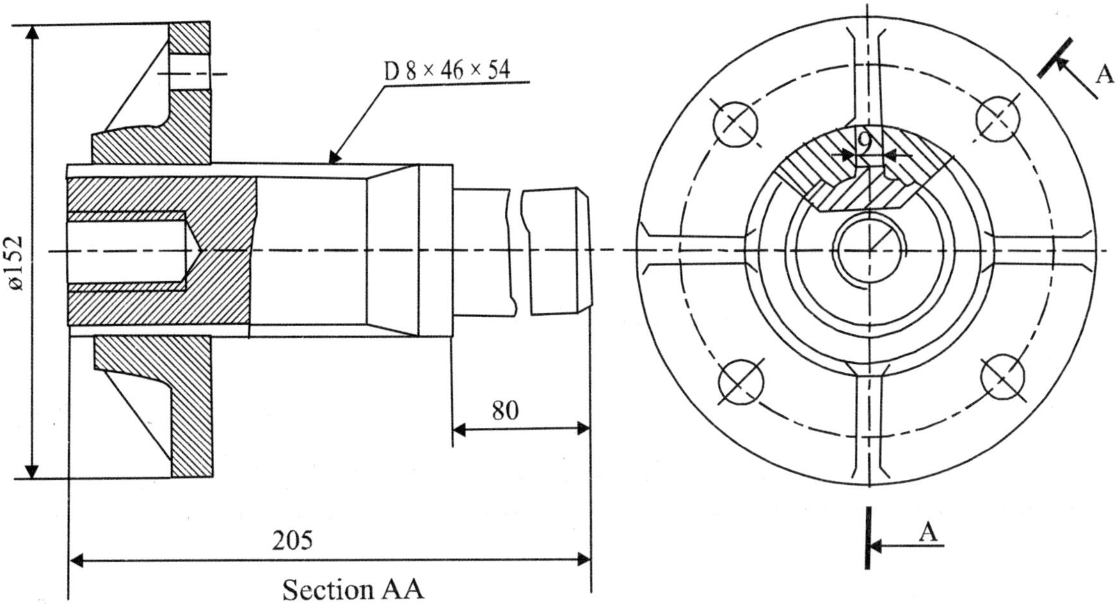

Section AA

8 nos., 8mm depth, 9mm width splin on ϕ 54 diameter

Prob. 12.13

CHAPTER – 13

DEVELOPMENT AND INTERSECTIONS OF SURFACE

Many objects of engineering practice for everyday use are made of cardboard, plastic and other metallic materials, which are thin, and normally not more than 3 mm thick referred to as sheet. Objects made from these sheets are obtained by cutting out the patterns drawn on the sheets. The sheet is folded, rolled or bent to get the desired shape. There are four basic shapes from which various types of work like bins, chutes, ducts for ventilation, hoppers, vessels, pipe bends *etc.* are designed. The basic shapes are :

(1) Prism (2) Pyramid (3) Cylinder (4) Cone

A layout of the complete surface of an object is called the **development**. The development of an object bounded by plane surfaces may be obtained by unfolding the object as shown in Fig. 13.1 so as to place the imaginary surface upon a plane. The surface is said to be developed when the successive surfaces are drawn in their true sizes with their common edges joined. The basic shapes stated above can be developed by three commonly used methods, namely:

1. **Parallel line development:** for prism and cylinders (Figs.13.1 & 13.3)

2. **Radial line development:** for pyramids and cones (Figs.13.2 & 13.4)

3. **Triangulation:** for transition pieces which are used for joining various shapes.

It must be remembered that a pattern (development) is folded to get the desired component (chutes, bins *etc.*) and in practice allowance must be provided for extra material required for joints, bends *etc.*.

1. Prism

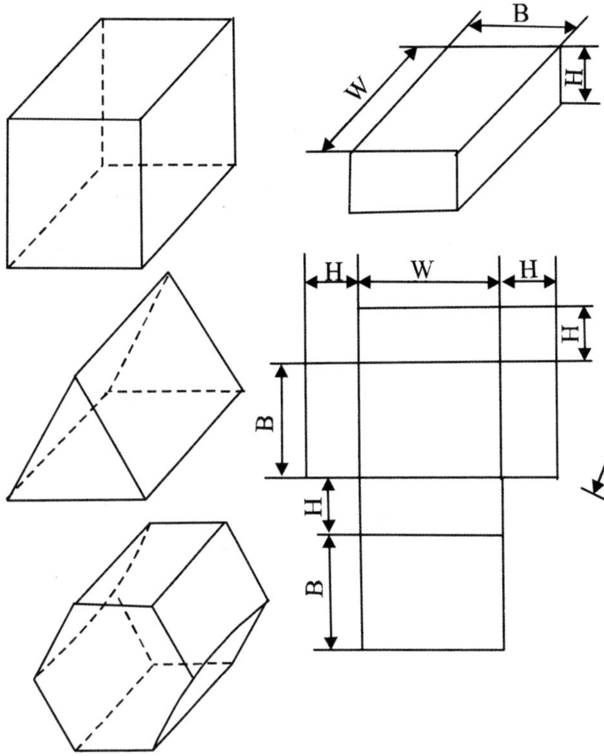

2. Pyramid

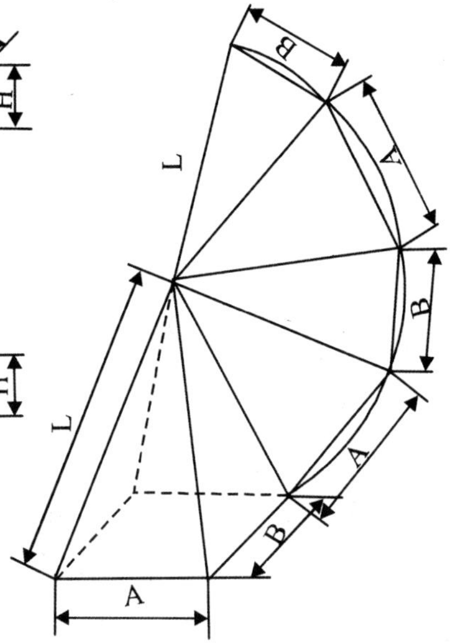

3. Cylinder

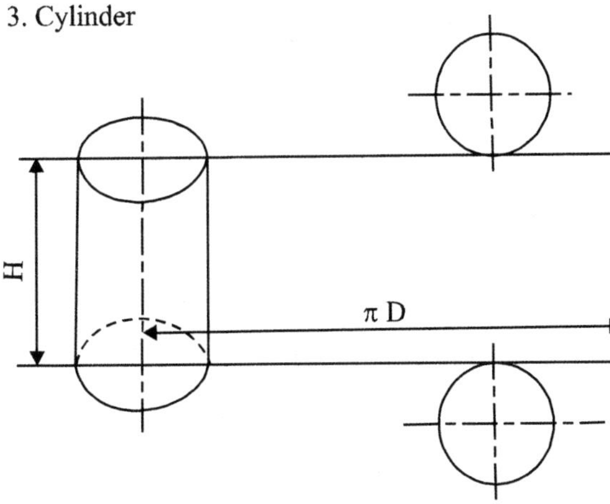

4. Cone

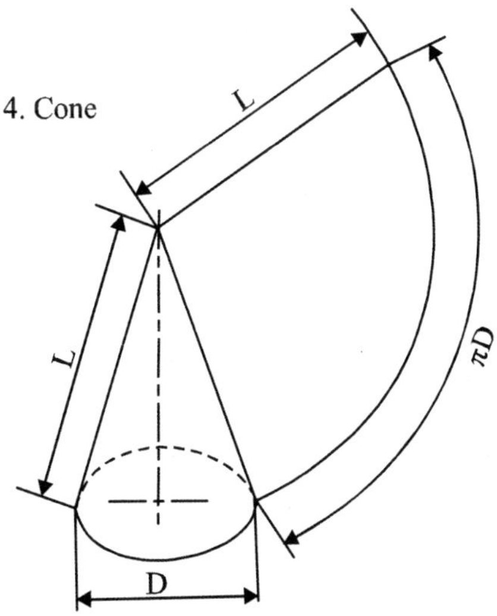

L = Slant height

Fig. 13.1 – 13.4

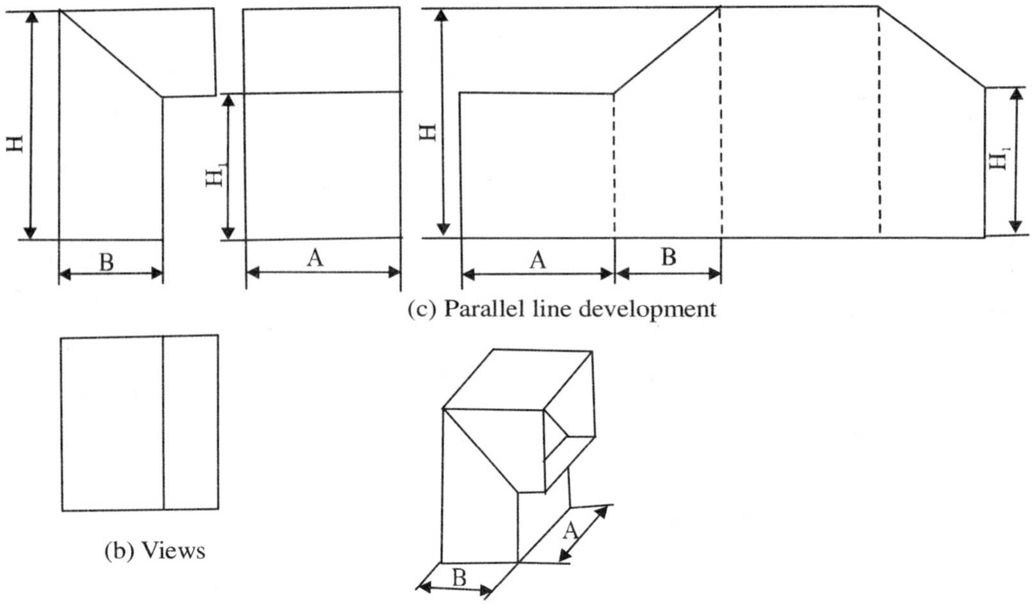

(c) Parallel line development

(b) Views

(a) Object

Figs. 13.5 (a), (b), (c) Elbow, rectangular section

TRUNCATED CYLINDER

A truncated cylinder is one in which all the lateral elements are intersected by a cutting plane making any angle other than 90° with its axis. In Fig. 13.6 the plan view of the cylinder is divided into 12 or any suitable number of equal parts and the points are projected as elements on the lateral surfaces seen in the front view. Then draw a line (known as stretch out line) equal to the circumference of the cylinder, πD and divide it into the same number of equal parts as the plan, here 12. The lengths of the successive lateral elements are measures or parts as the plan, here 12. The length of the successive lateral elements are measured or projected from the front view by drawing lines parallel to the stretch out line from the points marked (*e.g.* 1′, 2′, 3′ *etc.*). Erect perpendiculars from the stretch out line to intersect the corresponding projected lines at the required points on the profile to be developed. The auxiliary true shape projections may be attached to the developed lateral surface. It may be noted that the elliptical shape is obtained by taking an auxiliary view in a plane parallel to plane AA. Here the distances 6 to 8, 5 to 9, 4 to 10 *etc.* on the auxiliary view are measured equal to the distances 6 to 8, 5 to 9, 4 to 10 *etc.* on the plan.

Development of Elbows

The principle discussed in connection with the development of the truncated cylinder may be used to develop the five-piece elbow shown in Fig. 13.7 as elbows may be split into a number of truncated cylinders. In order to form the elbow from sheet metal the surface must be developable, *i.e.* they must flat on a plane surface. It is not possible to make a single developed surface for the elbow as a whole; the elbow must be divided into a convenient number of developable segments which may finally be joined by suitable means like welding. From Fig.13.7 it is clear how the development is achieved. It is to be noted that two pieces at the ends are similar as are the three intermediate pieces.

To develop the segments draw the inner radius r_1 and the outer radius r_2. Divide the base into 12 equal parts. Project these points as elements on the lateral surface of the five radially proportioned pieces of the elbow. Follow the procedure previously discussed in case of a truncated cylinder. To save materials the pattern for the elements may be laid as shown in Fig. 13.7 on top right position.

RADIAL LINE DEVELOPMENT

Development of a Truncated Pyramid

Development of a pyramid has been illustrated earlier and in fact it consists of drawing the true surface of each side and attaching them together at their edges which intersect at the apex.

A truncated pyramid is developed in the same manner, except that the true lengths the edges of the slant surfaces are obtained by revolving them parallel to the ground line in the plan view and projecting the revolved positions in the view. In the front view (vide Fig.13.8) the edge 0-1 is placed parallel to the vertical plane, therefore it represents the true length; with this length ($\neq$ H) as radius, draw an arc with center O. Mark of chords 1-2, 2-3 *etc.*, (equal B) on this arc which are equal to the true base size of the pyramid. Then draw the slant elements meeting at the apex 0 (*i.e.*, join 01, 02, 03 *etc.* on the development drawing just developed). Now set of the true lengths of the sides 11′, 22′, 33′, *etc.*, (figure on the right) equal to 11′, 12′, 13′ *etc.* (elevation of the truncated pyramid) as obtained by revolving the lines in the plan view and projecting to the front view. Join the points 1′, 2′, 3′ *etc.*, to get the development.

Development of the Lateral Surface of a Truncated Right Cone

Development of a right cone has been shown earlier. Now the development of a truncated cone can be obtained by following the method similar to that described for the truncated pyramid. First of all draw the two views as shown, then with the slant height 0-1 ($\neq$ H) as radius draw an arc of length equal to the circumference of the base of the cone. Take some suitable number of equally spaced elements, say 12, on the surface of the cone in both the views and mark them as shown in Fig. 13.9. Find the true lengths of the elements by revolving them parallel to 0-1 in plan view, when their projections will represent the true lengths along the extreme elements 0-1. Note that A, OB_1, OC_1 *etc.* are the true lengths of the elements OA, OB, OC *etc.* respectively, as shown in the plan view where these were foreshortened. Cut off these true lengths along the elements of the developed surface from the center of the arc to A, B_1, C_1, D_1 *etc.* Join the points by a smooth curve. The development of the inclined face can be obtained by taking an auxiliary view in a plane parallel to the XX – plane as shown.

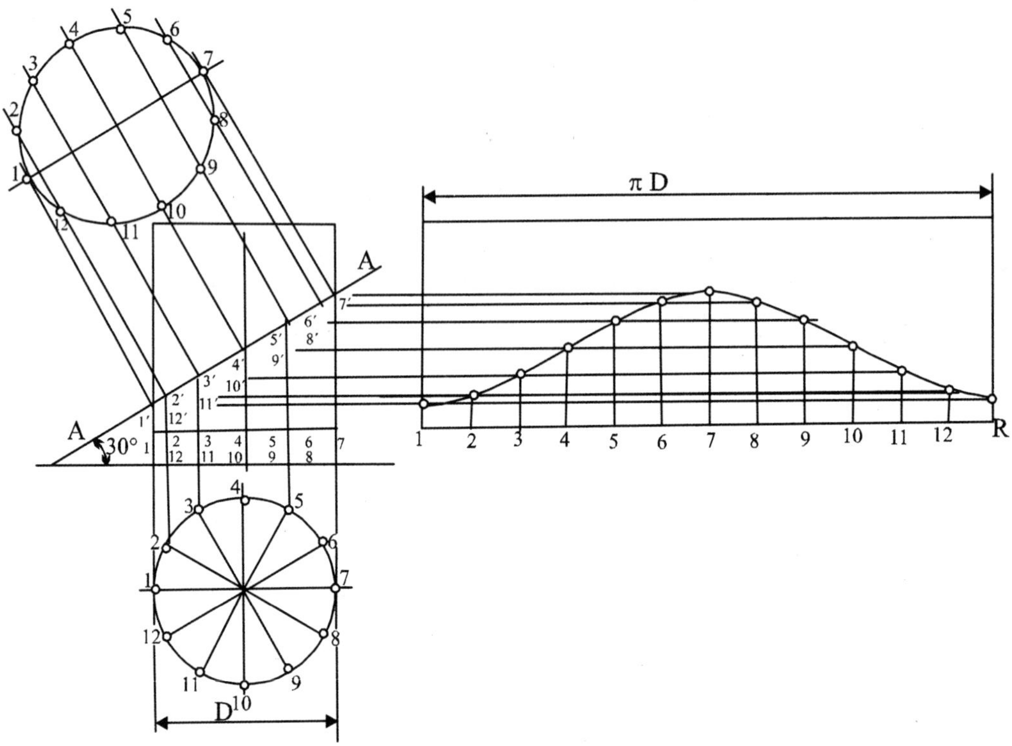

Fig. 13.6 Truncated cylinder

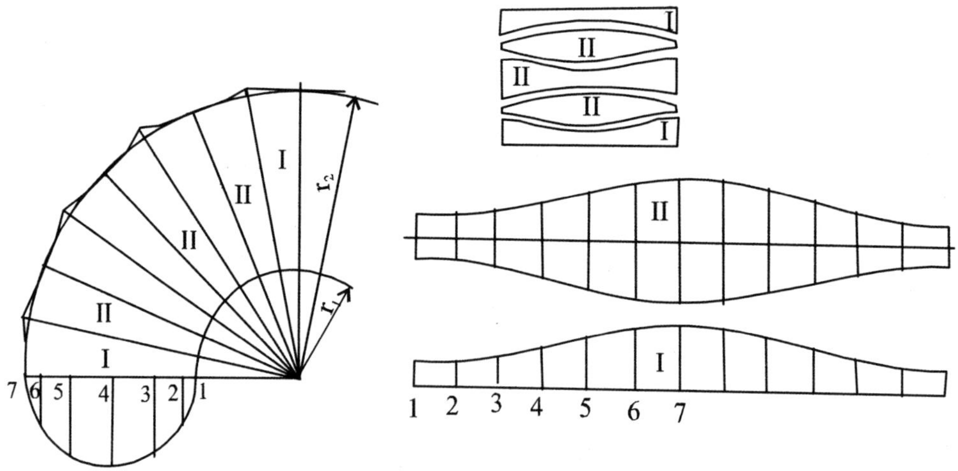

Fig. 13.7 Development of 5-piece elbow

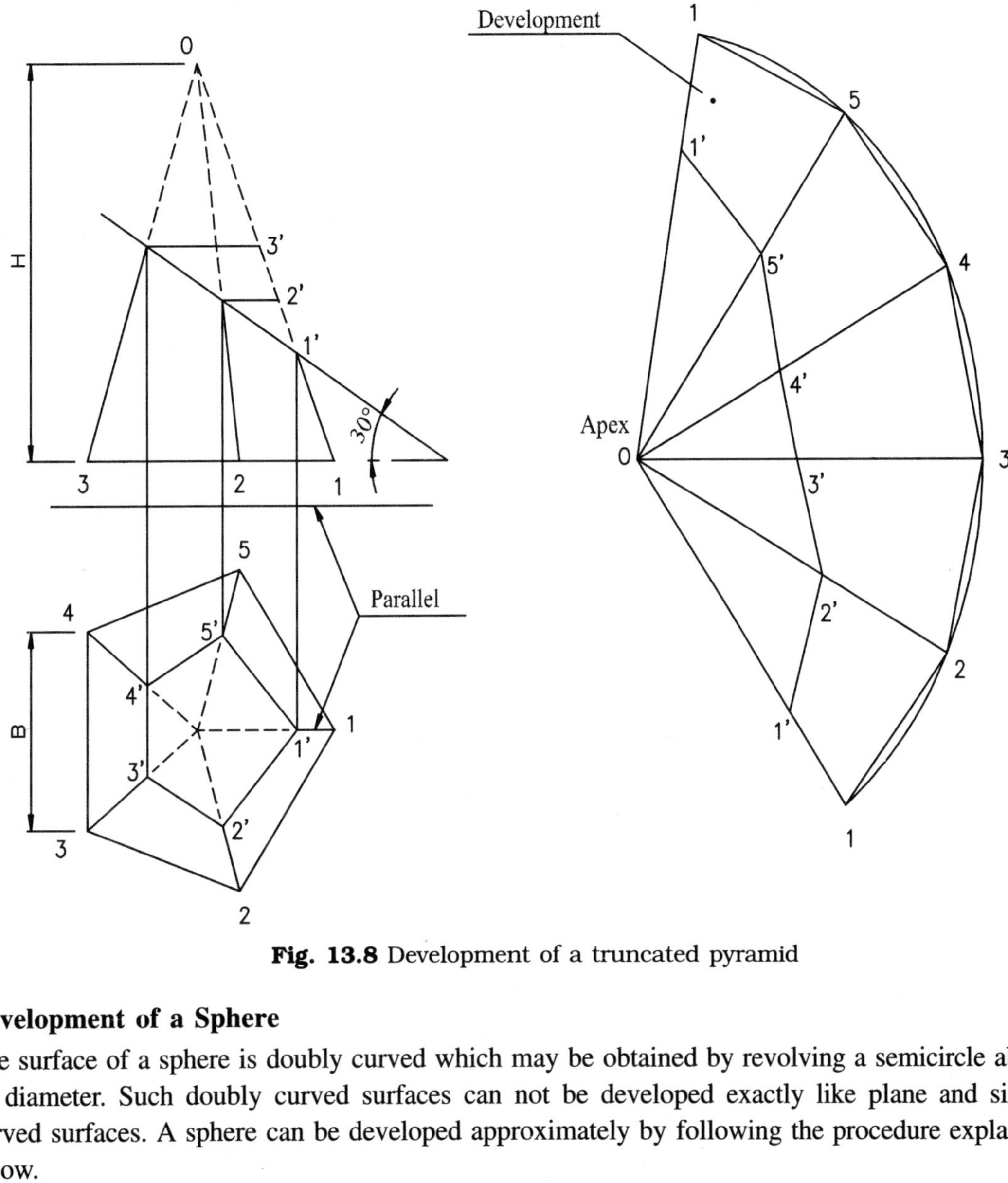

Fig. 13.8 Development of a truncated pyramid

Development of a Sphere

The surface of a sphere is doubly curved which may be obtained by revolving a semicircle about its diameter. Such doubly curved surfaces can not be developed exactly like plane and single curved surfaces. A sphere can be developed approximately by following the procedure explained below.

The surface of the sphere is sectioned into a number of parts by passing imaginary planes as shown in Fig. 13.10 which convert the sphere into several approximate frustums of cone *e.g.* 1, 2, 3, 4. The chords a b, c d of the sphere are the slant heights of the frustums 1, 2, 3 & 4 respectively. The cones are joined to their apices A, B, C and D respectively. Now each of these frustums of a cone can be isolated and developed by the radial line development method discussed earlier. Half the development of each frustum is shown.

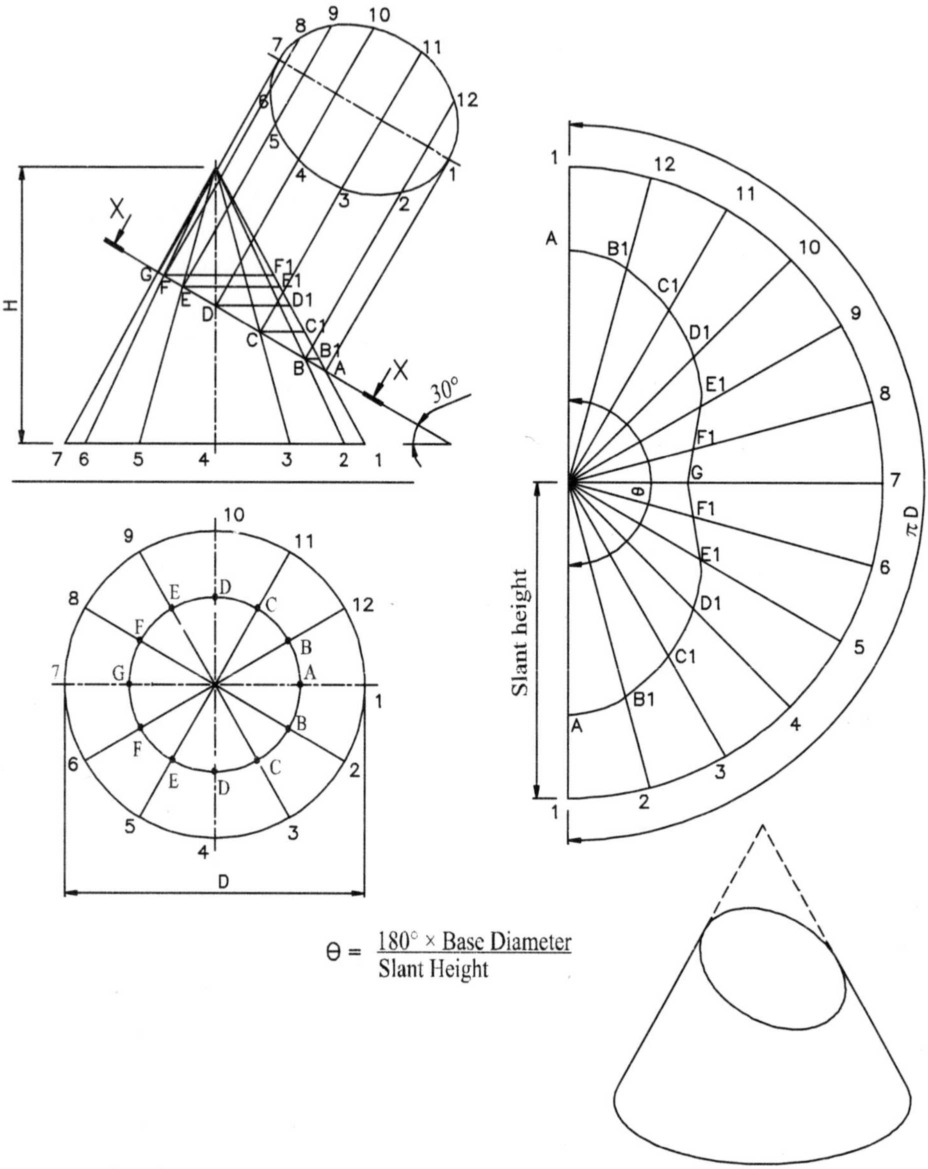

$$\Theta = \frac{180^\circ \times \text{Base Diameter}}{\text{Slant Height}}$$

Fig. 13.9 Development of a truncated cone

LINE OF INTERSECTIONS OF GEOMETRICAL SURFACES

The line of intersection of two surfaces is a line that is common to both. It may be considered as the line that would contain the points at which the elements of one surface would pierce the other. It may be recalled that every line on a practical orthographic representation is a line of intersection, therefore the following discussion is nothing but an extension of the same subject.

The line of intersection of two surfaces is found by determining a number of points common to both surfaces and drawing a line or lines through these points in order. The resulting line of intersection may be straight or curved or a combination of both.

To Find the Intersection of Two Cylinders:

If a series of elements are drawn on the surface of the small horizontal cylinder, as in Fig.13.11a, the points A, B, C and D in which they intersect the vertical cylinder will be points on the line of intersection (vide pictorial view – right hand top). These points, a^H, b^H, c^H & d^H in the top view, are located in the front view by projecting them upward to the corresponding elements in the front view shown as points a^V, b^V, c^V & d^V. The desired intersection is represented by the smooth curve passing through these points.

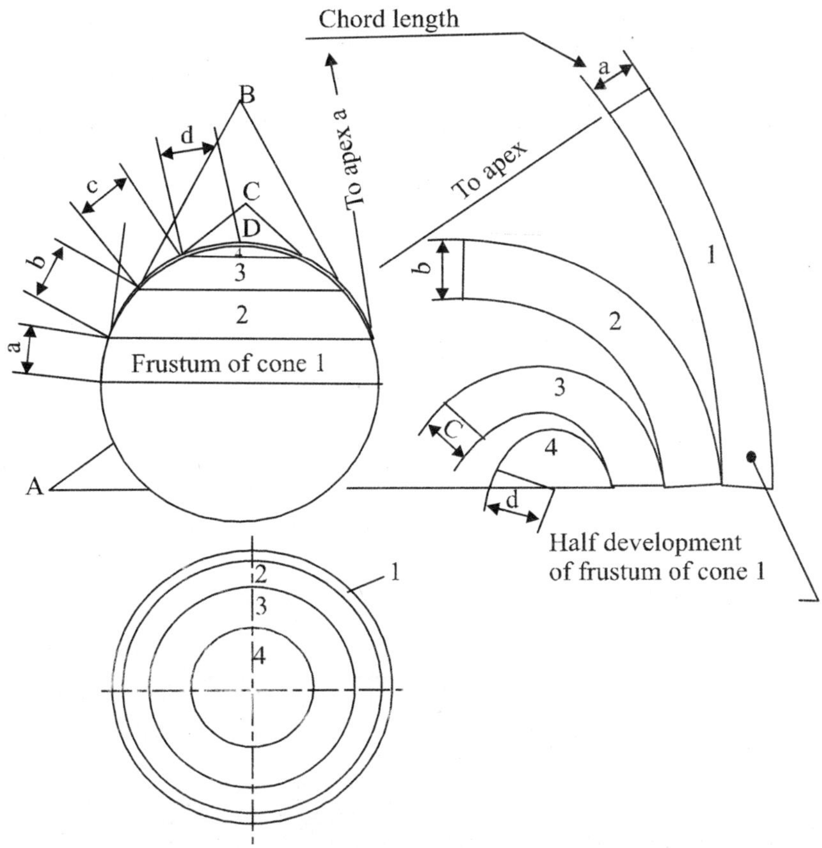

Fig. 13.10 Development of a sphere

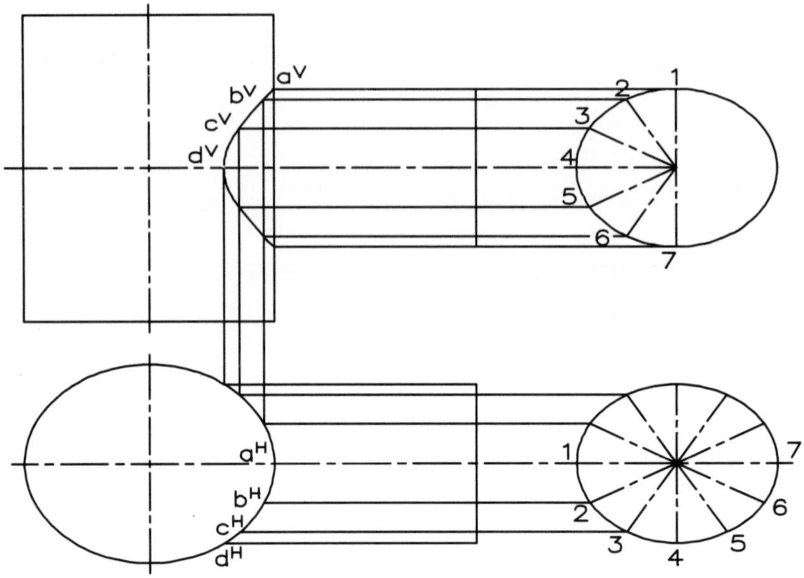

(a) Intersecting cylinders (Axes at 90°)

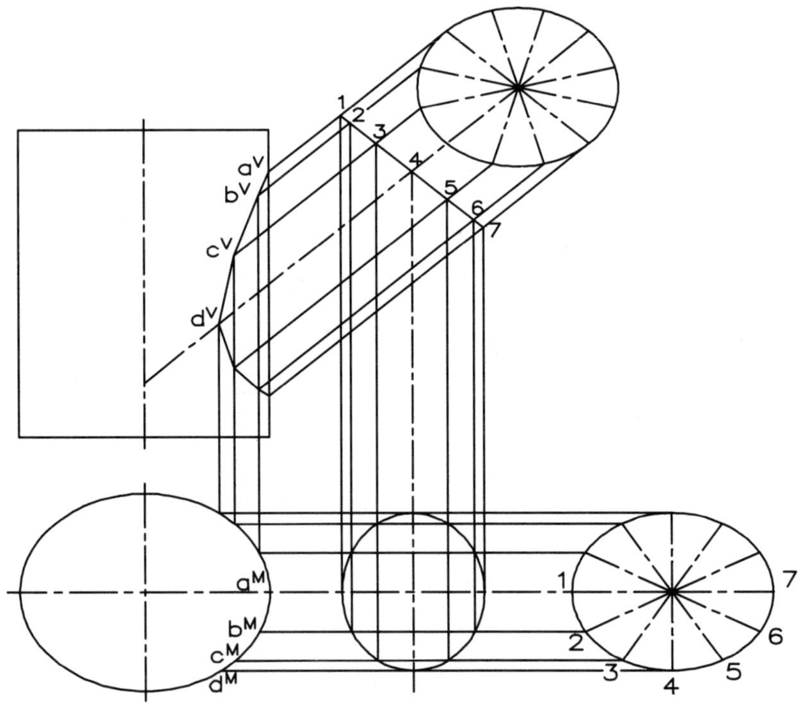

(b) Intersecting cylinders (Axes at 45°)

Fig. 13.11

To Find the Intersection of Two Cylinders Oblique to Each Other

Draw a revolved right section of the oblique cylinder in the front view of that cylinder (Fig. 13.11b). The circumference of the right section is divided into a number of equal divisions and elements are drawn through these division points; the points A,B,C & D in which the elements intersect the surface of the vertical cylinder are the points on the intersection line (vide pictorial view – right hand bottom). These points are found first in the top view and then are projected upward to the corresponding elements in the front view. The smooth curve drawn through these points is the desired line of intersection in the front view.

PROBLEMS

13.1 to 13.4: Make complete orthographic drawings of the following objects cut by a vertical projecting plane making an angle of 30° with the horizontal plane and passing through the mid- height of each object. Find the true shapes of the sections and draw isometric views of the cut objects. Construct the development of the surfaces.

(i) A regular hexagonal prism, height 60 mm, base side 20 mm.

(ii) A cylinder, height 80 mm, ϕ 50 mm base.

(iii) A regular pentagonal pyramid, height 60 mm, base side 20 mm.

(iv) A right cone, height 80 mm, ϕ 50 mm base.

13.5.1 to 13.5.5: Develop all the surfaces of the pyramids, the rectangular elbow and the hexagonal prism. Make construction lines light. Show construction for finding true length.

13.5.6 to 13.5.8: Draw the line of intersection of the intersecting cylinders as shown in Prob. Nos. 13.5.6 to 13.5.8. Develop the lateral surfaces of each cylinder.

13.5.9: A three piece elbow is shown in Prob. 13.5.9. The angles are 22.5°, 45° and 22.5° respectively. Make a layout of the surface to economise material.

Prob. 13.6

A right circular cone of base diameter 80 mm and height 80 mm is standing on its base on ground with 20 mm square hole drilled in it at the mid-height and axis parallel to the base. Draw complete development of the lateral surface of the cone showing the position of the hole.

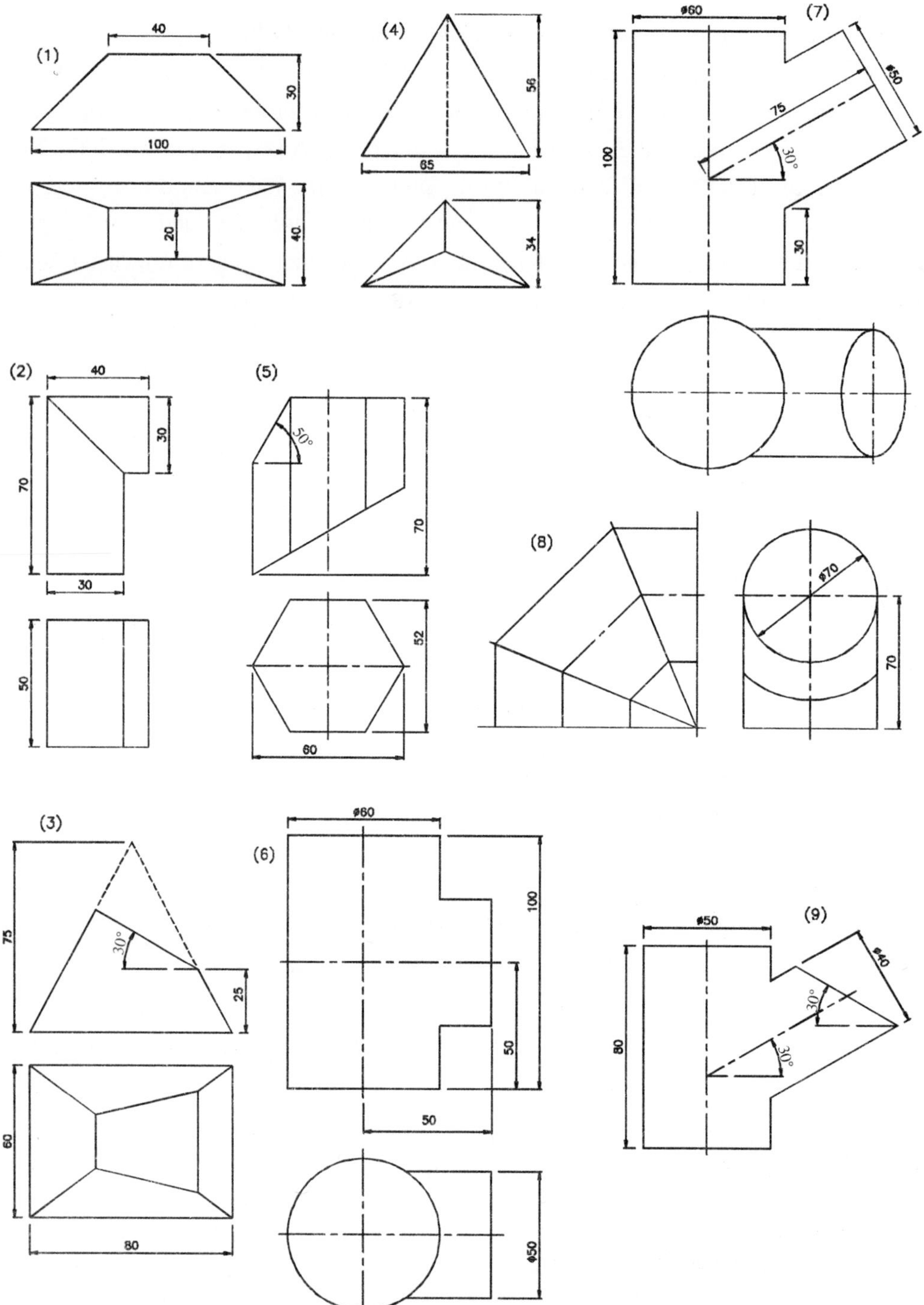

Probs. 13.5.1 - 13.5.9

CHAPTER – 14

BUILDING DRAWING

(For further details, reference may be made to the National Building Code of India)

As engineers, in general, are required to deal with various drawings of buildings, mills, factories *etc*. They should be familiar with the same. The present chapter has been designed to fulfill that end.

BUILDING

Any structure for whatsoever purpose and of whatsoever materials constructed and every part thereof whether used as human habitation or not and includes foundation, plinth, walls, chimneys, drainage works, verandah, balcony, cornice or projection, part of a building or anything affixed thereto and any wall enclosing or intended to enclose any land or space is a building.

MASONRY*

The form of construction composed of bricks, stones, structural steel, clay tiles, concrete blocks, gypsum or other similar building units or materials or a combination of these materials laid up unit by unit, bounded together and set in mortar.

---

* **Lime mortar –** mixture of quick lime and sand in the proportion of 1 lime : 3 sand (by volume) in addition to water, used for internal plastering.

Cement mortar – mixture of cement and sand in the proportion of 1 cement : 5 to 6 sand in addition to water.

Mix proportion – for R.C.C. work (Base, Columns, Lintel, Beams) 1 cement : 2 sand : 4 stone chips (20 mm maximum nominal size) Granite-Black or sand stone-rounded pebbles.)

Minimum size of column bar - ϕ 16 mm

Minimum size of link bars - ϕ 6 mm

Size of modular bricks – 190 mm × 90 mm × 90 mm

A building, in general, consists of :

(a) Foundation

The part of the building to transmit the superincumbent loads from walls and columns to mother earth. Foundation shall be designed to ensure safety of building in that the ultimate pressure on earth shall be well within the allowable pressure limits thus preventing the subsoil from spreading and avoiding unequal settlement of the structure.

In its widest sense the term foundation may be defined as an expanded base of a wall of pier in addition to the ground or subsoil which supports it. The ground which receives the building is known as a natural foundation, and the extended base which is constructed of concrete or masonry is called an artificial foundation.

An artificial foundation comprises

(i) a concrete bed only.

(ii) one or more courses of stone work called footings which are wider than the wall or pier they support and

(iii) a concrete bed together with footings – mostly used now-a-days.

For a two-storey residential building where the wall load is not likely to exceed 3300 kg per linear metre, the foundation widths for different subsoils are given below:*

Nature of Subsoil	*Width of Foundation*
Rock	Equal to the wall-width
Compact gravel and sand or stiff clay	300 mm to 310 mm
Loose sand	600 mm to 620 mm
Soft clay	680 mm to 700 mm
Very soft clay	925 mm to 950 mm

Minimum depth of foundation = 450 mm

(a) Plinth*:* the portion of a structure between the surface of the surrounding ground and surface of the floor, first above the ground.

(b) Brick*:* work or similar panel fitting.

(c)(i) Floor*:* (ii) Mezzanine floor(if any) an intermediate floor between two floors, above ground level.

(d) Lintel: a member of wood, brick or concrete above windows / large openings, which is fixed horizontally and used to support the structure above the opening. Most lintels now are of reinforced concrete.

* *Building Construction*, Vol. 1, W.B. Mckay, Longmans, 1963

(f) Wall above lintel level upto the ceiling level.

(g) Roofing materials: concrete slab/any other material, which may be used with equal effect.

(h) Roof terrace

(i) Parapet (if any): A low wall built along the edge of a roof or a floor not more than 1.2 m in height.

APPLICATION FOR BUILDING PERMIT

Persons intending to erect, re-erect or making materials alteration in any space in a building must submit to the appropriate authority the following:

1. Site Plans
2. Building Plans
3. Services Plans (water supply, sanitation, electricity, gas etc.)
4. Specification – general and detailed.

Site Plan

This must not be drawn to a scale of less than 1:100 and shall show:

(a) the boundaries of the site or any contiguous land belonging to the owner thereof.

(b) the position of the site in relation to neighbouring streets.

(c) the name of street in which the construction is proposed.

(d) all existing buildings standing on, over or under the site.

(e) free passage or way in front of the proposed building.

(f) the direction of north point relative to the plan of the buildings and

(g) other particulars as desired by the sanctioning authority.

Building Plan

This should be drawn to a scale of 1:100 with clear indication of north and should include;

(a) floor plans of all floors together with the covered area, accessory buildings and basement plans. Such drawings shall clearly indicate the size and spacing of all supporting members, sizes of rooms *etc.*

(b) exact location of essential services *e.g.* w.c., sink, bath *etc.*

Colouring of Drawing

Sl. No.	Item	Site Plan			Building Plan		
		White Plan	Blue Print	Ammonia Print	White Plan	Blue Print	Ammonia Print
1.	Existing work	Black (outline)	White	Blue	Black	White	Blue
2.	Proposed work	Red filled in	Red	Red	Red	Red	Red
3.	Drainage & sewerage work	Red dotted	Red dotted	Red dotted	Red dotted	Red dotted	Red dotted
4.	Water supply work	Black dotted thin	Black dotted thin	Black dotted thin	Black dotted thin	Black dotted thin	Black dotted thin
5.	Work proposed to be demolished	Yellow hatched	Yellow hatched	Yellow hatched	Yellow hatched	Yellow hatched	Yellow hatched
6.	Open spaces	No Colour	No Colour	No Colour	No Colour	No Colour	No Colour
7.	Plot lines	Thick black	Thick black	Thick black	Thick black	Thick black	Thick black
8.	Permissible building lines	Thick dotted black	Thick dotted black	Thick dotted black	×	×	×
9.	Existing streets(s)	Green	Green	Green	×	×	×
10.	Future street (s) if any	Green dotted	Green dotted	Green dotted	×	×	×

(c) sectional drawings showing clearly the sizes of foundation footings, thickness of basement wall and all roof slabs and floor slabs, wall construction, size and spacing of framing members, stairways, ceiling heights and parapet heights with their materials. The section should indicate the drainage and the slope of the roof. At least one section should be taken through the staircase.

(d) all street elevations.

(e) details of Service Privy, if any.

(f) terrace plan indicating the drainage and the slope of the roof.

Some recommended values to act as guide while making plans for building are given below:

(a) Habitable Rooms

(i) Minimum height = 2.75 m measured from floor – finish to the soffit of the ceiling (*i.e.* the lowest point of the ceiling).

For row housing, minimum height = 2.6 m

(This practice of reducing the height should be discarded).

For air-conditioned room, minimum height = 2.4 m measured from floor-finish to the lowest point on the air-conditioning duct.

(ii) Minimum area (size) = 9.5 m^2 with a minimum width of 2.4 m (when there is only one room).

For two rooms, minimum area of one room = 9.5 m^2

and minimum area of other = 7.5 m^2

(minimum width in each case = 2.4 m)

(b) Kitchen

(i) Minimum height = 2.75 m

(ii) Minimum area (size) = 5.5 m^2 with a minimum width of 1.8 m

Minimum area = 4.5 m^2 with a separate store space, minimum width being 1.8 m

Minimum area of Kitchen-cum-dining room = 9.5 m^2 minimum width being 2.4 m

(c) Bath Rooms and Water Closets

(i) Minimum height = 2.2 m (normally same height as that of the main building should be maintained).

(ii) Minimum size of a bath room = 1.5 m × 1.2 m (= 1.8m^2)

Minimum area for W.C. and bath combined = 2.8 m^2 with a minimum width of 1.2 m

Minimum area of Water Closet = 1.1 m^2

(d) Mezzanine Floor

Size — If used as a living room, minimum area = 9.5 m^2

The maximum aggregate area of such mezzanine floors in a building
= 1/3 × Plinth area of the building.

(e) Internal Staircase

Minimum width = 100 mm.

Minimum width of treads without nosing = 250 mm

Maximum height of risers = 190 mm

Minimum height of handrails = 1000 mm

Maximum number of steps (risers) per flight = 12

CHARACTERISTIC FEATURES OF ARCHITECTURAL ENGINEERING DRAWING

(1) Use standardized conventional signs for openings of windows, doors, gates, lifts, stairs etc.. The openings for windows and doors are represented by thinner lines compared to the wall outlines.

(2) Show the North line clearly.

(3) Draw the building plan to a scale of 1:100. Separate parts and details of buildings are usually drawn to larger scales – 1:50, 1:10, 1:5 – for clarity depending upon the size of the object and the purpose of the separate drawings.

(4) Instead of arrowheads, dashes inclined at 45° are generally used at the intersection of dimension lines with extension lines.

(5) The elevation or level, of the ground floor is taken to be the Datum Line. The elevations of floors, ceilings, staircase landings and other elements of a building are referred to from this Datum Line. A minus sign will indicate its position downwards. The onventional symbol for level marks is an isosceles triangle with its right hand half filled with ink.

(6) Cross-hatching is done in uniformly spaced thin parallel lines drawn at an angle of 90°, 45°or 0° to the axis of the element or its outline. The hatching is applied for only a short distance from the border of the area when a large area of building material is sectioned.

PROBLEMS

14.1 Read the building drawings represented in Figs. 14.1 and 14.2. Hence solve some problems assigned by the class-teacher.

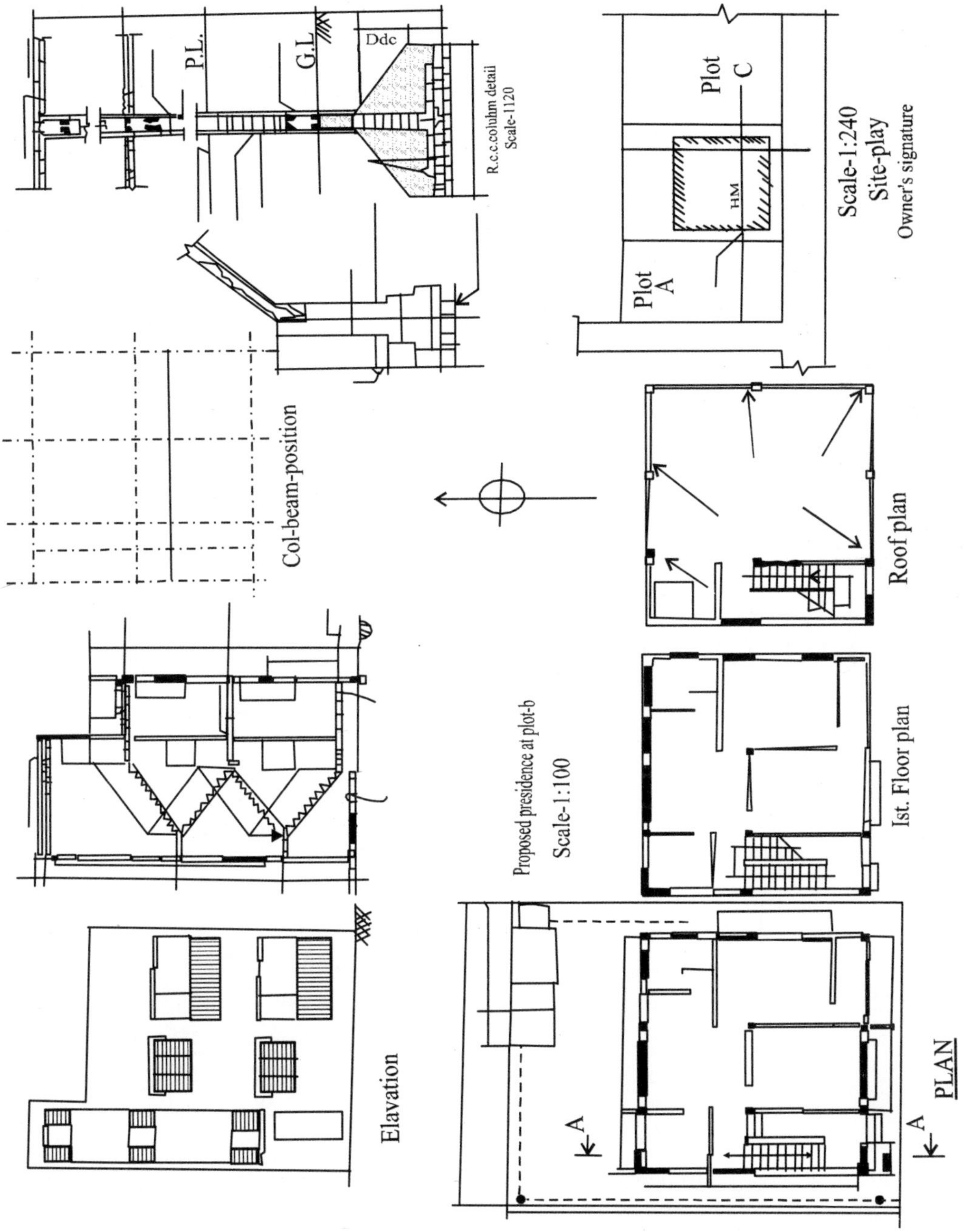

Fig. 14.1 R.C.C framed building

S.T. – Septic Tank – for further details please refer to IS: 2470 Pt.I & II
Dotted line shows ϕ 100 S. W. Pipe

CHL. CH-Chlorination chamber;
Separator bars in beam-ϕ 16 m.s.

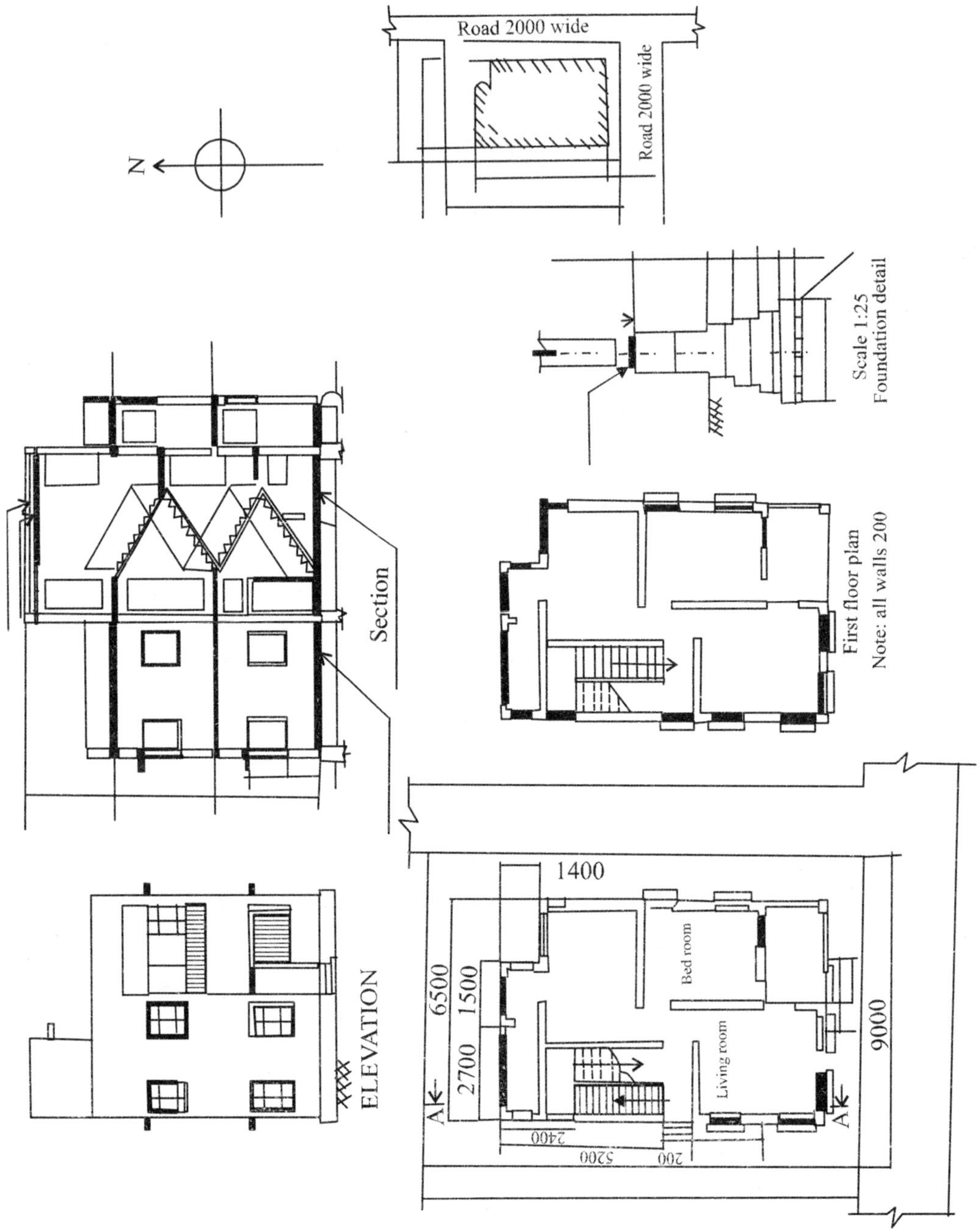

Fig. 14.2 Conventional building

(Terrace Plan not shown, Chujjah (sun shade) details not given and rammed earth not shown by conventional symbol)

CHAPTER – 15

COMPUTER-AIDED DRAWING

Introduction

Students have practiced Engineering drawing using sets squares, tee while fixing the drawing sheets, of various sizes in the wooden drawing board with clamps or board pins and instrument box. Now this a good practice for a learner of the Engineering Drawing. Now with changing time and in a globalised society we need to do it using computers. This is much faster and economic for production of large scale drawing of machine elements, components parts and machine itself. We don't have to use pencil and compass, eraser or any other drawing instrument.

Everything can be done using keyboard & mouse. The most popular software used for implementing such type of facilities one have to have a good knowledge of computer applications. AUTOCAD is the most popular user friendly windows based software being used over several decades. It is being improved and taught separately in different Institutes. The software provider has got user manual and help in its pull down menu. Every year it is being updated and having some customized facilities. One have to buy a licensed version of AUTOCAD and install it in his computer with appropriate configuration and start working with simple geometrical drawing, e.g., starting from lines of various type, then rectangle, polygons, circle. Ellipse etc. Then there are Toolbars and pull down menus which can be used very easily and the new learner can explore them and learn. The help menu also explains all the features which are necessary for drawing a particular object. You also need the help of a teacher at the learning stage.

Here we have given a point by point instruction to learn the computer aided drawing using autocad package of AUTODESK.

To Create a New Drawing

Command Line: New

Menu: File ➢ New

Standard Toolbar: New

To Start a New Drawing

1. If changes were made to your current drawing, the Save Changes to (file name) message box appears with options for **Yes, No,** and Cancel. If you previously have not invoked Save or Qsave, select Yes to save any edited or unsaved drawings before opening a new drawing. If you wish to save your current drawing and have previously not named it, the Save Drawing As dialog box opens and you may enter a name in the File Name Edit box.
2. When the Create **New** Drawing dialog box opens, specify an option by clicking one of four buttons described under Options below.

   If the Filedia system variable is set to 0, the following command line prompt appears instead of the dialog box:

   Enter template file (or . for none) <C:\ AutoCAD\.......\ template\ filename.dwt>: Specify a file name containing a .DWT or .DWG extension.

Options

Use a Wizard Click Use a Wizard and choose between Quick Setup or Advance Setup. The Quick Setup wizard contains two tabbed sections: Setup 1: Units and Step 2: Area. The Step 1 tab is the same for both wizards and allows you to select from Decimal, Engineering, Architectural, Fractional, and Scientific units of measure. The Step 2 in Quick setup and Step 5 in Advanced Setup contains Width and Length edit boxes to help you specify your working area. The Advanced Setup dialog box offers four additional steps: Angle Measure, Angle Direction, Title Block and Layout. Each tabbed section displays a preview window showing you the results of your selection.

Angle Measure Pick a radio button or enter a value to determine the direction in which AutoCAD measures angles.

Angle Directon Set a clockwise or counter-clockwise direction for AutoCAD to measure angles.

Title Block Click drop-down boxes for Title Block Description, choosing among Ansi, Arch/Eng, Din, Iso, or Jis sizes or Title Block File Name, selecting among the various drawing files. Use Add… to open the Select Title Block file dialog box and append drawings to the drop down box or the Remove button to delete a title block file.

Layout Radio buttons Yes or No allow you to use advanced paper space layout capabilities and offer you a way to start your drawing.

Use a Template Click this button to start a drawing based on a template. AutoCAD has created 36 template files that you can select from, serving as prototypes, some with predefined layers, dimension styles, and views for creating new drawings. AutoCAD stores template drawings as .DWT files in the C:\ ACADR13\ TEMPLATE directory. You can create your own templates by renaming the .DWG extension to .DWT and copying the file to the template directory. Double-click More files… in the list box to open the Select Template dialog box and locate a .DWG or .DWT template file from any directory or drive.

Start from Scratch Use the Start from Scratch button for a drawing setup similar to the Acad.DWG prototype offered in release 13. It allows you to start a drawing quickly with either English or metric settings.

Instructions To help you understand and decide which procedure best suites your needs, AutoCAD provides a brief description when you click the Instructions button.

Done Select Done to exit the dialog box and start AutoCAD using the specifications you selected.

Show at startup Toggle the Create New Drawing dialog box off if you do not want it to display when you start an AutoCAD session. To toggle it on, click Show the Startup dialog box in the Capability tab section of the Preferences dialog box.

See Also Exit, Open, Qsave, Quit, Save, System Variables: Filedia, Dwgname, Dwgprefix, Dwgtitled, Savename

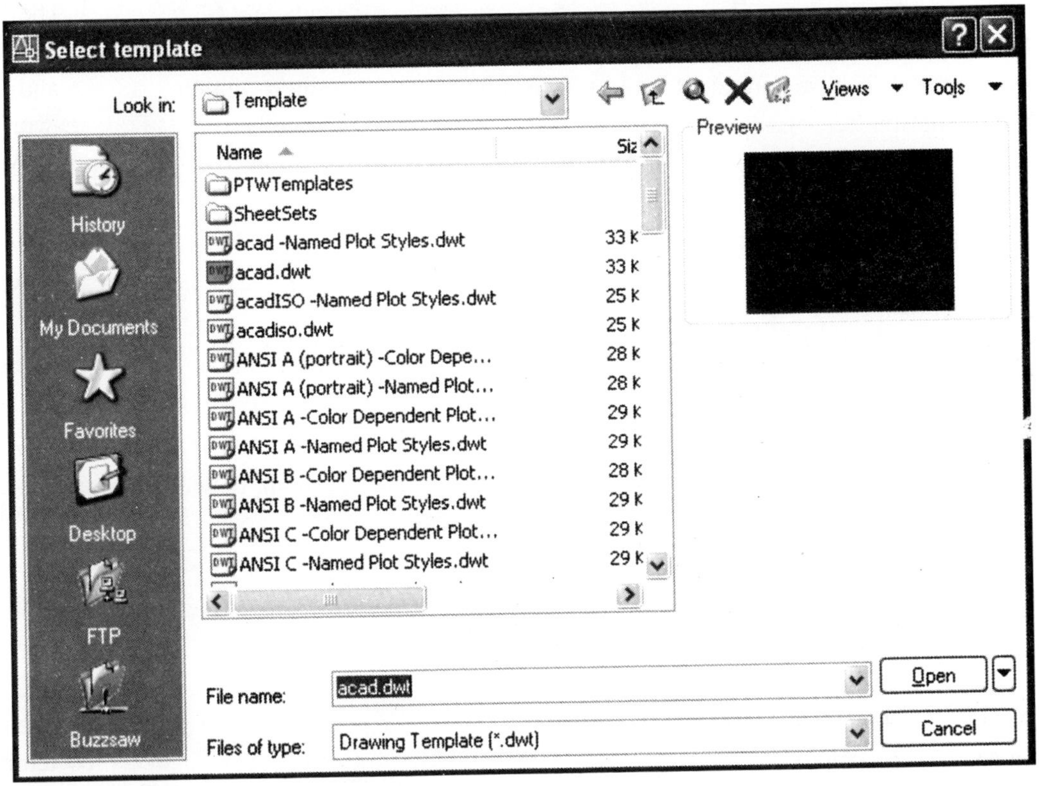

To Change Drawing Units

Command Line: Units (or 'Units to use transparently)

1. Report formats: Enter choice, 1 to 5 <2>: Enter the number corresponding to the desired unit system, as described below. With the exception of the Engineering and Architectural modes, you can use these modes with any basic unit of measurement.
2. Number of digits to the right of decimal point(0 to 8):<4>: Enter a number to specify the degree of precision.
3. System of angle measure: Enter choice, 1 to 5 <1>: Enter the number corresponding to the desired angle measure system, as described below.
4. Number of fractional places for display of angle (0 to 8) <0>: Enter a number to specify the degree of precision.
5. Direction for angle 0: Enter direction for angle 0 <0>: Enter the desired angle for the 0 degree direction, as described below.
6. Do you want angle measured clockwise?<N>: Enter Y if you want angles measured clockwise, otherwise, press.

Options

System of units Sets format of units that AutoCAD will accept as input:

Report Format	Example
1. Scientific	1.5500E+01
2. Decimal	15.5
3. Engineering	1'-3.50"
4. Architectural	1'3 1/2"
5. Fractional	15 1/2

System of angle measure Sets format of angle measurement AutoCAD will accept as input:

Measurement Format	Example
1. Decimal	45.0000
2. Degrees/minutes/	45d0'0" seconds
3. Grads	50.0000g
4. Radians	0.7854r
5. Surveyor's units	N 45d0'0"E

Direction for angle 0 Sets direction for the 0 angle:

East	3 o'clock = 0
North	12 o'clock = 90
West	9 o'clock = 180
South	6 o'clock = 270

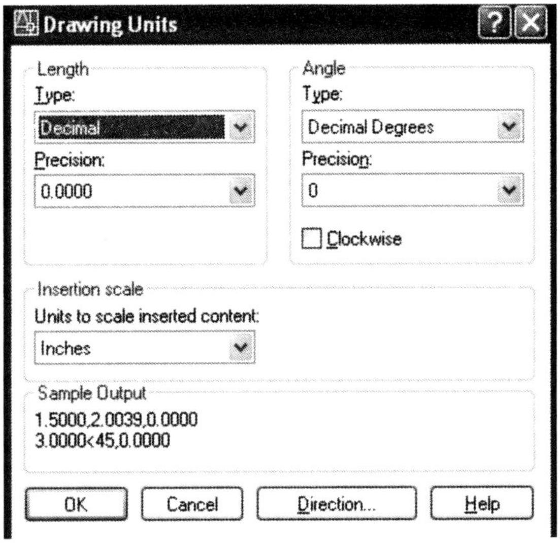

LAYER

Layer must be used at the command line to create new layers, assign colors and line types to layers, set the current layer, repress editing of layers, and allow you to control which layers are displayed. The dialog box equivalents are Ddlmodes and Layer.

To Create and Modify Layers

Command Line: -Layer (or '-Layer, to use transparently)

?/Make/Set/New/ON/OFF/Color/Ltype/Freeze/Thaw/LOck/Unlock: Enter an option.

Options

? Displays the list of existing layers. Wildcards are accepted.

Make Creates a new layer and makes it current.

Set Makes an existing layer the current layer.

New Creates a new layer.

On Turns on layers.

Off Turns off layers.

Color Sets color of a layer.

Ltype Sets line type of a layer.

Freeze Freezes one or more layers.

Thaw Unfreezes one or more layers.

Lock Prevents editing of visible layers.

Unlock Releases locked layers to allow editing.

Note All Layer options except Make, Set, and New allow you to enter wildcard characters (question marks and asterisks) for input. For example, if you want to turn off all layers whose names begin with G, enter G* at the prompt.

Note The option pairs Freeze/Thaw and On/Off both control whether or not a layer is displayed. However, unlike Off, Freeze makes AutoCAD ignore objects on frozen layers. This allows faster regenerations. Freeze also affects blocks differently than Off. Thawing layers requires a Regen if Regenauto is off.

Note Layer 0, the default layer when you open a new file, is white (number 7) and has the continuous line type. Layer 0 also has some unique properties. If you include objects on Layer 0 in a block, they take on the color and line type of the layer on which the block is inserted. The objects must be created with the Byblock option (see Color). The dimension layer Defpoints is also unique. When it is turned off, objects on this layer are still displayed and are not selectable, but will not appear on prints or plots. Defpoints is unique in that objects assigned to this layer will not plan or accept a color setting, everything remains white. This makes the Defpoints layer suitable for layout lines.

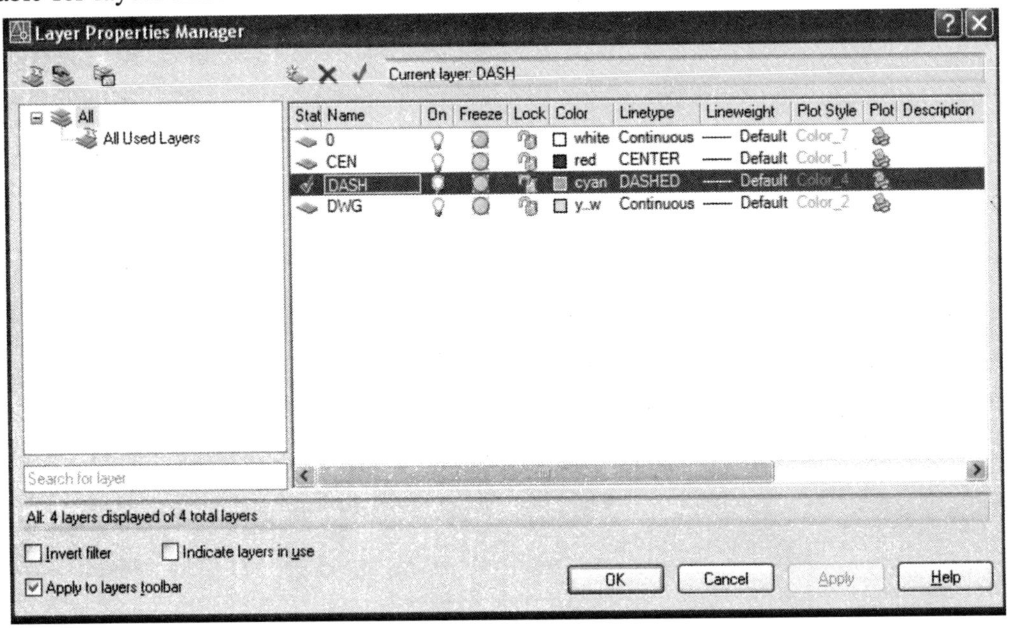

LINE

Line draws simple lines—either a single line or a series of line segments end-to-end.

To Draw a Line

Command Line: Line

Menu: Draw ➢ Line

Draw Toolbar: Line

1. From point: Select a point to begin the line.
2. To point: Select the line endpoint.
3. To point: Continue to select points to draw consecutive lines or press to exit the command.

Options

C Closes a series of lines, connecting the last start point and the last endpoint with a line.

At the "From point:" prompt, lets you continue a series of lines from a previously entered line, arc, point, or polyline. If the last object drawn is an arc, the line is drawn at a tangent from the end of the arc.

U At the "To point:" prompt, deletes the last line segment.

Note Convert lines to polylines using the Pedit command.

See Also AutoLISP, Mline, Osnap, Pedit, Pline, Ray, Xline

LINETYPE

Linetype enables you to control the type of line you can draw. The default line type is continuous Bylayer, but you can choose from several other types, such as a dotted or dashed line or a combination of the two (see Figure 13). Predefined line types are stored in a file called Acad.LIN. You can list these line types by entering a question mark at the "-Linetype" prompt, or using the Linetype command and clicking Load in the Layer and Linetype Properties dialog box. Ddltype is the dialog box equivalent.

To Change the Line Type

Command Line: -Linetype (or '-Linetype, to use transparently)

?/Create/Load/Set: Enter the option name.

Options

?	Lists available line types in a specified external line type file.
Create	Creates a new line type.
Load	Loads a line type from a specified line type file.
Set	Sets the current default line type.

The Create option first prompts you for a line type name. This name can be any alphanumeric string of 47characters or less (although the status line will display only the first eight characters). You are then prompted for the name of the file in which to store your line type. Next, you enter a description or graphic representation of the line type. Finally, you enter the line type pattern on the next line, where you will see an "A" (for pattern alignment) followed by a comma and the cursor. A-type alignments force lines and arcs to start and end with a dash. Enter a string of numeric values separated by commas. These values should represent the lengths of lines as they will be plotted. Positive values represent the "drawn" portion of the line; negative values represent the "pen up," or blank, portion of the line; and a zero indicates a dot. The following code produces a line type segment with a dash .3 drawing units long and three dots spaced .05 drawing units apart:

A, .3, –.05, 0, –.05, 0, –.05, 0, –.05

You can create complex line types only by editing the Acad.LIN file.

You can assign line types to layers or to individual objects. Use the Ltscale command to make the scale of the line types correspond with the scale of your drawing.

A line type may appear continuous even though it is a noncontinuous type. Several things can affect the appearance of line types. For example, if the drawing scale is not 1:1, the Ltscale must be set to correspond with your drawing scale. If the drawing scale is "1/4" equals 1', the Ltscale must be set to 48. A low Viewres value can also affect appearance, making line types appear continuous onscreen even though they plot as a noncontinuous line type. Regenerating exhibits the true appearance of line types. To display a list of line types currently loaded in your drawing, use Ddltype.

Use the Ltscale command to set the global linetype scale factor, Celtscale for individual linetype scaling per object and Psltscale (with a value of 1) for viewport scaling. For global editing, it is best to define linetypes by layer and not by a property

CIRCLE

Circle offers several methods for drawing circles, the default being to choose a center point and enter or pick a diameter or radius.

To Draw a Circle

Command line: Circle

Menu: Draw ➢ Circle ➢ Preset Options

Draw Toolbar: Circle

1. 3P/2P/TTR/<Center point>: Pick a center point or enter an option.
2. Diameter/<Radius> <current default>: Depending on the current default, provide a diameter or radius by dynamically dragging, entering the letter D or R, or enter a value.

   Selecting TTR prompts appear to enter two tangent osnap points and a radius.

1. Enter Tangent spec: Pick first osnap point
2. Enter second Tangent spec: Pick second osnap point
3. Radius <current value>: Specify a radius.

The Draw ➢ Circle ➢ Tan, Tan, Tan pulldown menu option offers a method similar to the three point, allowing you to draw a circle by specifying three tanget points and automatically displaying Deferred Tangent autosnap points. Prompts for First point:, Second point: and Third point: appear to set the circle's size.

Options

3P (3 Point) Allows you to define a circle based on three points. Once you select this option, you are prompted for a first, second, and third point. The circle will be drawn to pass through these points.

2P (2 Point) Allows you to define a circle's diameter based on two points. Once you select this option, you are prompted to select the first and second point. The two points will be the opposite ends of the diameter.

TTR (Tangent, Tangent, Radius) Allows you to define a circle based on two tangent points and a radius. The tangent points can be on lines, arcs or circles.

ARC

Arc allows you to draw an arc using a variety of methods. The system prompts shown below will vary with the options chosen.

To Draw an Arc

Command Line: Arc

Menu: Draw ➢ Arc ➢ Preset Options

Draw Toolbar: Arc

1. Center/<Start point>: Use the mouse to pick the start point of the arc or select C for more options.
2. Center/End/<Second point>: Pick the second point of the arc or select C for Angle/Length of Chord/End point: options or E for Angle/Direction/Radius/<Center point> options.

   Depending on the option selected, prompt 3 or 4 will appear:
3. Angle/Length of Chord/End point: Pick the end point of the arc.
4. Angle/Direction/Radius/<Center point>: Pick the end point of the arc or specify a value for selected option.

Options

Angle Enters an arc in terms of degrees or current angular units. The "Included angle:" prompt appears. You can enter an angle value or use the cursor to select angle points on the screen.

Center Enters the location of an arc's center point. At the prompt "Center:", enter a coordinate or pick a point with your cursor.

Direction Enters a tangent direction from the start point of an arc. At the prompt "Direction from start point:", either enter a relative coordinate or pick a point with your cursor.

End Enters the end point of an arc. At the prompt "End point:", enter a coordinate or pick a point with your cursor.

Length Enters the length of an arc's chord. At the prompt "Length of chord:", enter a length or drag and pick a length with your cursor.

Radius Enters an arc's radius. At the prompt "Radius:", enter a radius or pick a point that defines a radius length.

Start point Enters the beginning point of an arc.

If you press at the first prompt of the Arc command, AutoCAD uses the most recent point entered for a line or arc as the first point of the new arc. It then prompts you for a new end point. An arc is drawn at a tangent to the last line or arc drawn.

If you select from the pull-down menu, the Arc Cascading menu appears, with ten preset arc options. For example, S, E, D allows you to select the start point, the end point, and the direction of the arc. Figure 1 illustrates how these options draw arcs.

You can convert arcs to light weight polyline arcs with the Pedit or Bpoly command. You can also lengthen existing arcs using the Lengthen command.

See Also Change, Elev, Ellipse, Lengthen, UCS

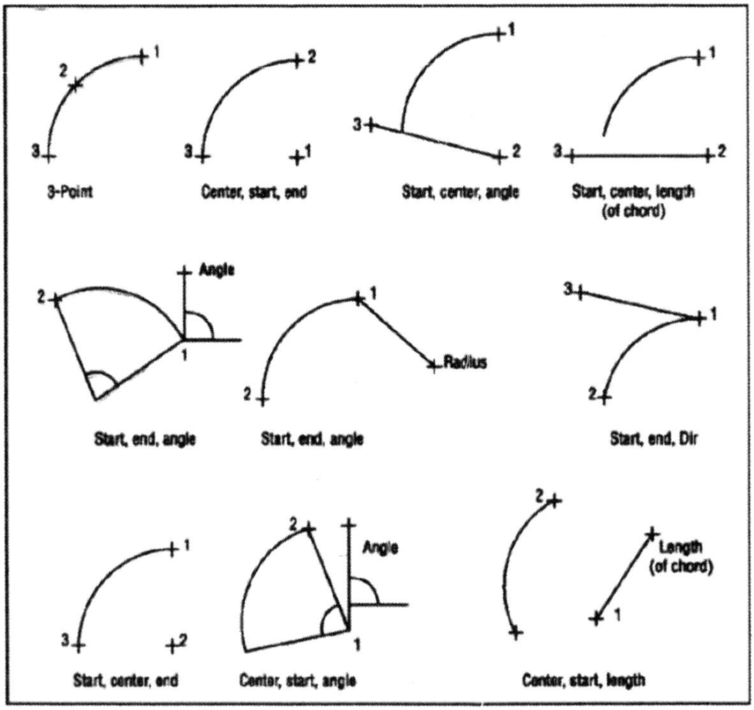

Figure 1: The Arc menu options and their meanings

MOVE

Move displaces a single object or a set of objects.

To Move Objects

Command Line: Move

Menu: Modify ➢ Move

Modify Toolbar: Move

1. Select objects: Select the objects to be moved.
2. Base point or displacement: Pick the reference or "base" point for the move.
3. Second point of displacement: Pick the distance and direction in relation to the base point orenter the displacement value.

Alternate: Grips

If you have enabled grips for moving, the prompts for Steps 1 and 2 and the Grips Options are the same as for Mirror, except for the following:

1. Cycle through the grip mode commands by pressing or the spacebar, or by entering Move or M a sufficient number of times until you see the following prompt:

   MOVE

   <Move to point>/Base point/Copy/Undo/eXit:
2. To move the object(s) with the selected grip as the base, drag your cursor and pick the second or displacement point. To move the object(s) using a new base point, enter B to pick your first reference point or enter coordinate values, then pick the second point of displacement.

If you chose the displacement option rather than picking a base point, you can enter the total amount of movement you want the object to make (in the X and Y directions) and press Enter twice.

AutoCAD assumes you want to move objects within the current UCS. However, you can move objects in 3D space by entering XYZ coordinates or using the Osnap overrides to pick objects in 3D space.

If you press at the "Second point:" prompt without entering a point value, the objects selected may be moved to a position completely off your drawing area. Use the U or Undo command to recover.

To make multiple copies of the selected object(s), follow the steps above to define your base point, then hold the Shift key while picking the first copy point (to set copy mode on), and continue by picking additional points. (Entering C after selecting the base point will provide the same results.) Exit the command by pressing .

Pressing the Shift key in the last step to copy the first object from its source point to a destination point will set an automatic snap mode based on these two points. To apply Multiple copy using the snap mode, hold the Shift key down while copying additional objects.

COPY

Copy copies a single object or a set of objects.

To Copy

Command line: Copy

Menu: Modify Ø Copy

Modify Toolbar: Copy

1. Select objects: Select objects to be copied.
2. <Base point or displacement>/Multiple: Pick the reference or "base" point for copy or enter M for multiple copies. The M option instructs AutoCAD that you wish to make several copies and prompts you for the base point again.
3. Second point of displacement: Pick the copy distance and direction in relation to the base point or enter the displacement value.

Options

Multiple Allows you to make several copies of the selected objects. The second point is repeated until you press ø or Esc.

AutoCAD assumes you want to make copies within the current user coordinate system (UCS). However, you can make copies in 3D space by entering 0,0,0 as the base point and the desired X,Y,Z coordinates as the second point, or by using the Osnap overrides to pick objects in 3D space.

If you press ø at the "Second point:" prompt without entering a point value, the selected objects may be copied to an area that is off your current drawing. To recover, use the Undo command or Zoom entents. For multiple copying using grips.

TRIM

Trim shortens an object to meet another object or objects. One or more objects may also be trimmed to a point of implied intersection—that is, to the point at which they would intersect with the cutting edge if the cutting edge were extended out. Additionally, objects may be trimmed to the current UCS plane or along the current view direction—that is, to an apparent intersection in the current view. These options are controlled by the system variables Projmode and Edgemode.

To Trim an Object

Command Line: Trim

Menu: Modify ➢ Trim

Modify Toolbar: Trim

1. Select cutting edge(s)(Projmode - UCS, Edgemode - No extend: Choose object(s) that define the cutting edge(s) at which you want to trim other objects.
2. <Select objects to trim>: Pick the objects you want to trim, one at a time or using the Fence selection set option, or select an option:

- If you type P, you will be prompted None/Ucs/View/<current value>:.
- If you type E, you will be prompted Extend/ No extend <current value>:.

Options

Project Specifies the projection mode for AutoCAD to use when trimming objects: None specifies that only objects that actually intersect the cutting edge will be trimmed; UCS specifies projection onto the XY plane of the current UCS; View trims all selected objects which intersect with the cutting edge in the current view. The system variable Projmode settings control the projection mode.

Edge Allows cutting edges to be extended to the point where they would intersect with an option. You may select either Extend or No Extend (system variable Edgemode settings 1 and 0). If you choose No extend, only objects that actually intersect the cutting edge will be trimmed.

EXTEND

Extend lengthens an object to meet another object. Objects that can be extended include arcs, elliptical arcs, lines, open polylines, and rays. You can extend objects to an "implied" as well as to an actual boundary.

To Extend an Object

Command Line: Extend

Menu: Modify ➢ Extend

Modify Toolbar: Extend

1. Select boundary edges: (projmode - <current value>, Edgemode - <current value> Select objects: Select the object or use fence selection to designate the boundary objects. If you press at the "Select objects:" prompt, AutoCAD selects all objects in your drawing, then displays the prompt shown in step 2.
2. <Select objects to extend>Project/Edge/Undo: Select objects to be extended by picking an object, using fence selection, or entering U to undo the last extend operation. You may also enter P or E to reset the Project or Edge settings.

- If you type P, you will be prompted None/Ucs/View/ <current value>:.
- If you type E, you will be prompted Extend/No extend <current value>:.

Options

Project Specifies the projection mode for AutoCAD to use when extending objects: with None: only selected objects which actually intersect with the boundary are extended; UCS: specifies projection onto the XY plane of the current UCS; View: extends all selected objects that intersect with the boundary in the current view. This option is controlled by the system variable Projmode. Projmode settings are zero = "None"; 1 = "UCS"; 2 = "View".

Edge Allows you to extend objects to implied as well as actual boundaries; boundary edges do not have to be exactly in the path of the objects to be extended. This option is controlled by the system variable Edgemode: a setting of 0 "No extend" will not extend objects to the implied boundary; a setting of 1 "Extend" will allow objects to extend to an implied boundary.

Undo Restores the previously executed option. You cannot extend objects within blocks or use blocks as boundary edges.

ERASE

Erase deletes one or several selected objects from a drawing.

To Erase Objects

Command Line: Erase

Menu: Modify ➢ Erase

Modify Toolbar: Erase

Select objects: Select the objects to be erased, by picking or using any selection method.

Options

As well as the standard selection options, the following are useful with the Erase command:

Single Lets you pick a single object only. From the keyboard you can also enter Si at the "Select objects:" prompt.

Last Lets you erase the last entity drawn. From the keyboard you can enter L at the Select objects: prompt. You can also enter Previous or P from the keyboard to erase the previous selection set.

Oops Lets you unerase the last erase. (See Oops.) (The Undo command will reverse the entire Erase sequence but not the last object erased.)

OOPS

Oops restores objects that have been accidentally removed from a drawing.

To Restore an Erased Object

Command Line: Oops

Oops retrieves all objects erased by the last Erase and after executing Bmake, Block, or Wblock.

See Also U, Undo

UNDO

Undo allows you to undo parts of your editing session. This can be useful if you accidentally execute a command that destroys part of all of your drawing. Undo also allows you to control how much of a drawing is undone.

To Reverse Commands

Command Line: Undo

Menu: Edit ➢ Undo

Auto/Control/BEgin/End/Mark/Back/<default number>: Enter an option to use or the number of commands to undo.

Options

Auto Makes AutoCAD view menu macros as a single command. If Auto is set On, the effect of macros issued from a menu will be undone regardless of the number of commands the macro contains.

Control Allows you to turn off the Undo feature to save disk space or to limit the Undo feature to single commands. You are prompted for All, None, or One. All fully enables the Undo feature, None disables Undo, and One restricts the Undo feature to a single command at a time.

Begin, End Begin marks the beginning of a sequence of operations. All edits after that point become part of the same group. End terminates the group. This allows you to mark a group of commands to be undone together.

Mark, Back Allows you to experiment safely with a drawing by first marking a point in your editing session to which you can return. Once a mark has been issued, you can proceed with your experimental drawing addition. Then, you can use Back to undo all the commands back to the place that Mark was issued.

VIEWPORTS

Vports or Viewports displays multiple views, or tiled viewports, of your drawing at one time. This command is disabled when the Tilemode system variable is set to 0.

To Display Multiple Viewports

Command Line: Vports or Viewports

Menu: View ➢ Tiled Viewports

Save/Restore/Delete/Join/SIngle/?/2/<3>/4: Enter the desired option or press for the "Horizontal/Vertical/Above/Below/Left/<Right>:" default prompt.

Options

Save	Saves the current viewport arrangement.
Restore	Restores a previously saved viewport arrangement.
Delete	Deletes a previously saved viewport arrangement.
Join	Joints two adjacent viewports of the same size to make one larger viewport.
Single	Changes the display to a single viewport.

1 Displays a list of saved viewport arrangements along with each viewport's coordinate location.

2 Splits the display to show two viewports. You are prompted for a horizontal or vertical split.

3 Changes the display to show three viewports.

4 Changes the display to show four equal viewports.

Layout If you are working from the pull-down View menu, you may select Layout... to open the Tiled Viewport Layout dialog box. This displays an icon menu of predefined viewport arrangements from which you may select the desired layout.

Each viewport can contain any type of view you like. For example, you can display a perspective view in one viewport and a plan view of the same drawing in another viewport.

You can only work in one viewport at a time. To change active viewports, pick any point inside the desired viewport. The border around the selected viewport will thicken to show that it is active. The standard cursor appears only in the active viewport. (When you move the cursor into an inactive viewport, it changes into an arrow.) Any edits made in one viewport are immediately reflected in the other viewports.

Each viewport has its own virtual display within which you can pan and zoom at redraw speeds. For this reason, the Regen and Redraw commands affect only the active viewport. To regenerate or redraw all the viewports at once, use the Regenall and Redrawall commands.

ZOOM

Zoom controls the display of your drawing dynamically in real time.

To Use Zoom

Command Line: Zoom

Menu: View ➢ Zoom ➢ Realtime/Previous/Window/Dynamic/Scale/In/Out/All/Extents

Zoom Toolbar: Zoom Window Zoom Dynamic

Zoom Scale Zoom Center Zoom In Zoom Out

Zoom All Zoom Extents

All/Center/Dynamic/Extents/Previous/Scale(X/XP)/Window/<Realtime>:

Press Esc or Enter to exit, or right-click to activate pop-up menu. Press for real time Zoom or enter the desired option.

Options

In/Out Thae Zoom In and Zoom Out icon tools on the Zoom toolbar or the Zoom flyout on the Standard toolbar allow you to zoom in and out by a predefined scale factor. Zoom In doubles the size of the drawing display; Zoom Out shrinks the display to half of the current size so that you can view double the drawing are.

Realtime Displays a magnifying glass with a plus and minus sign for real-time pan and zoom. Combines Rtzoom and Rtpan into a single command allowing you to switch between pan and zoom. Right-clicking your mouse displays a cursor menu to alternate among Pan, Zoom, Zoom Window, Zoom Previous, Zoom Extents options or Exit the Zoom command.

All Displays the area of the drawing defined by the drawing's limits or extents, whichever are greater (see Limits).

Center Displays a view based on a selected point. You are first prompted for a center point for your view and then for a magnification or height. A value followed by an X is read as a magnification factor; a lone value is read as the desired height in the display's drawing units.

Dynamic Displays the virtual screen and allows you to use a view box to select a view. The drawing extents, current view, and the current virtual screen area are indicated as a solid white box, a dotted green box, and red corner marks, respectively. You can pan, enlarge, or shrink the view by moving the view box to a new location, adjusting its size, or both. When the view box appears, press the pick button to adjust the view box size, then pick again to restore the X in the view box. Press enter to set the new view.

Extents Displays a view of the entire drawing centered on the screen. In Release 14, this option no longer causes a regen.

Previous Displays the last view created by a Zoom, Pan or View command.

Window Enlarges a rectangular area of a drawing, based on a defined window.

Scale(X) Expands or shrinks the drawing display. If an X follows the scale factor, it will be in relation to the current view. If no X is used, the scale factor will be in relation to the area defined by the limits of the drawing. A value of .5x displays a view half the size of the current view.

Scale (XP) Sets a viewports scale in relation to the Paperspace scale. For example, if you have set up a title block in Paperspace at a scale of 1"=1" and your full-scale Modelspace drawing is to be at a final plot scale of 1/4"-1', you can enter 1/48xp at the Zoom prompt to set the viewport at the appropriate scale for Paperspace. You must be in floating Modelspace to use this option.

PAN

Pan allows you to view different parts of your drawing by moving the cursor, displayed as an open hand, in any direction. In Release 14, real-time pan is the default option for interactive panning.

To Move around in Your Drawing

Command Line: Pan (or 'Pan to use transparently)

Menu: View ➢ Pan ➢ Preset Options

Standard Toolbar: Pan Real-time

1. Press Esc or Enter to exit, or right-click to activate pop-up menu. Enter the command and the cursor changes to a hand cursor. Press the pick button on your pointing device to lock the cursor in its current location relative to the viewport coordinate system and the display screen moves in the same direction as the cursor. As you drag the hand your drawing moves around the graphics screen. To exit the command, press the Escape key or .

   To stop panning, release the pick button; to restart the process, press it again.
2. Right click the mouse button to display a pop-up menu for switching between Exit, Pan, Zoom, Zoom Window, Zoom Previous, and Zoom Extents.

DIMENSIONING COMMANDS

In earlier releases of AutoCAD, dimensioning was accomplished by first using the Dim command to put the command line in dimensioning mode (Dim: mode), then issuing a dimensioning subcommand. To maintain compatibility with previous versions of AutoCAD, all of the existing Dimensioning mode subcommands are still available and may be used in the same way.

AutoCAD generally uses the same types of dimensions and dimension label components as standard drafting. Figure 2 gives examples of the five types of dimensions possible in AutoCAD drawings: linear, angular, diametric, radial, and ordinate. Dimension labels consist of the elements illustrated in Figure 3.

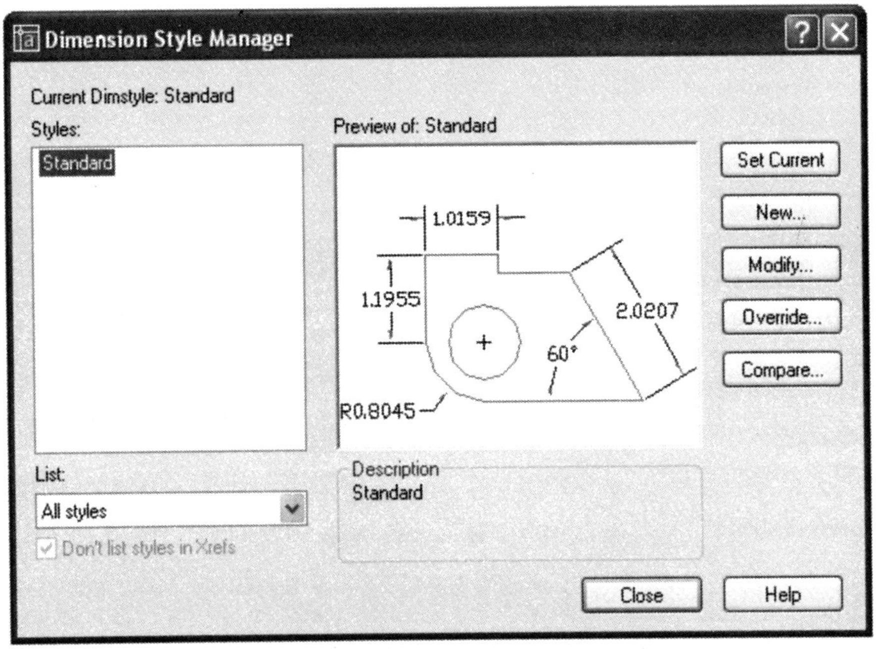

Figure 2: Types of dimensions

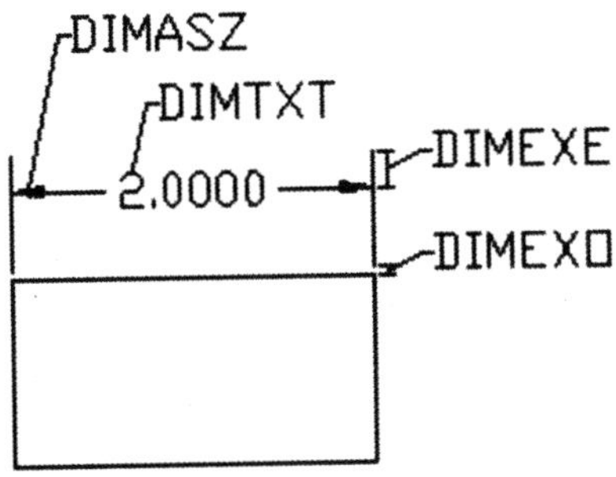

Figure 3: Components of dimension labels

Variables for Controlling Dimensions

Table 4 describes variables that control the way AutoCAD draws dimensions. These variables control extension line and text location, tolerance specifications, arrow styles and sizes, and much more.

Table 4: The Dimension Variables

Variable	Description
Dimalt	When on, dimension texts for two measurement systems are inserted simultaneously (**alt**ernate). Dimaltf and Dimaltd must also be set appropriately. The alternate dimension is placed within brackets. Angular dimensions are not affected. This variable is commonly used when inches and metric units must be displayed at the same time in a dimension. The default setting is off.
Dimaltd	When Dimalt is on, Dimaltd controls the number of decimal places the alternate dimension will have (**alt**ernate **d**ecimal places). The default value is 2.
Dimaltf	When Dimalt is on, Dimaltf controls the multiplication factor for the alternate dimension (**alt**ernate **f**actor). The value held by Dimaltf will be multiplied by the standard dimension value to determine the alternate dimension. The default value is 25.4, the number required to display metric units.
Dimapost	When Dimalt is on, you can use Dimapost to append text to the alternate dimension (alternate **post**). For example, if Dimapost is given the value "mm," the alternate dimension will appear as value mm instead of just value. The default value is null. To change a previously set value to null, enter a period for the Dimapost new value.
Dimaso	When on, dimensions will be associative (**asso**ciative). When off, dimensions will consist of separate drawing entities with none of the associative dimension properties. The default is on.
Dimasz	Sets the size of dimension arrows or Dimblks (**a**rrow **s**i**z**e). If set to 0, a tick is drawn in place of an arrow. (See Dimblk.) The default value is .18 units.

Dimblk You can replace the standard AutoCAD dimension arrow with one of your own design by creating a drawing of your symbol and making it a **block.** You then give Dimblk the name of your symbol block. This block must be drawn corresponding to a one-by-one unit area and must be oriented as the right side arrow. The default value is null.

Dimblk1 With Dimsah set to on, you can replace the standard AutoCAD dimension arrows with two different arrows using Dimblk1 and Dimblk2. Dimblk1 holds the name of the block defining the first dimension arrow, while Dimblk2 holds the name of the second dimension arrow **block**.

Dimblk2 See Dimblk1.

Dimcen Sets the size of **cen**ter marks used during the Center, Diameter, and Radius dimension subcommands. A negative value draws center lines instead of the center mark cross, while a 0 value draws nothing. The default value is 0.09 units.

Dimclrd Lets you specify **colors** for **d**imension lines, arrowheads, and dimension leader lines.

Dimclre Sets **color** for dimension **e**xtension lines.

Dimclrt Sets **color** for dimension **t**ext.

Dimdle With Dimtsz given a value greater than 0, dimension lines can extend past the extension lines by the amount specified in Dimdle (**d**imension **l**ine **e**xtension). This amount is not adjusted by Dimscale. The default value is 0.

Dimdli Sets the distance at which dimension lines are offset when you use the Baseline or Continue dimension subcommands (**d**imension **l**ine **i**ncrement). The default is 0.38 units.

Dimexe Sets the distance the extension lines are drawn past the dimension lines (**ex**tension line **e**xtension). The default value is 0.18 units.

Dimexo Sets the distance between the beginning of the extension line and the actual point selected at the "Extension line origin:" prompt (**ex**tension line **o**ffset). The default value is 0.0625 units.

Dimgap Sets the distance or **gap** between the text and the dimension line and lets you enclose text within a box by assigning it a negative value. The default value is 0.09.

Dimlfac Sets the global scale factor for dimension values (**l**ength **fac**tor). Linear distances will be multiplied by the value held by Dimlfac. This multiple will be entered as the dimension text. The default value is 1.0. This can be useful when drawings are not drawn to scale.

Dimlim When set to on, dimension text is entered as two values representing a dimension range rather than a single value. The range is determined by the values given to Dimtp (**p**lus **t**olerance) and Dimtm (**m**inus **t**olerance). The default value is off.

Dimpost Automatically appends text strings to dimension text. For example, if Dimpost is given the value "inches," dimension text will appear as value inches instead of just value. The default value is null. To change a previously set value to null, enter a period for the Dimpost new value. If you use Dimpost in conjunction with appended dimension text, the Dimpost value is included as part of the default dimension text.

Dimrnd Sets the amount to which all dimensions are **r**ou**nd**ed. For example, if you set Dimrnd to 1, all dimensions will be integer values. The number of decimal places affected depends on the precision value set by the Units command. The default is 0.

Dimsah When set to on, allows the separate arrow blocks Dimblk1 and Dimblk2 to replace the standard AutoCAD arrows (**s**eparate **a**rrow **h**eads). If Dimtsz is set to a value greater than 0, Dimsah has no effect.

Dimscale Sets the **scale** factor for dimension variables that control dimension lines and arrows and text size (unless current text style has a fixed height). If your drawing is not full scale, you should set this variable to reflect the drawing scale. For example, for a drawing whose scale is ~q~" equals 1~', you should set Dimscale to 48. The default value is 1.0.

Dimse1 When set to on, the first dimension line extension is not drawn (**s**uppress **e**xtension **1**). The default is off.

Dimse2 When set to on, the second dimension line extension is not drawn (**s**uppress **e**xtension **2**). The default is off.

Dimsho When set to on, dimension text in associative dimensions will dynamically change to reflect the location of a dimension point as it is being moved (**sho**w dimension). The default is off.

Dimsoxd When set to on, dimension lines do not appear outside of the extension lines (**s**uppress **o**utside e**x**tension **d**imension lines). If Dimtix is also set to on and the space between the extension lines prohibits the display of a dimension line, no dimension line is drawn. The default is off.

Dimstyle Identifies the current dimension **style**. Use the Save and Restore dimensioning commands for alternate Dimstyles. The default is *UNNAMED.

Dimtad When set to on and Dimtih is off, dimension text in linear dimensions will be placed above the dimension line (**t**ext **a**bove **d**imension line). When off, the dimension line will be split in two and text will be placed in line with the dimension line. The default value is off.

Dimtdec Controls **t**olerance value **dec**imal place for primary units dimension. Values can be selected from the Precision popup list under Primary Units in the Annotation dialog box.

Dimtfac Sets **t**ext scale **fac**tor for text height of tolerance values based on Dimtxt, the dimension text height variable. Use Dimtfac to display the plus and minus characters when Dimtol is on and Dimtm does not equal Dimtp, or when Dimlim is on. The default is 1.0.

Dimtih When set to on, dimension text placed between extension lines will always be horizontal (**t**ext **i**nside **h**orizontal). When set to off, text will be aligned with the dimension line. The default value is on.

Dimtix When set to on, dimension text will always be placed between extension lines (**t**ext **i**nside e**x**tension). The default value is off.

Dimtm When Dimtol or Dimlin is on, Dimtm determines the minus tolerance value of the dimension text (**t**olerance **m**inus).

Dimtofl With Dimtofl on, a dimension line is always drawn between extension lines even when text is drawn outside (**t**ext **o**utside—**f**orced **l**ine). The default is off.

Dimtoh With Dimtoh on, dimension text placed outside extension lines will always be horizontal (**t**ext **o**utside—**h**orizontal). When set to off, text outside extension lines will be aligned with dimension line. The default is on.

Dimtol With Dimtol on, tolerance values set by Dimtp and Dimtm are appended to the dimension text (**tol**erance). The default is off.

Dimtp When Dimtol or Dimlim is on, Dimtp determines the plus tolerance value of the dimension text (**t**olerance **p**lus).

Dimtsz Sets the size of tick marks drawn in place of the standard AutoCAD arrows (**t**ick **s**i**z**e). When set to 0, the standard arrow is drawn. When greater than 0, tick marks are drawn and take precedence over Dimblk1 and Dimblk2. The default value is 0.

Dimtvp When Dimtad is off, Dimtvp allows you to specify the location of the dimension text in relation to the dimension line (**t**ext **v**ertical **p**osition). A positive value places the text above the dimension line, while a negative value places the text below the dimension line. The dimension line will split to accommodate the text unless the Dimtvp value is greater than 1.

Dimtxt Sets the height of dimension **text** when the current text style height is set to 0. The default value is 0.18.

Dimzin Determines the display of inches when Architectural units are used. When set to 0, zero feet or **z**ero **in**ches will not be displayed. When set to 1, zero feet and zero inches will be displayed. When set to 2, zero inches will not be displayed. When set to 3, zero feet will not be displayed.

Dimensioning and Drawing Scales

Take care when dimensioning drawings at a scale other than one-to-one. If the Dimscale dimension variable is not set properly, arrows and text will appear too small or too large. In extreme cases, they may not appear at all. If you enter a dimension and arrows or text do not appear, check the Dimscale setting and make sure it is a value equal to the drawing scale. For instance, a drawing at 1 to 1 that is going to be plotted at _" =1' the dimscale should be set to 48.

Starting the Dimensioning Process

Depending upon your system configuration or your preferences, you may access the dimensioning commands in a number of different ways.

- Toolbar From the Dimensioning toolbar, click on the appropriate Dimensioning button.
- Pull-down menu From the Dimensioning Menu, select the appropriate command on the menu.
- Command line directly Each dimensioning mode command has an equivalent AutoCAD command that can be invoked directly at the command prompt; for example, Dimaligned.
- Command line with Dim/Dim1 All of the dimensioning subcommands from previous releases of AutoCAD are still available for compatibility purposes. To use them, enter Dim or Dim1 at the command prompt. At this point, the prompt changes to "Dim:" and you can enter any dimensioning subcommand. These and the transparent commands (see Transparent) are the only AutoCAD commands you can enter while in the dimensioning mode. When you have finished entering dimensions under the Dim command, issue the Exit command or press Escape to return to the standard command prompt. If you want to enter only a single dimension, use Dim1. The 1 tells AutoCAD to return you automatically to the command prompt after you complete one dimension. You can invoke any dimension command by entering just its first three letters. For example, you can enter Dia instead of Diameter. Alternative short command forms are shown with each subcommand as appropriate.

Dimaligned

Dimaligned aligns a dimension with two points or an object. The dimension text appears in the current style. Figure 4 illustrates the difference between aligned and rotated dimensions. Dimension settings are modified with the Ddim dialog box. The Dimensioning mode equivalent is Dimaligned/Al.

To Align a Dimension

Command Line: Dimaligned

Menu: Dimension ➤ Aligned

Dimension Toolbar: Aligned Dimension

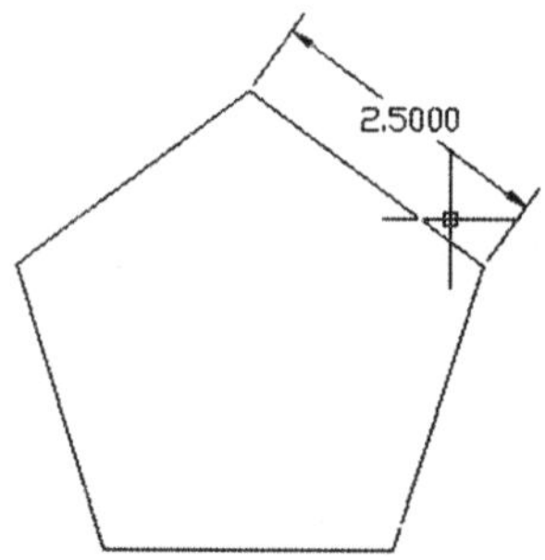

Figure 4: Aligned and rotated dimensions

1. First extension line origin or press ENTER to select:Pick one end of object to be dimensioned. If you entered coordinates for an extension line, you will be prompted for the following: Second extension line origin: Pick the other end of the object.
2. Dimension line location (Mtext/Text/Angle): If you wish to modify the text or text angle, enter an option (M, T, A). To position the dimension line dynamically, drag your cursor and pick a point; or enter a coordinate for the location of the dimension line.
3. Dimension text <default dimension>: Press to accept the default dimension, or enter dimension value.

Options

Mtext Enter (M) to open the Multiline Text Editor dialog box and display the text as angle brackets <>. Accept the default and pick OK, or enter a new dimension text. Additional text can be placed before or after the angle brackets as described in the Text option.

Text This option (T) allows you to edit and customize the dimension. AutoCAD displays the dimension measurement at the command prompt Dimension text <measured length>: and allows you to modify it. If you wish to enter a different value, enter it at the command prompt and press . If you wish to insert additional text before the dimension measurement, or append text after it, place angle bracket signs (<>) to represent the default dimension value and add the desired text before or after. For example, if you wish to add "inches" after the value, type <> inches at the "Dimension text:" prompt.

Angle This option (A) allows you to change the default angle of the text. Text will be placed horizontally (0 degrees) unless you specify a dimension text orientation to the "Enter text angle:" prompt.

Dimangular

Dimangular creates a dimension label showing the angle described by an arc, circle, or two lines, or by a set of three points. An arc with dimension arrows at each end is drawn and the angle value is placed using current text style. Dimension settings can be modified with the

Ddim dialog box, or by selecting each individual Dim variable. The Dimensioning mode equivalent is Dim ø Angular/An.

To Create an Angle Dimension Label

Command Line: Dimangular

Menu: Dimension ➤ Angular

Dimension Toolbar: Angular Dimension

1. Select arc, circle, line, press ENTER: Pick an object as indicated by the prompt or press to indicate angles using your cursor.
   - If you selected an arc, or a circle, continue with Step 2.
   - If you selected a line, you will be prompted Second Line:. Select another line, and then continue with step 2.
   - If you pressed, you may specify the desired angle by giving three points: the Angle vertex:, the First angle endpoint:, and the Second angle endpoint:. Then continue with step 2.
2. Dimension arc line location (Mtext/Text/Angle): If you wish to modify the text or text angle, enter an option (M, T, A). Otherwise, drag your cursor and pick a point, or enter a coordinate for the location of the dimension line as the extension line with text appears.

Options

For a description of the Mtext, Text and Angle options, see the Dimaligned command.

Dimbaseline

Dimbaseline continues a dimension string using the first extension line of the most recently inserted dimension as its first extension line. You are prompted only for the second extension line origin. Each new dimension is offset by an increment set in the Dimdli system variable from the previous dimension line. Prompts vary depending on the type of dimension that was last created. The Dimensioning mode equivalent is Dim Baseline/B.

To Continue a Dimension from the Baseline

Command Line: Dimbaseline

Menu: Dimension ➤ Baseline

Dimension Toolbar: Baseline Dimension

1. Specify a second extension line origin or (<select>/Undo): Pressing prompts to Select base dimension: if you wish to pick a new dimension for the start point. Otherwise, select as many points as you wish along the measured object. AutoCAD responds Dimension text =default dimension and draws a series of dimensions, each starting from the second point extension of the last dimension in the chain. To end the command, press or use the Escape key.
   - If no dimensions exist in the current session, the prompt appears: Select base dimension:, Linear, Ordinate, or Angular associative dimension required.

- If the previous dimension was linear, the prompt in step 1 appears. Otherwise, previous angular or ordinate dimensions display the prompt: Specify a point on an object or (<select>/ Undo):. Pick a point, use an object snap mode option to select, or press to select a base dimension.

Option

Undo Removes the last dimension string. AutoCAD allows you to repeat the U option until all dimension strings are undone, then displays the prompt to Select base dimension.

See Also Dimension Variable: Dimdli

Dimcenter

Dimcenter places a cross at the center point of a selected arc or circle. To choose center lines instead of center marks, and to change the size of the center mark, use the Dimcen dimension variable or reset them in the Geometry dialog box of the Ddim command. The Dimensioning mode equivalent is Dim Center/Ce.

To Mark the Center of an Arc or Circle

Command Line: Dimcenter

Menu: Dimension ➢ Center Mark

Dimension Toolbar: Center Mark

Select arc or circle: Pick an arc or circle and the mark will appear.

Dimcontinue

Dimcontinue draws a series, or "chain," of related dimensions by using the second extension line of the most recently inserted dimension as its first extension line. All of the "chained" dimension measurements add up to a total measurement of the object. You are prompted only for the second extension line origin. If the original dimension is linear, the related dimensions are placed in line with, and parallel to, this dimension. The Dimfit system variable controls the positioning of the dimension lines and text between the selected points. The Dimensioning mode equivalent is Dim Continue/Co.

To Continue a Dimension

Command Line: Dimcontinue

Menu: Dimension ➢ Continue

Dimension Toolbar: Continue Dimension

Specify a second extension line origin or (<select>/Undo): Select as many points as you wish along the measured object. AutoCAD responds with Dimension text=default dimension and draws a series of dimensions, each starting from the second extension line of the last dimension in the chain.

Note Pressing displays the prompt "Select continued dimension:". Pick a new dimension for the start point or to exit Dimcontinue.

Option

Undo Removes the last dimension string. AutoCAD allows you to repeat the U option until all dimension strings are undone, then displays the prompt to "Specify a second extension line origin or (<select>/Undo):".

Dimdiameter

Dimdiameter draws different types of diameter dimensions depending upon the size of the arc or circle and upon the dimension settings created in Ddim Format dialog box, and stored in the dimensioning system variables as a dim style. The Dimensioning mode equivalent is Dim ø Diameter/Di.

To Dimension an Arc or Circle Diameter

Command Line: Dimdiameter.

Menu: Dimension ➢ Diameter

Dimension Toolbar: Diameter Dimension

1. Select arc or circle: Pick the arc or circle as appropriate.
2. Dimension line location<Mtext/Text/Angle>: As the leader line extends in a dragging mode from the arc or circle, pick a point that indicates the location of the text or enter an option (M,T, A). AutoCAD displays the value with Dimension text =default dimension. The direction of the leader is toward the center point. A center mark is also placed at the center of the arc or circle.

Options

For a description of the Mtext, Text and Angle options, see the Dimaligned command.

Note Unless you modify the text, the dimension value will be preceded by a diameter symbol. The point at which you pick the arc or circle determines one end of the dimension arrow. If you want the dimension to be in a horizontal or vertical orientation, use the Quadrant Osnap override. Pick the left or right quadrant for a horizontal dimension; pick the top or bottom quadrant for a vertical dimension. If the Dimtix dimension variable is set to on, the dimension is placed inside the circle starting at your pick point. A center mark is also placed at the center of the arc or circle.

See Also Dimension Variables: Dimcen, Dimtix.

Dimedit

Dimedit allows you to change and manipulate selected text and extension lines. Dimedit can be used to change the same elements on several dimension objects at once. Using this command, you can Rotate text; shift it back to its original "home" position; and edit dimension text. The Oblique option allows you to skew existing dimension extension lines to an angle other than 90 degrees to the dimension line. This is used for isometrics. The Dimedit command combines the dimensioning subcommands Dim: Hometext, Dim: Newtext, Dim: Trotate, and Dim: Oblique.

To Edit Dimension Text

Command Line: Dimedit

Menu: Dimension ➢ Oblique

Dimension Toolbar: Dimension Edit

1. Dimension Edit (Home/New/Rotate/Oblique) <Current>: Select an option or press.
2. Select Objects: Select dimensions and dimension text to edit.

Options

Home Repositions all selected dimension text objects in their default positions. This has the same effect as the Dim: Home dimensioning subcommand.

New Opens the Multiline Text Editor, allowing you to replace associative dimension text on several dimensions at once. Replace the angle brackets with new dimension text, then pick OK. Entering text outside the brackets appends it to the existing dimension string. All associative dimensions selected will be changed to the new text. This has the same effect as the Dim: Newtext dimensioning subcommand.

Rotate Allows you to rotate dimension text to a specified angle. At the prompt Enter text angle, enter the required angle. This has the same effect as the Dim: Trotate dimensioning subcommand.

Oblique Allows you to change all selected extension lines to a specified angle. At the prompt "Enter obliquing angle (press ENTER for none):", enter the required angle. This has the same effect as the Dim: Oblique dimensioning subcommand.

Dimlinear

Creates a linear dimension. Dimlinear allows you to position linear dimensions in either a horizontal or a vertical direction, as well as allowing you to rotate the dimension line and text to a specified angle. This command combines the dimensioning subcommands Dim: Horizontal, Dim: Vertical, and Dim: Rotate.

To Create Linear Dimensions

Command Line: Dimlinear

Menu: Dimension Ø Linear

Dimension Toolbar: Linear Dimension

1. First extension line origin or press ENTER to select: Pick a point on the object to be dimensioned or press.
2. Second extension line origin: Pick a second point on the object.

- If you pressed , you will be prompted to Select object to dimension:. When you select an object, AutoCAD automatically locates the origin points for the extension lines.

3. Dimension line location (Mtext/Text/Angle/Horizontal/Vertical/Rotated): Enter an option, pick a location or enter a coordinate for the location of the dimension line as the extension line with mtext or text appears.

Options

For a description of the Mtext, Text and Angle options, see the Dimaligned command.

Horizontal Allows you to specify a horizontal dimension at "the Dimension line location (Mtext/Text/Angle):" prompt, or enter an option. You may also do this dynamically. If you drag your cursor up or down the screen, the horizontal dimension will appear.

Vertical Allows you to specify a vertical dimension at "the Dimension line location (Mtext/Text/Angle):" prompt, or enter an option. You may also do this dynamically. If you drag your cursor to the left or right across the screen, the vertical dimension will appear.

Rotated Allows you to create rotated dimensions. If you enter R to select the Rotated option, you will be prompted "Dimension line angle <default angle>:". Enter an angle of rotation for the dimension line.

Note If you select a line or an arc, its end points are used as origins for the dimension lines; if you select a circle, the diameter end points are used as the origin points. Selecting polylines and other explodable objects only dimensions the individual line or arc segments. Objects in a non-uniformly scaled block reference are not selectable.

Dimordinate

Dimordinate creates an ordinate dimension string based on a datum or origin point. The Dimensioning mode equivalent is Dim Ordinate/Or. To set the origin point use the UCS command and specify an origin point for the part, i.e., the lower left corner of the part.

To Create Ordinate Dimensioning

Command Line: Dimordinate

Menu: Dimension ➢ Ordinate

Dimension Toolbar: Ordinate Dimension

1. Select Feature: Pick the location of the feature to be dimensioned.
2. Leader endpoint (Xdatum/Ydatum/Mtext/Text): Specify an endpoint for the leader. AutoCAD will determine automatically whether to measure the X or Y coordinate. Alternatively, you may enter X or Y to specify the axis along which the dimension is to be taken. Enter T or M if you wish to modify the dimension text. Then press to display the value.
   - Dimension text = <dimension default >

Options

For a description of the Mtext and Text options, see the Dimaligned command.

Note If you pick a point at the first Leader endpoint: prompt, then AutoCAD selects the dimension axis based on the angle defined by the points you pick during the Select Feature: and Leader endpoint: prompts.

See Also UCS

Dimoverride

Dimoverride changes an individual dimension's properties, such as its arrow style, colors, scale, text orientation, etc. It allows you to override current dimension variable settings for selected objects only, without changing the current Dimension Style. The Dimensioning mode equivalent is Dim Override/Ove.

To Override Dimension Variable Settings

Command Line: Dimoverride

Menu: Dimension ➢ Override

1. Dimension variable to override (or Clear to remove overrides): Enter the name of a dimension variable, or type C to clear.
2. Current value <(variables value)> New value: Enter new value or setting for the dimension variable.
3. Dimension variable to override: Enter another dimension variable name and repeat the process or press to continue.
4. Select objects: Select dimension object(s) to change.

Options

Clear Allows you to clear any dimension variable overrides. AutoCAD clears any overrides and the dimension object reverts to the setting defined by its dimension style

See Also Dimstyle, Restore, Ddmodify

Dimradius

Dimradius adds a radius dimension to arcs and circles in essentially the same way that Dimdiameter adds diameter dimensions. The Dimensioning mode equivalent is Dim Radius/Ra.

To Add a Radius Dimension

Command Line: Dimradius

Menu: Dimension ➢ Radius

Dimension Toolbar: Radius Dimension

1. Select arc or circle: Pick the arc or circle. The point at which you pick the arc or circle determines one end of the dimension arrow.

2. Dimension line location (Mtext/Text/Angle): If you wish to modify the mtext, text or text angle, enter an option (M, T, A). Or drag the cursor to position the mtext or text (and leader). The Dimension text = default radius value is displayed above the prompt.

Options

For a description of the Mtext, Text and Angle options, see the Dimaligned command.

Note Unless you modify the mtext or text, the dimension value will be preceded by the letter R. If a dimensioned circle or arc is large enough, the radius dimension will be positioned within it. Otherwise it is positioned outside the circle or arc, with a leader line pointing to the center. A center mark is also placed at the center of the arc or circle. To drag the dimension text outside the circle or arc turn on dimupt or use Ddim and click User defined in Format. In Ddim it is in Format..., user defined.

Dimstyle

Dimstyle allows you to create and modify dimension styles at the command line. The Ddim dialog boxes offer a more detailed and graphical approach to creating dimension styles, but Dimstyle allows a quick method of saving and creating dimension styles on the fly. The Dimensioning mode equivalent is Dim Dimstyle.

To Create/Modify a Dimension Style

Command Line: Dimstyle

Menu: Dimension ➢ Dimstyle

Dimension Toolbar: Dimension Style

1. Dimension Style Edit (Save/Restore/STatus/Variables/Apply/?) <Restore>: Type ? to list the dimension styles in the current drawing, or select an option. AutoCAD also reports the current dimension style along the prompt.
2. Depending upon the option chosen, the prompts will vary. Enter the required information.

Options

Save Displays the prompt?/Name for new dimension style: when you type S. Enter a name to save the current dimension variable settings as a dimension style that can later be restored using the Restore option. You can save multiple dimension styles. You can also redefine an existing style by saving it back as the same name. This updates all the dims that have that style.

Restore Typing R returns the prompt "?/Enter dimension style name or press ENTER" to select dimension: Enter a name to makes an existing dimension style the current default style. You can either name a specific style or select an object that uses the style. Restore immediately changes the current dimension style to reflect any differences.

Status Select this option by entering Sta to display the current dimension variable settings and descriptions. Once the dimension variables have been listed, Dimstyle ends.

Variables To list the dimension variable settings of a dimension style, enter V, then press . You can either name a specific style or select an object that uses the style at the "Select dimension:" prompt. Once the dimension variables have been listed, Dimstyle ends.

Apply Typing A displays the "Select objects:" prompt allowing you to update selected dimension objects with the current dimension style, including any overrides.

? Lists the named dimension styles in the current drawing. Pressing displays the "Dimension style(s) to list <*>:" prompt and reports the Named dimension styles: that are stored in the current drawing.

See Also Restore, Save, Ddim

Dimtedit

Modifies the placement, justification, and rotation angle of a single dimension text object. The Dimensioning mode equivalent is Dim ø Dimtedit.

To Modify Dimension Text

Command Line: Dim ø Tedit/Te

Menu: Dimension ➢ Align Text ➢ Home/Angle/Center/Right

Dimension Toolbar: Dimension Text Edit

1. Select dimension: Pick a single dimension.
2. Enter text location (Left/Right/Home/Angle): Enter an option or pick a new location for the dimension text. You can drag the text to the next location with the cursor. If you move the text to a position in line with the dimension line, the dimension will automatically join to become a continuous line.

Options

Left Justifies text to the left within the extension lines on linear, radius, and diameter dimensions.

Right Justifies text to the right within the extension lines on linear, radius, and diameter dimensions.

Home Places text in its default dimension. This has the same effect as the Hometext subcommand.

Angle Changes the angle for text. At the prompt "Text angle:", enter an angle value, or indicate an angle by picking two points.

Note When Tedit is used on a dimension, the style is updated to the current dimension style setting. If no style is associated with the dimension, then current dimension variable settings are used

See Also Dim:, Hometext; System Variables: Dimaso, Dimsho

Dimensioning Mode Subcommands

Most of the dimensioning commands are still available as subcommands in Dimensioning mode (Dim: mode). Dim: mode equivalents, where available, are identified with each command reviewed above. A number of dimensioning variables remain as Dim subcommands only.

Table 5: The DIM Subcommands and the Dimensioning Commands

DIM subcomand	Dimensioning Commands
Dim: Aligned/AL	DIMALIGNED
Dim: Angular/AN	DIMANGULAR
Dim: Baseline/B	DIMBASELINE
Dim: Center/CE	DIMCENTER
Dim: Continue/CO	DIMCONTINUE
Dim: Diameter/D	DIMDIAMETER
Dim: Hometext/HOM	DIMEDIT Home
Dim: Horizontal/HOR	DIMLINEAR Horizontal
Dim: Leader/L	LEADER
Dim: Newtext/N	DIMEDIT New
Dim: Oblique/OB	DIMEDIT Oblique
Dim: Ordinate/OR	DIMORDINATE
Dim: Override/OV	DIMOVERRIDE
Dim: Radius/RA	DIMRADIUS
Dim: Restore/RES	DIMSTYLE Restore
Dim: Rotated/RO	DIMLINEAR Rotated
Dim: Save/SA	DIMSTYLE Save
Dim: Status/STA	DIMSTYLE Status
Dim: Tedit/TE	DIMTEDIT
Dim: Trotate/TR	DIMEDIT Angle
Dim: Variables/VA	DIMSTYLE Variables
Dim: Tedit/TE	DIMTEDIT
Dim: Vertical/VE	DIMLINEAR Vertical
Dim: Exit/E	Exits Dim mode and returns to command mode
Dim: Redraw/RED	Redraws the display
Dim: Style/STY	Switches to a new text style
Dim: Undo/U	Erases the most recent dimension objects
Dim: Update/UP	Redraws dimensions to match the current settings

Dim: Exit/E

This command exits the Dim command and returns you to the standard command prompt.

To Exit Dim

Command Line: Dim Exit/E

You can press the Escape key or type E in place of Exit. If you entered Dim1 to begin dimensioning, Exit is not needed.

Dim: Leader/L

Dim: Leader adds leadered dimensions to drawings based on the current dimension style. You can replace the default dimension text with a single line of text.

To Add Dimensions/Text with Leaders

Command Line: Dim Leader/L

1. Leader start: Pick a point to start the leader. This is the point where the arrow will be placed.
2. To point: Pick the next point along the leader line or continue to pick points as you would in drawing a line. When you are finished, press. You can enter U to Undo the last drawn line segment.
3. Dimension text <default dimension>: Press to accept the default, enter new dimension text or a single line of text.

The distance between the Leader start point and the next point must be at least twice the length of the arrow; otherwise an arrow will not be placed. If the last line segment of the leader is not horizontal, a horizontal line segment is added.

Leadered text does not take on the properties of associative dimensioning. You can append text to the default dimension value. The default value is usually the last dimension entered.

See Also Dimstyle, Leader, Mtext, Tolerance, Spline

Dim: Oblique/Ob

Dim: Oblique skews existing dimension extension lines to an angle other than 90 degrees to the dimension line.

To Skew Extension Lines

Command Line: Dim Oblique/Ob

Menu: Dimension ➢ Oblique

1. Select objects: Pick the dimensions to be edited.
2. Enter obliquing angle (press ENTER for none): Enter the desired angle for extension lines.

If the dimension being edited has a dimension style setting, then this setting is maintained. If no style is associated with the dimension, the current dimension variable settings are used to update the obliqued dimension.

See Also Dimedit, Dimstyle

Dim: Redraw/Red

Redraw behaves like a transparent command and allows you to refresh the screen in the current viewport.

To Refresh the Screen

Command Line: Dim Redraw/Red

See Also Redraw

Dim: Restore/Res

Makes an existing dimension style the current default style.

To Use Restore

Command Line: Dim Restore/Res

?/Enter dimension style name or press ENTER to select dimension: Enter ? to list available dimension styles, enter the name of a known dimension style (wildcard characters are accepted), or press to pick a dimension whose dimension style you want to make current.

To select a dimension style, either pick an associative dimension that is associated with the desired style, or enter the name of the style. The tilde (~) can also be used to compare styles. To find the differences between the current dimension style and another style, enter a dimension style name preceded by a tilde (~) at the prompt "Dimension style(s) to list <*>:". Differences are displayed in a list of dimension variable settings.

See Also Dimoverride, Dim: Save

Dim: Save/Sav

Dim: Save saves the current dimension variable settings as a dimension style that can later be restored using the Dim: Restore subcommand. You can save multiple dimension styles.

To Save Dimension Settings

Command Line: Dim Save/Sa

?/Name for new dimension style: Enter a new dimension style name. If you enter ?, the following prompt appears:

Dimension style(s) to list <*>: or enter a name specification using wildcards.

Note The tilde (~) can also be used to compare styles. To find the differences between the current dimension style and another style, enter a dimension style name preceded by a tilde (~) at the prompt "Dimension style(s) to list <*>:". Differences are displayed in a list of dimension variable settings.

See Also Ddim, Dimoverride, Restore

Dim: Status/Sta

Dim: Status displays the current dimension variable settings and descriptions.

To Display Dimension Settings

Command Line: Dim Status/Sta

The text screen appears listing dimension settings, their status or values and a brief description.

Dim: Style/Sty

Dim: Style specifies a text style for the dimension text. Once the text style is changed, any subsequent dimensions will contain text in the new style. Existing dimension text is not affected.

To Specify a Text Style

Command Line: Dim: Style/Sty

1. New text style <current style nane>: Press to accept the current style, or enter another previously created style name to set as the current text style.

- If you enter the name of a style that does not exist, you receive the message No such text style: Use the main STYLE command to create it.

See Also Ddstyle, Style

Dim: Undo/U

Undo rescinds a dimension you decide you do not want, as long as you are still in the dimension mode.

To Undo a Dimension

Command Line: Dim Undo/U

Note If you issue Undo during the Leader command, the last leader line segment drawn will be undone.

Dim: Update/Up

Update changes old dimensions to new dimension variable settings. Update works only on associative dimensions that have not been exploded.

Command Line: Dim Update/Up

Menu: Dimension ➢ Update

Dimension Toobar: Update

Select objects: Pick dimension string to update.

Dim: Variables/Va

Variables lists the dimension variable settings of a dimension style.

To List the Dimension Variable Settings

Command Line: Dim Variables/Va

?/Enter dimension style name or press ENTER to select dimension: Enter? to view available dimension styles, enter the name of a known dimension style (wildcards are accepted), or press to pick a dimension whose dimension style you wish to list.

The tilde (~) can also be used to compare styles. To find the differences between the current dimension style and another style, enter a dimension style name preceded by a tilde (~) at the prompt "Dimension style(s) to list <*>:". Differences are displayed in a list of dimension variable settings.

See Also Ddim

Additional Dimensioning Options

The two processes described below can be used in conjunction with both Dimensioning commands as well as in Dim: mode.

At the "Dimension text:" prompt, you can append text to the default dimension text.

To Append Dimension Text

Start the required dimensioning command, then provide the following information:

Dimension text <default text>: Place <> signs where you want the text to appear, then enter the text.

At the "First extension line origin or press ENTER to select:" prompt or the "Select arc, circle, line, or press ENTER:" prompt, you can let AutoCAD dimension a line, arc, or circle automatically.

To Use Automatic Dimensioning

Start the required dimensioning command, then provide the following information:

1. First extension line or press ENTER to select: Press .
2. Select arc, circle, line, or press ENTER: Pick the object to be dimensioned.

Note After selecting a line, circle, or arc, you may drag your cursor and the dimension text and then pick a location point.

SAVE

Save stores your currently open file to disk. Use Save to save changes to your drawing under the same name.

To Save Your Drawing

Command Line: Save

Your drawing has not yet been named, the Save Drawing As dialog list box opens. Enter a drawing name in the File edit box. The next time you enter Save, the Save Drawing As dialog box reappears. Click the OK button to save the drawing to the same file name or specify a new file name, directory, and/or path.

Note Use the Savetime command to set a time interval for automatically saving your drawing. The current drawing is saved to the default file name Auto.sv$

SAVEAS

Saveas saves an unnamed drawing with a file name or allows you to rename your drawing, making it the current drawing.

To Save An Existing Drawing as the Current Drawing

Command Line: Saveas

Menu: File ➢ Saveas

Your drawing has not yet been named, the Save Drawing As dialog list box opens. Press to save the drawing as the same name in the File edit box or enter a new file name, directory and/or path.

PROBLEMS

1. How to Draw a Rectangle (80mm x 40mm)?

Command Line: line ↵
LINE Specify first point:
Specify next point or [Undo]: @80<0 ↵
Specify next point or [Undo]: @40<90 ↵
Specify next point or [Close/Undo]: @80<180 ↵
Specify next point or [Close/Undo]: c ↵

Command Line: dimlinear ↵
Specify first extension line origin or <select object>: end
of
Specify second extension line origin: end
of
Specify dimension line location or
[Mtext/Text/Angle/Horizontal/Vertical/Rotated]:
Dimension text = 80 ↵

Command Line: dimlinear
Specify first extension line origin or <select object>: end
of
Specify second extension line origin: end
of

Specify dimension line location or
[Mtext/Text/Angle/Horizontal/Vertical/Rotated]:
Dimension text = 40 ↵

Command Line: qsave ↵
Wright down the file name : <Drawing.dwg>

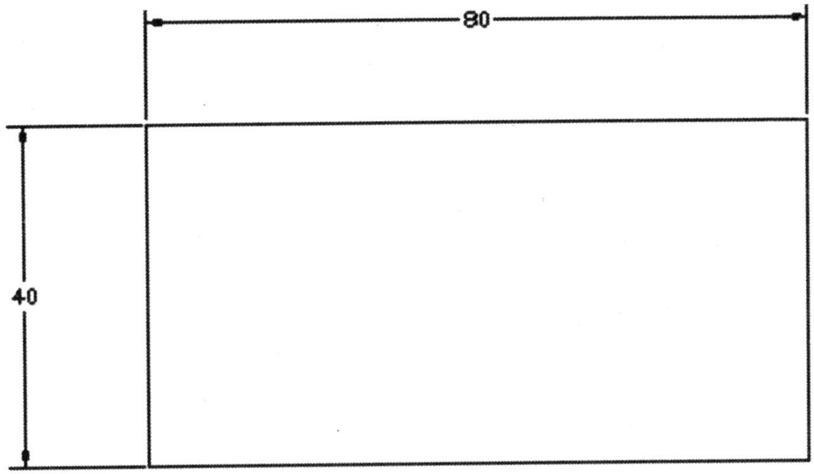

2. How to Draw a Hexagon?

Command Line: polygon Enter number of sides <4>: 6 ↵

Specify center of polygon or [Edge]:
Enter an option [Inscribed in circle/Circumscribed about circle] <I>: I ↵

Specify radius of circle: 10 ↵

Command Line: _polygon Enter number of sides <6>: ↵

Specify center of polygon or [Edge]:
Enter an option [Inscribed in circle/Circumscribed about circle] <I>: C ↵

Specify radius of circle: 10 ↵

Command Line: polygon Enter number of sides <6>: 6 ↵

Specify center of polygon or [Edge]: E ↵
Specify first endpoint of edge: Specify second endpoint of edge: @10<0 ↵

Command Line: qsave ↵
Wright down the file name : <Drawing.dwg>

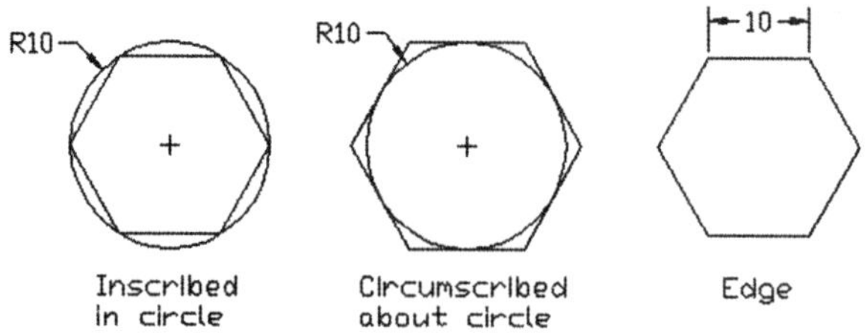

3. How to Draw a Isometric View?

(Page 72, Prob-7.1)

Command Line: line ↵
LINE Specify first point:
Specify next point or [Undo]: @600<30 ↵

Specify next point or [Undo]: ↵

Command Line: line ↵
LINE Specify first point:
Specify next point or [Undo]: @600<150 ↵

Specify next point or [Undo]: ↵

Command Line: copy ↵
COPY
Select objects: Specify opposite corner: 1 found

Select objects:
Specify base point or displacement, or [Multiple]: end
displacement or <use first point as displacement>: end

Command Line: copy ↵
COPY
Select objects: Specify opposite corner: 1 found

Select objects:
Specify base point or displacement, or [Multiple]: end
displacement or <use first point as displacement>: end

Command Line: copy ↵
COPY
Select objects: Specify opposite corner: 1 found

Select objects:

```
Specify base point or displacement, or [Multiple]: end
displacement or <use first point as displacement>: @100<270 ↵

Command Line: line ↵
LINE Specify first point:
Specify next point or [Undo]: @100<90 ↵

Specify next point or [Undo]:

Command Line: line ↵
LINE Specify first point:
Specify next point or [Undo]: @100<90 ↵

Specify next point or [Undo]:

Command Line: line ↵
LINE Specify first point:
Specify next point or [Undo]: @100<90 ↵

Specify next point or [Undo]:

Command Line: line ↵
LINE Specify first point: mid
of
Specify next point or [Undo]: mid
of
Specify next point or [Undo]: mid
of
Specify next point or [Close/Undo]: mid
of
Specify next point or [Close/Undo]: mid
of
Specify next point or [Close/Undo]: ↵

Command Line: offset ↵
OFFSET
Specify offset distance or [Through] <Through>: 50 ↵

Select object to offset or <exit>:
Specify point on side to offset:
Select object to offset or <exit>:
Specify point on side to offset:
Select object to offset or <exit>:
Specify point on side to offset:
```

```
Select object to offset or <exit>:
Specify point on side to offset:
Select object to offset or <exit>:

Command Line: trim ↵
TRIM
Current settings: Projection=UCS, Edge=None
Select cutting edges ...
Select objects: Specify opposite corner: 1 found
Select objects: Specify opposite corner: 1 found, 2 total
Select objects: Specify opposite corner: 1 found, 3 total
Select objects: Specify opposite corner: 1 found, 4 total
Select objects:
Select object to trim or shift-select to extend or [Project/Edge/Undo]:
Select object to trim or shift-select to extend or [Project/Edge/Undo]:
Select object to trim or shift-select to extend or [Project/Edge/Undo]:
Select object to trim or shift-select to extend or [Project/Edge/Undo]:
Select object to trim or shift-select to extend or [Project/Edge/Undo]:
Select object to trim or shift-select to extend or [Project/Edge/Undo]:
Select object to trim or shift-select to extend or [Project/Edge/Undo]:
Select object to trim or shift-select to extend or [Project/Edge/Undo]:
Select object to trim or shift-select to extend or [Project/Edge/Undo]:
Select object to trim or shift-select to extend or [Project/Edge/Undo]:
Select object to trim or shift-select to extend or [Project/Edge/Undo]:
Command Line: copy ↵
COPY
Select objects: Specify opposite corner: 1 found
Select objects: Specify opposite corner: 1 found, 2 total
Select objects: Specify opposite corner: 1 found, 3 total
Select objects: Specify opposite corner: 1 found, 4 total
Select objects:
Specify base point or displacement, or [Multiple]: end
displacement or <use first point as displacement>: @100<90
Command Line: line ↵
```

LINE Specify first point:
Specify next point or [Close/Undo]: end
Specify next point or [Close/Undo]: end
Specify next point or [Close/Undo]: end

Command Line: copy ↵
COPY
Select objects: Specify opposite corner: 2 found

Select objects: 1 found, 3 total

Select objects: 1 found, 4 total

Select objects:
Specify base point or displacement, or [Multiple]: end
displacement or <use first point as displacement>: @100<90

Command Line: offset ↵
OFFSET
Specify offset distance or [Through] <50.0000>: 50

Select object to offset or <exit>:
Select object to offset or <exit>:
Select object to offset or <exit>:
Specify point on side to offset:
Select object to offset or <exit>:
Specify point on side to offset:
Select object to offset or <exit>:
Specify point on side to offset:
Select object to offset or <exit>:
Specify point on side to offset:
Select object to offset or <exit>:

Command Line: trim ↵
TRIM
Current settings: Projection=UCS, Edge=None
Select cutting edges ...
Select objects: Specify opposite corner: 2 found

Select objects: Specify opposite corner: 2 found, 4 total

Select objects:

Select object to trim or shift-select to extend or [Project/Edge/Undo]:
Select object to trim or shift-select to extend or [Project/Edge/Undo]:

Select object to trim or shift-select to extend or [Project/Edge/Undo]:
Select object to trim or shift-select to extend or [Project/Edge/Undo]:
Select object to trim or shift-select to extend or [Project/Edge/Undo]:
Select object to trim or shift-select to extend or [Project/Edge/Undo]:
Select object to trim or shift-select to extend or [Project/Edge/Undo]:
Select object to trim or shift-select to extend or [Project/Edge/Undo]:
Select object to trim or shift-select to extend or [Project/Edge/Undo]:
Select object to trim or shift-select to extend or [Project/Edge/Undo]:
Select object to trim or shift-select to extend or [Project/Edge/Undo]:
Select object to trim or shift-select to extend or [Project/Edge/Undo]:
Select object to trim or shift-select to extend or [Project/Edge/Undo]:

Command Line: copy ↵
COPY
Select objects: Specify opposite corner: 4 found

Select objects:
Specify base point or displacement, or [Multiple]: Specify second point of displacement or <use first point as displacement>: @100<90 ↵

Command Line: line ↵
LINE Specify first point: end
of
Specify next point or [Undo]: @100<90 ↵

Specify next point or [Undo]:

Command Line: line ↵
LINE Specify first point: end
Specify next point or [Undo]: @100<90 ↵

Specify next point or [Undo]:

Command Line: ↵
LINE Specify first point: end
Specify next point or [Undo]: @100<90 ↵

Specify next point or [Undo]:

Command Line: ↵
LINE Specify first point: end
of
Specify next point or [Undo]: @100<90 ↵

Specify next point or [Undo]:

Command Line: trim ↵
TRIM
Current settings: Projection=UCS, Edge=None
Select cutting edges ...
Select objects: Specify opposite corner: 23 found

Select objects:

Select object to trim or shift-select to extend or [Project/Edge/Undo]:
Select object to trim or shift-select to extend or [Project/Edge/Undo]:
Select object to trim or shift-select to extend or [Project/Edge/Undo]:
Select object to trim or shift-select to extend or [Project/Edge/Undo]:
Select object to trim or shift-select to extend or [Project/Edge/Undo]:
Select object to trim or shift-select to extend or [Project/Edge/Undo]:
Select object to trim or shift-select to extend or [Project/Edge/Undo]:

Command: qsave
Wright down the file name : <Drawing.dwg>

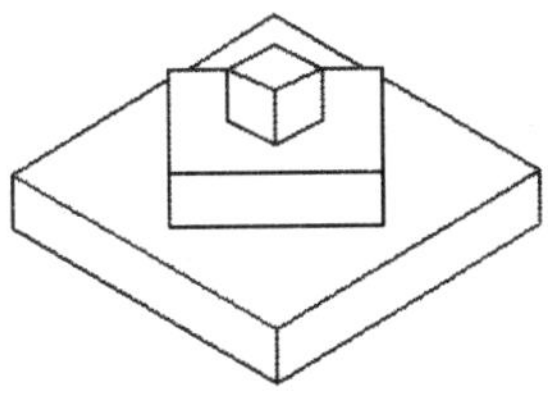

4. How to Draw a Shaft Coupling?

(Page 129, Prob-12.9)

Command: line ↵
LINE Specify first point:

Specify next point or [Undo]: @60<30 ↵

Specify next point or [Undo]: *Cancel*

Command: line ↵
LINE Specify first point:
Specify next point or [Undo]: @45<45 ↵

Specify next point or [Undo]:

Command Line: line ↵
LINE Specify first point: *Cancel*

```
Command: trim ↵
TRIM
Current settings: Projection=UCS, Edge=None
Select cutting edges ...
Select objects: Specify opposite corner: 4 found
Select objects: pick ↵
Select object to trim or shift-select to extend or [Project/Edge/Undo]:
Select object to trim or shift-select to extend or [Project/Edge/Undo]:
Select object to trim or shift-select to extend or [Project/Edge/Undo]:
Select object to trim or shift-select to extend or [Project/Edge/Undo]:
Select object to trim or shift-select to extend or [Project/Edge/Undo]:
Select object to trim or shift-select to extend or [Project/Edge/Undo]:
Select object to trim or shift-select to extend or [Project/Edge/Undo]:
Select object to trim or shift-select to extend or [Project/Edge/Undo]:

Command: mirror ↵
MIRROR
Select objects: Specify opposite corner: 1 found
Select objects: Specify opposite corner: 1 found, 2 total
Select objects:
Specify first point of mirror line: Specify second point of mirror line:
Delete source objects? [Yes/No] <N>:
Command: Specify opposite corner:
Command: Specify opposite corner:
Command: Specify opposite corner:
Command: _.erase 1 found

Command: trim ↵
TRIM
Current settings: Projection=UCS, Edge=None
Select cutting edges ...
Select objects: 1 found
Select objects: Specify opposite corner:
Invalid window specification.
Select objects: Specify opposite corner: 0 found
```

Select objects:

Select object to trim or shift-select to extend or [Project/Edge/Undo]:
Select object to trim or shift-select to extend or [Project/Edge/Undo]:
Select object to trim or shift-select to extend or [Project/Edge/Undo]:

Command: offset ↵
OFFSET
Specify offset distance or [Through] <10.0000>: 65

Select object to offset or <exit>:
Specify point on side to offset:
Select object to offset or <exit>: *Cancel*

Command: move ↵
MOVE
Select objects: Specify opposite corner: 1 found

Select objects:
Specify base point or displacement: Specify second point of displacement or <use first point as displacement>:

Command: stretch ↵
STRETCH
Select objects to stretch by crossing-window or crossing-polygon...
Select objects: Specify opposite corner: 3 found

Select objects:

Specify base point or displacement:
Specify second point of displacement or <use first point as displacement>:
Command: *Cancel*

Command: *Cancel*

Command: trim ↵
TRIM
Current settings: Projection=UCS, Edge=None
Select cutting edges ...
Select objects: Specify opposite corner: 1 found

Select objects:

```
Select object to trim or shift-select to extend or [Project/Edge/Undo]:
Select object to trim or shift-select to extend or [Project/Edge/Undo]:
Select object to trim or shift-select to extend or [Project/Edge/Undo]:
Select object to trim or shift-select to extend or [Project/Edge/Undo]:

Command:
DIST Specify first point: Specify second point:
Distance = 65.0000, Angle in XY Plane = 270, Angle from XY Plane = 0
Delta X = 0.0000, Delta Y = -65.0000, Delta Z = 0.0000

Command: move ↵
MOVE
Select objects: Specify opposite corner: 9 found
Select objects: Specify opposite corner: 3 found, 12 total
Select objects: Specify opposite corner: 2 found, 14 total
Select objects:
Specify base point or displacement: Specify second point of displacement or
<use first point as displacement>:

Command: mirror ↵
MIRROR
Select objects: Specify opposite corner: 9 found
Select objects:
Specify first point of mirror line: Specify second point of mirror line:
Delete source objects? [Yes/No] <N>:
Command: _dimlinear ↵
Specify first extension line origin or <select object>:
Specify second extension line origin:
Specify dimension line location or
[Mtext/Text/Angle/Horizontal/Vertical/Rotated]:
Dimension text = 10.0000
Command: mirror ↵
MIRROR
Select objects: *Cancel*
```

```
Command: line ↵
LINE Specify first point:
Specify next point or [Undo]:
Specify next point or [Undo]:
Command: _.erase 1 found
Command: mi
MIRROR
Select objects: Specify opposite corner: 6 found

Select objects:
Specify first point of mirror line: Specify second point of mirror line:
Delete source objects? [Yes/No] <N>:

Command: _.erase 1 found

Command: tr
TRIM
Current settings: Projection=UCS, Edge=None
Select cutting edges ...
Select objects: Specify opposite corner: 1 found

Select objects: Specify opposite corner: 2 found, 3 total

Select objects:

Select object to trim or shift-select to extend or [Project/Edge/Undo]:
Select object to trim or shift-select to extend or [Project/Edge/Undo]:
Select object to trim or shift-select to extend or [Project/Edge/Undo]:

Command: m
MOVE
Select objects: Specify opposite corner: 28 found

Select objects:
Specify base point or displacement: Specify second point of displacement or
<use first point as displacement>:
Command: Specify opposite corner:
Command: _.erase 1 found

Command: Specify opposite corner:
Command: _.erase 1 found
```

Command: _qsave

Wright down the file name : <Drawing.dwg>

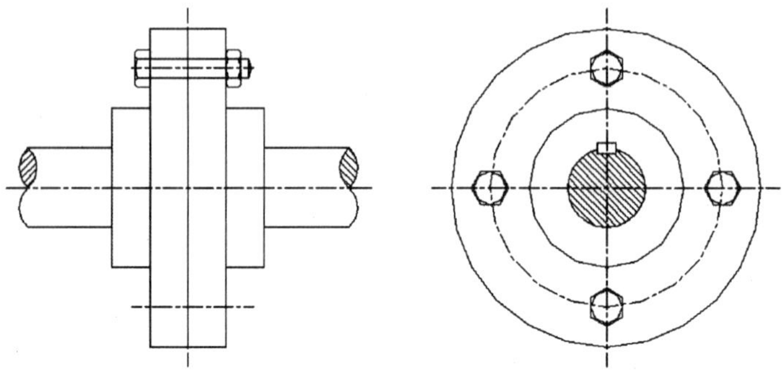

PART – B

CHAPTER – 1

POINTS TO REMEMBER

In this Part of the book we will discuss and present various advanced aspect of machine drawing e.g., tolerancing, surface finish, specifications of materials and common methods of manufacture and form design *etc.*

- Suitable scale should be selected and stated on the drawing.
- Letterings are as important as the drawing.
- Minimum number of views consistent with the clarity of the object should be given.
- Check up the appropriateness of dimensions given. Mating parts should be toleranced, whenever required.
- Sharp corners should be avoided as far as practicable.
- Accuracy demanded should be determined only from functional standpoint.
- More than one parallel surface of a component should not mate with corresponding surfaces of a mating one.
- Bolts preferably should not be used for locating purposes.
- Materials specified should be commonly available in the market.
- Any special feature, for which no standard convention might have been framed, should be clearly stated on the drawing.
- Space between views should be so given as not to overcrowd the dimensions.

METHODS OF PROJECTION

First angle projection and third angle projection, should be used.

CHOICE OF VIEWS

The number of views should be limited to the minimum necessary and sufficient to represent the object without ambiguity.

SPACE BETWEEN VIEWS

Sufficient space should be provided between views to permit placing of dimensions without crowding and to preclude the possibility of notes pertaining to one view overlapping or crowding the other views.

CONVENTIONAL REPRESENTATION

In view of the cases where complete delineation of the part would involve unnecessary drawing line or space and variety of materials are used, it is desirable to resort to conventional representations. An explanatory table is given in page 3.

DIMENSIONING

All dimensions, tolerances *etc.* required for correct functioning of the parts should be placed directly on the drawing to describe the part in its finished form with due consideration to manufacturing and inspection requirements.

CHAPTER – 2

FITS AND TOLERANCES

FITS

When two parts are to be assembled, the relation resulting from the difference between their sizes before assembly is called a fit.

TOLERANCE

The inevitable inaccuracy of manufacturing methods renders it impossible to make a part precisely to a given dimensions and so it is made to lie between two limits, maximum and minimum, the difference between which is the tolerance. Tolerance (dimensional deviation) is a function of size and the grade which determines the manufacturing process. Fig.2.1 illustrates the same with respect to a shaft and a hole.

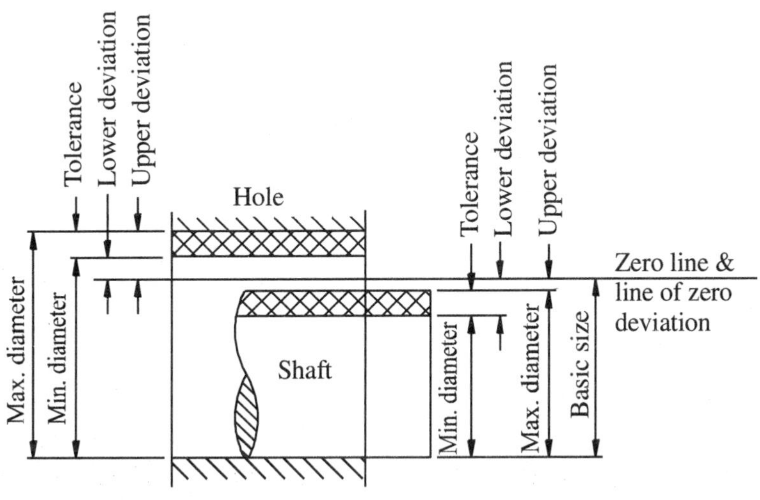

Fig. 2.1

Indian standards provide for

(i) 18 Grades of Standard Tolerances (*e.g.* IT 01, IT 0 and IT 1 to IT 16 in order of decreasing accuracy) and

(ii) 25 Fundamental Deviations indicated by letter symbols which determine the position of the tolerance zones with respect to the zero line, being a function of the basic size (capital letters for holes and small letters for shafts as shown in Fig.2.2).

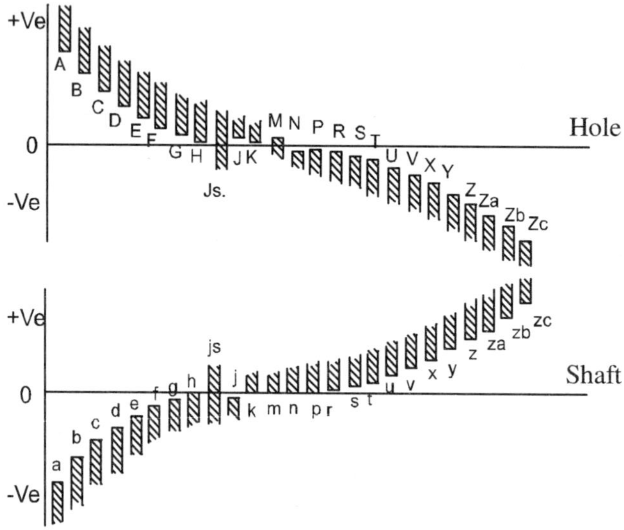

Fig. 2.2

THERE ARE TWO METHODS OF TOLERANCING

(i) Hole basis – Different types of fits are obtained in associating various shafts with a single hole having zero lower deviation (Symbol h).

(ii) Shaft basis – Different types of fits are obtained in associating various holes with a single shaft having zero upper deviation (Symbol h).

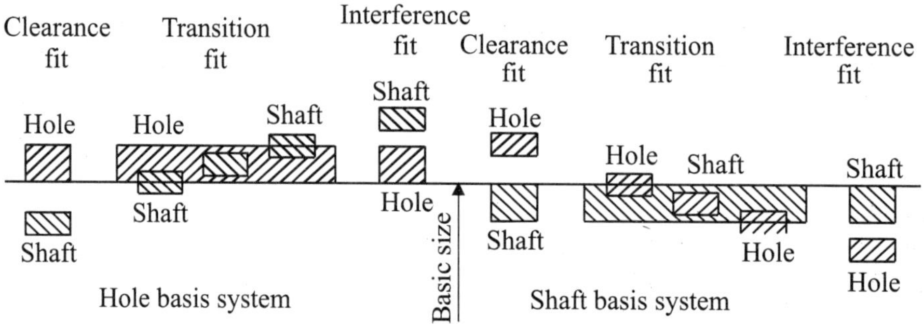

Fig.2.3 Examples illustrating hole basis and shaft basis system

Illustrative example = 50 H7/r6
Basic size of Hole and Shaft = 50 mm
Fundamental Deviation of Hole = H
Tolerance Grade of Hole = 7
Fundamental Deviation of Shaft = r
Tolerance Grade of Shaft = 6

Shafts mean all external features of a part and holes mean internal features of a part including non-cylindrical ones.

STANDARD TOLERANCE UNIT, i

$$i \text{ (microns)} = 0.45\ D^{1/3} + 0.1D \text{ (D in mm)}$$

where D is the geometric mean dimension in mm of standard stepping as given below.
3 – 6 – 10 – 18 – 30 – 50 – 120 – 180 – 250 – 315 – 400 – 500

Table 2.1 Tolerance grade and machining methods

Tolerance grade	Relative magnitude	Matchining methods
5	7i	Precision boring/turning followed by find grinding or honing.
6	10i	
7	16i	Boring/turning followed by grinding.
8	25i	Boring followed by reaming, Finish Turning.
9	40i	
10	64i	Drilling, Turning.
11	100i	

For economic manufacture, tolerance should be as wide as possible consistent with satistactory functioning. As the holes are more difficult to be produced than shafts, the tolerance grade of the holes, in general, is one grade coarser than that of the mating shafts *e.g.* H8-g7. Now the total tolerance can be calculated as

$$T = \text{Grade Tolerance} \times \text{Standard Tolerance}$$

Tabke 2.2 Recommended fits

Type of fit	Fit	Typical applications
	H7-c8	I.C engine exhaust valve in guide.
	H7-c8	Gland seals, loose pulleys, idle gears.
Clearance	H8-e8	Widely separated bearings.
	H7-f7	Bush bearings for gear shafts, small electric motor shafts, pump shafts etc.
	H9-d9	Plummer block bearings.
	H8-g7	Connecting rod cap bolt, bush bearing for accurate link work, piston.
	H11-h11	Rough machines, Transport equipment, cranes, lift
	H7-j6	Worm wheel keyed to shaft.
	H6-k5	Couplings, handwheels, gears on shafts, handles, gudgeon pin on piston bosses.
Transition	H8-j7	Centering spigot on shaft couplings.
	H6-m5	Gears, pulleys, flywheels.
	H7-m6	Mounting bush in housings.
	H7-r6	Press-in bearing bushes, pump impeller on shaft.
	H6-r5	Small end bush in connecting rod.
Interfence	H7-r6	Cylinder liner fitting.
	H6-u5	Locomotive wheel on axie.

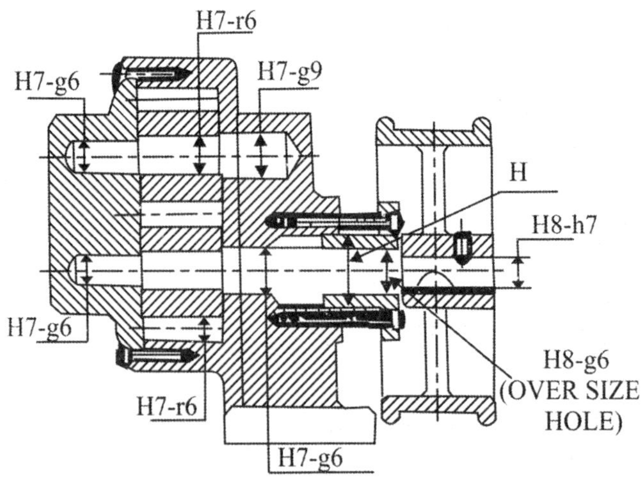

Fig. 2.4 Self driven gear pump

ILLUSTRATION

The fitting of the piston rod in the gland bush in a stuffing box is given to be 50 H8-g7. Show the mating dimensions of the components and find out the extrems of fit.

An excerpt of tables from IS:919-1963 is given (Vide Tables 2.3, 2.4, & 2.5)

Table 2.3 Fundamental deviatons for shafts

	Hole	Shaft
Tolerance	39 Microns	25 Microns
Deviations	39 Microns	–9 Microns
Lower deviations	0	–34 Microns
Dimensions (mm)	50.039 50.000	49.991 49.966
Extremes of fit	+73 Microns +9Microns	i.e. clearance fit

Table 2.4 Fundamental deviatons for shafts

Fundamental deviations in Microns												
Diameter steps in mm		Upper deviations (es)				Lower deviations (ei)						
		d	f	g	h	j		k	m	r	s	u
Over	To	All Grades				5, 6, 7		4 to 7	All Grades			
10	18	–50	–16	–6	0	–3	–6	+1	+7	+23	+28	+33
	18	24										+41
24	30	–65	–20	–7	0	–4	–8	+2	+8	+28	+35	+48
30	40											+60
		–80	–25	–9	0	–5	–10	+2	+9	+34	+43	
40	50											+70
50	65									+41	+53	+87
		–100	–30	–10	0	–7	–12	+2	+11			
65	80									+43	+59	+102
80	100									+51	+71	+124
		–120	–36	–12	0	–9	–15	+3	+13			
100	120									+54	+79	+144
120	140									+63	+92	170
140	160	–145	–43	–14	0	–11	–18	+3	+15	+65	+100	190
160	180									+68	+108	201

GENERAL TOLERANCES FOR LINEAR DIMENSIONS

This is applicable for all machining processes with chips, like turning, milling etc. without chips, like drawing, embossing, bending *etc.* and is NOT applicable for production methods like casting, forging, welding, flame cutting *etc.*

Deviations for radii and chamfers and angular dimensions are also given in SP:46-198. Now instead of calculation it is easier to consult tables to find out the various tolerances on shafts and holes. The appropriate type of fit can also be chosen from the figure on Approximate Designation of Type of Fit (IS-2079-64) which gives a qualitative idea of the various types of fit used.

Table 2. 5 Fundamental tolerances

Value of tolerances in microns (1 microns = 0.001)

Diameter steps in mm	Tolerance grades																	
	01	0	1	2	3	4	5	6	7	8	9	10	11	12	13	14	15	16
Over 10 to 18	0.5	0.8	1.2	2	3	5	8	11	18	27	43	70	110	180	270	430	700	1100
Over 18 to 30	0.6	1	1.5	2.5	4	6	9	13	21	33	52	84	130	210	330	520	840	1300
Over 30 to 50	0.6	1	1.5	2.5	4	7	11	16	25	39	62	100	160	250	390	620	1000	1600
Over 50 to 80	0.8	1.2	2	3	5	8	13	19	30	46	74	120	190	300	460	740	1200	1800
Over 80 to 120	1	1.5	2.5	4	8	10	15	22	35	54	87	140	220	350	540	870	1400	2200
Over 120 to 180	1.2	2	3.5	5	8	12	18	25	40	63	100	160	250	400	630	1000	1600	2500

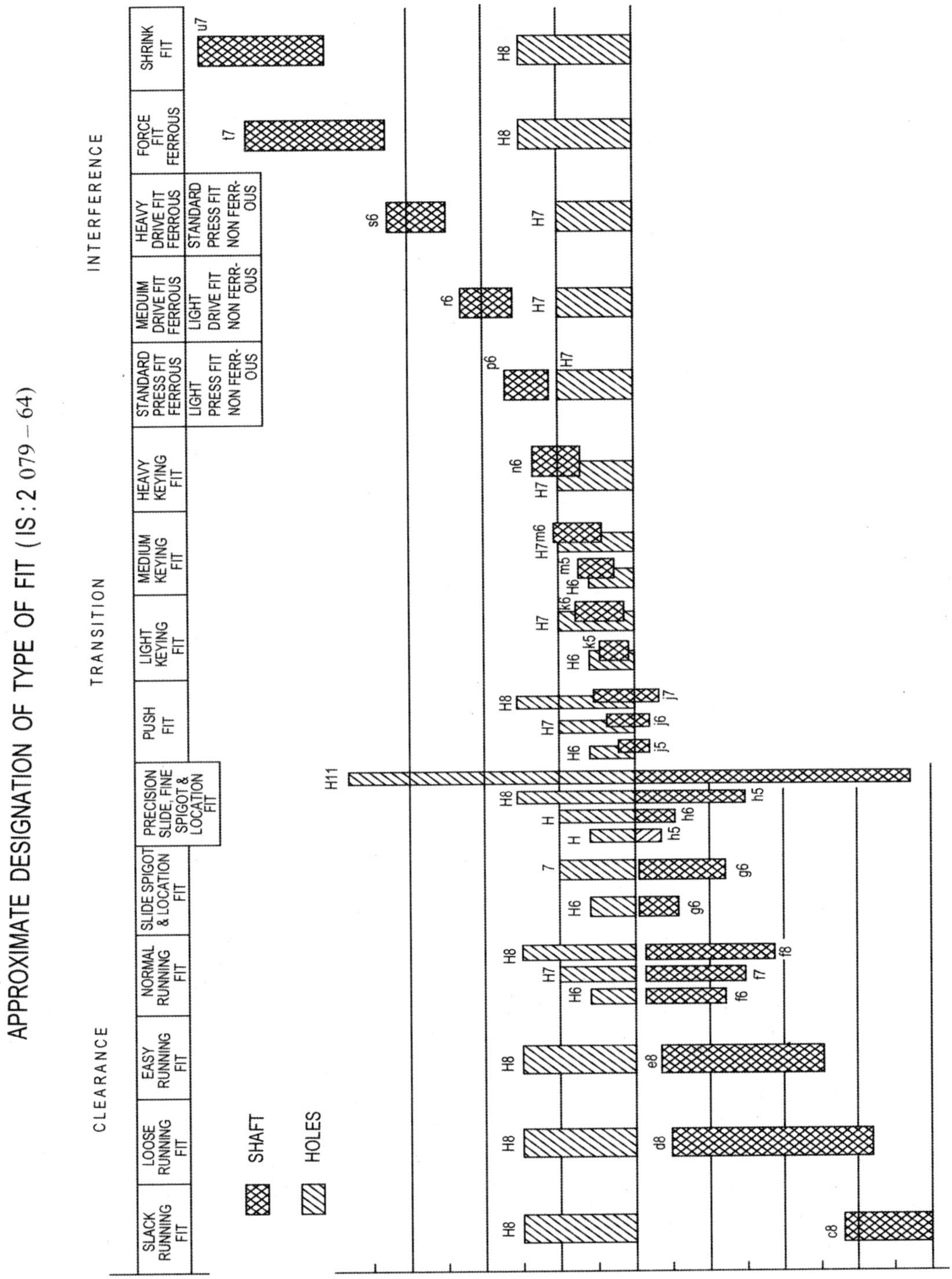

TABLE FOR LIMITS & FITS COMMONLY USED FOR DIFFERENT ASSEMBLY

Table 2.6 Tolerance on shafts and holes

Nominal size, mm		Tolerance on holes Unit 0.001 mm						Tolerance on shafts, microns				
Over	to	H6	H7	H8	H9	H10	H11	c8	c9	d8	d9	e8
-	3	+6	+10	+14	+25	+40	+60	-60	-62	-20	-20	-14
		0	0	0	0	0	0	-74	-85	-34	-45	-28
3	6	+8	+12	+18	+30	+48	+75	-70	-70	-30	-30	-20
		0	0	0	0	0	0	-88	-100	-48	-60	-38
6	10	+9	+15	+22	+36	+58	+90	-80	-80	-40	-40	-25
		0	0	0	0	0	0	-102	-116	-62	-76	47
10	14	+11	+18	+27	+43	+70	+110	-95	-95	-50	-50	-32
14	18	0	0	0	0	0	0	-143	-162	-98	-117	-73
30	40	+16	+25	+39	+62	+100	+160	-120	-120	-80	-80	-50
								-159	-182			
								-130	-130			
40	50	0	0	0	0	0	0	+169	-192	-119	-142	-89
50	65	+19	+30	+46	+74	+120	+190	-140	-140	-100	-100	-60
								-186	-214			
								-150	-150			
65	80	0	0	0	0	0	0	-96	-224	-146	-174	-106
80	100	+22	+35	+54	+87	+140	+220	-170	-170	-120	-120	-72
								-22	-257			
								-180	-180			
100	120	0	0	0	0	0	0	-234	-267	-174	-207	-126
120	140							-200	-200			
		+25	+40	+63	+100	+160	+250	-263	-300	-145	-145	-85
140	160							-210	-210			
		0	0	0	0	0	0	-273	-310	-208	-245	-148
160	180							-230	-230			
180	200							-240	-240			
		+29	+46	+72	+115	+185	+250	-312	-335	-170	-170	-100
200	225							-250	-250			
		0	0	0	0	0	0	-322	-365	-242	-285	-172
225	250							-280	-280			
								-352	-395			
250	280	+32	+52	+81	+130	+210	+320	-300	-300	-190	-190	-110
								-381	-430			
								-330	-330			
280	315	0	0	0	0	0	0	-411	-460	-271	-320	-191
315	355	+36	+56	+89	+140	+270	+360	-360	-360	-210	-210	-125
								-449	-500			
								-400	-400			
355	400	0	0	0	0	0	0	-489	-540	-299	-350	-214
400	450	+40	+63	+97	+155	+250	+400	-440	-440	-230	-230	-135
								-537	-595			
								-480	-480			
450	500	0	0	0	0	0	0	-577	-635	-327	-385	-232

Table 2.6 Contd.

Nominal size, mm		**Tolerance on shafts** Unit 0.001 mm									
Over	To	c9	f9	f7	f8	g5	g6	h5	j5	j6	j7
-	3	-14	-6	-6	-6	-2	-2	0	+2	+4	+6
		-39	-12	-16	-20	-6	-8	-4	-2	-2	-4
3	6	-20	-10	-10	-10	-4	-4	0	+3	+6	+8
		-50	-18	-22	-28	-9	-12	-5	-2	-2	-4
6	10	-25	-13	-13	-13	-5	-5	0	+4	+7	+10
		-61	-22	-28	-35	-11	-14	-6	-2	-2	-5
10	14	-32	-16	-16	-16	-6	-6	0	+5	+8	+12
14	18	-75	-27	-34	-43	-14	-17	-8	-3	-3	-6
18	24	-40	-20	-20	-20	-7	-7	0	+5	+9	+13
24	30	-92	-33	-41	-53	-16	-20	-9	-4	-4	-8
30	40	-50	-25	-25	-25	-9	-9	0	+6	+11	+15
40	50	-112	-41	-50	-64	-20	-25	-11	-5	-5	-10
50	65	-60	-30	-30	-30	-10	-10	0	+6	+12	+18
65	80	-134	-49	-60	-76	-23	-29	-13	-7	-7	-12
80	100	-72	-36	-36	-36	-12	-12	0	+6	+13	+20
100	120	-159	-58	-71	-90	-27	-34	-15	-9	-9	-15
120	140	-85	-43	-43	-43	-14	-14	0	+7	+14	+22
140	160	-185	-68	-83	-106	-32	-39	-18	-11	-11	-18
160	180										
180	200	-100	-50	-50	-50	-15	-15	0	+7	+16	+25
200	225	-215	-79	-96	-122	-35	-44	-20	-13	-13	-21
225	250										
250	280	-110	-56	-56	-66	-17	-17	0	+7	+16	+26
280	315	-240	-88	-108	-147	-40	-49	-23	-16	-16	-26
315	355	-125	-62	-62	-62	-18	-18	0	+7	+18	+29
355	400	-265	-98	-119	-151	-43	-54	-25	-18	-18	-28
400	450	-135	-68	-68	-68	-20	-20	0	+7	+20	+31
450	500	-290	-108	-131	-165	-47	-60	-27	-20	-20	-32

Table 2.6 Contd.

Nominal Size, mm		Tolerance on shaft, microns (0.01mm)							
Over	To	k5	k6	k7	m5	m6	n6	n7	p6
-	3	+4	+6	+10	+6	+8	+10	+14	+12
		0	0	0	+2	+2	+4	+4	+6
3	6	+6	+9	+13	+9	+12	+16	+20	+20
		+1	+1	+1	+4	+4	+8	+8	+12
6	10	+7	+10	+16	+12	+15	+19	+25	+24
		+1	+1	+1	+6	+6	+10	+10	+15
10	14	+9	+12	+19	+15	+18	+23	+30	+29
14	18	+1	+1	+1	+7	+7	+12	+12	+18
18	24	+11	+15	+23	+17	+21	+28	+36	+35
24	30	+2	+2	+2	+8	+8	+15	+15	+22
30	40	+13	+18	+27	+20	+25	+33	+42	+42
40	50	+2	+2	+2	+9	+9	+17	+17	+26
50	65	+15	+21	+32	+24	+30	+39	+50	+51
65	80	+2	+2	+2	+11	+11	+20	+20	+32
80	100	+18	+25	+38	+28	+35	+45	+58	+59
100	120	+3	+3	+3	+13	+13	+23	+23	+37
120	140								
		+21	+28	+43	+33	+40	+52	+67	+68
140	160								
		+3	+3	+3	+15	+15	+27	+27	+43
160	180								
180	200								
		+24	+33	+50	+37	+46	+60	+77	+79
200	225								
		+4	+4	+4	+17	+17	+31	+31	+50
250	280	+27	+36	+56	+43	+52	+66	+86	+88
280	+315	+4	+4	+4	+20	+20	+34	+34	+56
315	355	+29	+40	+61	+46	+57	+73	+94	+98
400	+450	32	+45	+68	+50	+63	+80	+103	+108
450	500	+5	+5	+5	+23	+23	+40	+40	+68

Table 2.6 Contd.

Nominal size, mm		Tolerance on shafts Unit 0.01 mm								
Over	To	p7	r6	r7	s6	s7	t6	t7	u6	u7
–	3	+16	+16	+20	+20	+24	–	–	+24	+28
		+6	+10	+10	+14	+14			+18	+18
3	6	+24	+23	+27	+27	+31	–	–	+31	+35
		+12	+15	+15	+19	+19			+23	+23
6	10	+30	+28	+34	+32	+38	–	–	+37	+43
		+15	+19	+19	+23	+23			+28	+28
10	14	+36	+34	+41	+39	+46	–	–	+44	+51
14	18	+18	+23	+23	+28	+28			+33	+33
18	24	+43	+41	+49	+48	+56	–	–	+54	+51
									+41	+41
24	30	+22	+28	+28	+35	+35	+54	+62	+57	+65
							+41	+41	+44	+44
							+64	+73	+76	+85
30	40	+51	+50	+59	+59	+68	+48	+48	+60	+60
		+70	+79	+86	+95					
							+54	+54	+70	+70
			+60	+71	+72	+83	+85	+96	+106	+117
		+62	+41	+41	+53	+53	+66	+66	+87	+87
			+62	+73	+78	+89	+94	+105	+121	+132
			+43	+43	+59	+59	+75	+75	+102	+102
			+73	+86	+93	+106	+113	+126	+146	+159
80	100	+72	+51	+51	+71	+71	+91	+91	+124	+124
100	120		+76	+89	+101	+114	+126	+139	+166	+179
			+54	+54	+79	+79	+104	+104	+144	+144
			+88	+103	+117	+132	+147	+162	+195	+210
120	140	+83	+63	+63	+92	+92	+122	+122	+170	+170
			+90	+105	+125	+140	+159	+174	+215	+230
140	160		+65	+65	+100	+100	+134	+134	+190	+190
160	180		+93	+108	+135	+148	+171	+186	+236	+250
			+68	+68	+108	+108	+146	+146	+210	+210
			+106	+123	+151	+168	+85	+212	+265	+282
180	200	+96	+77	+77	+122	+122	+156	+156	+236	+236
			+109	+126	+159	+176	+209	+226	+287	+304
			+80	+80	+130	+130	+180	+180	+258	+258
			+113	+130	+169	+86	+224	+241	+313	+330
			+84	+84	+140	+140	+195	+195	+284	+284
250	280		+126	+146	+190	+210	+250	+270	+347	+367
			+94	+94	+158	+158	+218	+218	+315	+315
280	315	+56	+130	+150	+202	+222	+272	+292	+382	+402
			+98	+98	+170	+170	+240	+240	+350	+350
			+144	+165	+226	+247	+304	+325	+426	+447
315	355	+119	+108	+108	+190	+190	+268	+268	+390	+390
355	400		+150	+171	+244	+265	+330	+351	+471	+492
			+114	+114	+208	+208	+294	+294	+435	+435
400	450		+166	+189	+272	+395	+370	+393	+530	+553
450	500	+131	+126	+126	+232	+232	+330	+330	+490	+490
			+172	+195	+292	+315	+400	+423	+580	+603
			+132	+132	+252	+252	+360	+360	+540	+540

CHAPTER – 3

SURFACE ROUGHNESS

Surface roughness refers to the finer irregularities on the apparently smooth surface which results from the inherent action of production processes and is expressed by its CLA (center line average) value, Ra in microns (IS : 3073-1967).

CLA value is the average height of the ordinates from the centre line of the surface profile. The ordinates are summed without considering their algebraic signs.

$$R_a = 1/L \int /y/dl,$$

where, L = Sampling length of the surface and

y = Height of the ordinate at a distance l.

The recommended sampling values are given in Table 3.1.

Same R_a values may be obtained for surfaces having different surface, textures which are characteristics of different manufacturing processes. For this reason R_a value specified on the drawing should be associated with the relevant manufacturing process, (Table 3.1).

R_a values obtained for the same surface are different if measured over different lengths; for this reason specified length known as ' Sampling Length' (standardized for different processes) should be chosen to find out the R_a value and it should be specified on the drawing.

Table 3.1: Recommended sampling lengths

Process	Sampling lenghts, mm				
Turning	—	0.8	2.5	—	—
Shaping	—	0.8	2.5	8	10
Milling	—	0.8	2.5	8	10
Boring	—	0.8	2.5	8	10
Grinding	0.25	0.8	2.5	—	—
Lapping	0.25	0.8	—	—	—

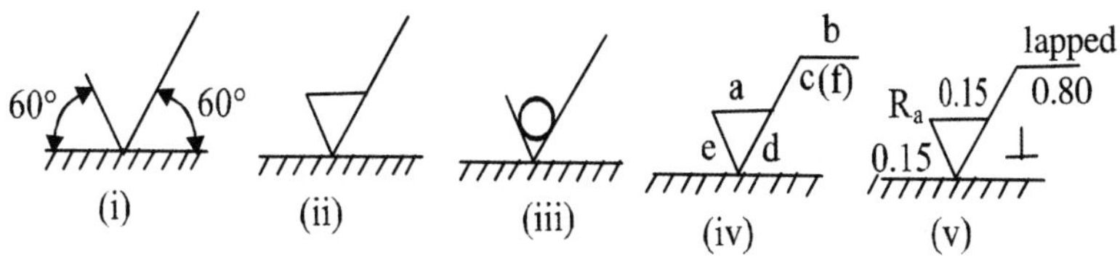

Fig. 3.1

METHOD OF INDICATING SURFACE ROUGHNESS

The basic symbol consists of two legs of unequal lengths implying at approximately 60° to the line representing the surface under consideration Fig. 3.1 (i). Machining symbol is shown in Fig. (ii) and when machining is not allowed symbol Fig. (iii) is used. In Fig. 3.1 (iv), in place of 'a' the R_a value grade Nos. N12, N11, N10, ……N1 are also put. The values of roughness grades are given in Table 3.4.

Usually the finer the surface finish, the lesser the surface is subject to wear and corrosion, and higher is the stress resistance of the component.

The production of a carefully finished surface costs more than that of a rough surface. Therefore, the quality of surface finish should always be specified with due consideration for the conditions under which the part will operate. The symbols for surface finish are given in Table 3.2.

Specifications of surface roughness is denoted in Fig. 3.1 (iv).

a = Roughness value R_a in microns.

b = Production method, treatment or coating.

c = Sampling length.

d = Direction of lay, Table 3.2

e = Machining allowance.

f = Other roughness values.

and Fig. 3.1 (v) shows an example.

Table 3.2: The symbols used for surface finish

SYMBOL			INTERPRETATION
=	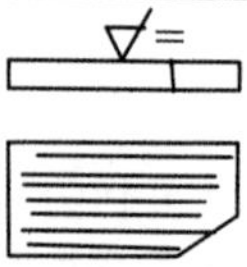	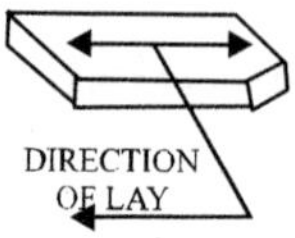	Parallel to the plane of projection of the view in which the symbol is used.
⊥	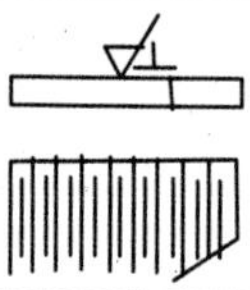	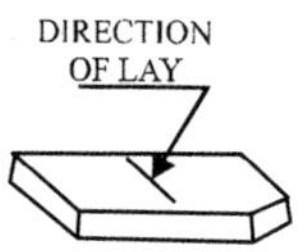	Perpendicular to the plane of projection of the view in which the symbol is used.
X	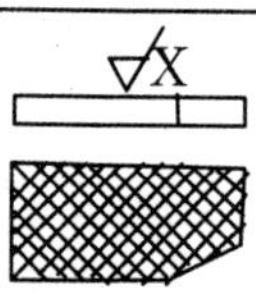	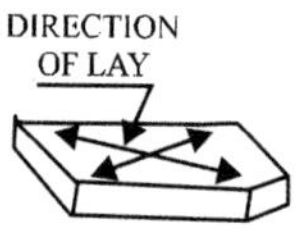	Crossed in two slant directions relative to the plane of projection of the view in which the symbol is used.
M	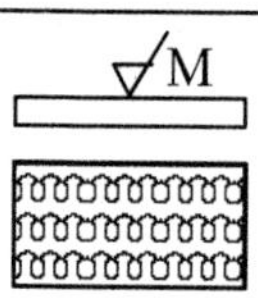	—	Multi-directional
C	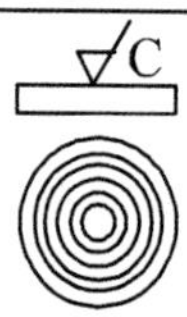	—	Approximately circular relative to the centre of the surface to which the symbol is applied.
R		—	Approximately radial relative to the center of the surface to which the symbol is used.

Table 3.3 Surface roughness obtained from different manufacturing processes

Manufacturing Process	Ra, Microns
Sand casting	6 – 25
Die casting	0.8 – 3.2
Hot rolling	2.5 – 50
Forging	1.6 – 25
Flame cutting, sawing, chipping	6.3 – 100
Hand grinding	6.3 – 25
Filling	0.25 – 25
Turning & milling	0.32 – 25
Shaping	1.6 – 25
Boring	0.4 – 6.3
Surface grinding	0.063 – 5
Lapping	0.012 – 0.16

Table 3.4 Roughness values and Grade No.

Roughness Values Ra, μm (microns)	Roughness Grade Nos.
50	N 12
25	N 11
12.5	N 10
6.3	N 9
3.2	N 8
1.6	N 7
0.8	N 6
0.4	N 5
0.2	N 4
0.1	N 3
0.05	N 2
0.025	N 1

CHAPTER – 4

SELECTION OF MATERIALS

Selection of materials is based on service requirements which dictate

(i) Strength – Ultimate, yield, proof, endurance and creep.

(ii) Ductility

(iii) Toughness

(iv) Wear-resistance

(v) High temperature properties

(vi) Corrosion resistance

The relative importance of the above six-factors influences the selection.

The rational selection also takes the following factors into considerations:

(i) Availability

(ii) Processability, *e.g.* casting, forming, machining, welding, heat-treatment *etc.*

(iii) Cost and

(iv) Standards

In the subsequent sections for the convenience of the user of the book several tables of common engineering materials and their important physical properties and typical uses in engineering practice have been illustrated. Designation of the materials as per Indian standards are also included. The specifications of the materials under different standard of various countries are explained subsequently. This is very important as the market and industries are being globalised in recent years and exports and imports are becoming very popular activities.

Table 4.1: Typical uses of cast iron

Designation of Cl		Typical uses
Grey C.I. (IS : 1210)	Grade 150	Brackets, Covers, Arms, Housings.
	Grade 200	Machine bed, column, saddle head-stock, tailstock, friction surface
	Grade 250	For components subjected to severe loading *e.g.* columns of presses chucks for Lathes, gear wheels, surface hardened guideways, hydraulic cylinders, high pressure pumps *etc.*
Mechanite (Alloy Cast Iron)		Pump-body, cylinders, valves, pistons, machine tool parts, machine frames, heavy brake drums, clutch plates, cylinder dies.
SG-iron (graphite precipitates as spheroids by addition of Mg/Ca)		Mechanical properties of SG-iron is comparable to low and medium carbon steels. It is suitable for 3 or 4 jaw chucks, gear, slides, tool-post body *etc.*
Malleable iron		Hand wheels, hand cranks, shifting levers.

Note: Grade means 150 MN/m^2 is the average UTS of this grade cast iron.

Table 4.2 Table of physical properties

Material	Specific Gravity Kg m/l	Melting Point °C	Modulus of Elasticity E, GPa	Modulus of Rigidity Gpa	Thermal Conductivity cals/scm °C	Coefficient of Linear expansion /oc	Poisson's Ratio
Aluminium	2.70	660	67.5	26.0	0.530	12.8×10^{-6}	0.34
Berylium	1.82	1280	292.8	–	0.380	12.3×10^{-6}	
Brass	8.45	900 - 950	97.0	35.0	0.310	1.67×10^{-6}	0.30 - 0.40
Bronze	8.73	910 - 1040	111.0	–	0.160	17.3×10^{-6}	–
Cast Iron	7.20	1160 - 1300	100.0	35.0	0.130	9.0×10^{-6}	0.23
Copper	8.96	1083	123.0	39.0	0.94	16.2×10^{-6}	0.26
Lead	11.34	327	16.0	7.6	0.83	28.3×10^{-6}	0.45

Table 4.2 contd.

Material	Specific Gravity Kg m/l	Melting Point °C	Modulus of Elasticity E, GPa	Modulus of Rigidity GPa	Thermal Conductivity cals/scm °C	Coefficient of Linear expansion/oc	Poisson's Ratio
Monel Metal	8.58	1315–s1350	159.0	67.0	0.060	14.0×10^{-6}	0.32
Steel C 15	7.85	1510	208.0	79.0	0.120	11.1×10^{-6}	0.30
Steel C 35	7.84	1490	206.0	to	0.120	11.1×10^{-6}	0.30
Steel C 60	7.83	1470	204.0	89.0	0.110	11.1×10^{-6}	0.30
Titanium	4.54	1800	105.0	–	0.041	11.8×10^{-6}	0.30
Tungesten	19.30	341	415.3	177.0	0.480	4.5×10^{-6}	0.17
Zirconium	6.50	1850	69.7		0.040	10.0×10^{-6}	

Table 4.3: Typical uses of plain carbon steel

Steel Designation	Typical Uses
C 07, C 14	Used for cold forming and deep drawing. Rimming quality used for automobile bodies, cold heading wires and rivets. Killed quality used for forging and heat treating applications.
C 10 & C 14	Case hardening steels used for making camshafts, cams, light duty gears, worms, gudgeon pins, selector forks, spindles, pawls, ratchets, chain, wheels, tappets, *etc.*
C 15	Used for lightly stressed parts. The material, although easily machinable is not designed specifically for rapid cutting, but is suitable where cold work, such as bending and riveting may be necessary.
C 15 Mn 75 C 20 C 25 & 35 Mn 75	General purpose steels for low stressed components
C 30	Used for cold formed levers, hardened and tempered tie rods, cables, sprockets, hubs and bushes, steel tubes.
C 35	Steel for making low stressed parts, automobile tubes & fasteners.
35 Mn 75	Steel for making low stressed parts in machine structures, cycle and motorcycle frame tubes, fish plates for rails and fasteners.
C 40	Steel for crankshafts, shafts, spindles, automobile axle beams, push rods, Connecting rods, studs, bolts, lightly stressed gears, etc.
C 45	Steel for spindles of machine Tools, bigger gears, bolts & shafts

C 50	Steel for making keys, shafts, cylinders, machine components requiring moderate wear resistance. In surface hardened condition it is also suitable for large-pitch and gears.
C 50 Mn 1	Rail steel. Also used for making spike bolts, gear shafts, rocking levers and cylinder liners.
C 55 & C 55 Mn 75	Steel used for making gears, cylinders, cams, keys, crank shafts, sprockets & machine parts, requiring moderate wear importance resistance for which touchness is not primary.
C 60	Steel for making spindless for machine tools, hardened screws & nuts, coupling, crank shafts, axles & pinions.
C 65	High tensile structural steel for making locomotive carriage & wagon tyres. Typical uses of this steel in the spring industry include engine valve springs, small washers & thin stamped parts.

Table 4.4 Typical uses of alloy steels

Designation	Uses
20 Mn 2 & 27 Mn 2	Welded structures, crank shafts, steering, levers, shafting, spindles, *etc.*
37 Mn 2	Axes, shafts, crank shafts, connecting rods *etc.*
35 Mn 2 Mo 28	Crank shafts, bolts, levers, connecting rods etc.
35 Mn 2 Mo 45	
40 Cr 1	Gears, connecting rods, wear resistant plates *etc.*
40 Cr 1 Mo 28	Shafts, gears, high tensile bolts and studs.
15 Cr 3 Mo 55	Components requiring medum of high tensile strengh in the nitrided
25 Cr 3 Mo 55	Condition for parts requiring medum of high surface hardness & wear resistance.
40 Cr 3 Mo 1 V 20	Components requiring high tensilie strength.
40 Cr 2 Al 1 Mo 18	Nitrided-components requiring high surface hardness & core strength.
40 Ni 3	Cold though steel-used at low temperatures such as refrigerators & strength of the order of 160 kgf/mm^2& where minimum distortion in heat treatment is essential.
35 Ni 1 Cr 60	Aircrafts & heavy vehicle components.
30 Ni	Highly stressed gears & other comopnents requiring ultimate tensile strength of the order of 160 kgf/mm^2 & where minimum distortion in heat treatment is essential.
40 Ni 1 Cr 1 Mo 15	General machine parts like bolts, gears, ets.
40 Ni Cr 1 Mo 28	High strength machine parts like collets, spindles, bolts, gears *etc.*

30 Ni Cr 65 Mo 55	Highly stressed bolts, shafts, gears, mandrels, aircraft power units and for low temperature service.
14 Mn 1 S 14	Parts where good machinability & finish are important.
40 Mn 2 S 12	Heat treated axels & shafts (not recommended for forgings in which transverse properties are important.)
17 Mn 1 Cr 95	Small gear wheels, shafts, & chain wheels.
15 Ni 2 Cr 140	Heavy duty components, gears & super charge gears.
13 Ni 3 Cr 80	Heavy duty gears, pinions etc. Very highly stressed machine parts which are to be hardened for high strength & toughness in core after hardening.
55 Si 2 Mn 90	Used for spring in the hardened & tempered condition.

Table 4.5: Typical characteristics and applications

Designation	Characteristics	Typical Applications
AUSTENTIC*		
302 C < 0.15,, Si < 1 Mn < 2 Cr 17-19 Ni 8-10	General purpose stainless steels	Widely used fro severely formed parts, *e.g.*, utensils, appliances and decorative work.
304 C < 0.08 Si < 1 Mn < 2 Cr 18-20 Ni 8-12	General puropse stainless steel specially for welded construction	For welded construction & heavy walled vessels formed in a number of draws and also used as type 302.
347 C < 0.08 Si < 1 Mn < 2 Cr 17-19 Ni 9-13	Stabilized 18/8 type with Columbium. Corrosion resistance comparable to type 304	Used in oxidising & corrosive environments within 420 to 870° C for welded construction. No annealing required after welding. Cryogenic equipment for storage and transportaion of liquid Nitrogen, Oxygen etc.
310 C > 0.25 Si < 1.5 Mn < 2 Cr 24-26 Ni 19-22	Excellent high temp. properties with ductility atmosphere & weldability	Used upto 1150°C for continuous service & upto 1050°C intermitted service in oxidizing
FERRITIC		
430 C < 12 Si < 1 Mn < Cr 14-18	Good combination of corrosion resistance, useful mechanical properties & good formability. Not hardenable by heat treate-	Utensils, hardwares, fasteners, applianes, furance parts, processing equipment in chemical, fertilizer industries & Nitric acid plant.
MARTENSITIC		
403 C < 0.15 Si < 0.5 Mn < 1 Cr 11.5-13	ment 12% Cr ensuring high degree of cleanliness, hardenable by heat treatment	Reactor components, turbine blades & rings & other stressed parts.

*Austenitic Stainless Steel is not hardenable by heat treatment. It hardens when cold worked.

Table 4.6 Typical use of aluminium alloys

A–1	Medium stressed casting such as timing gears, general purpose machine parts & pattern castings—possesses moderate strength, Non-heat treatable alloy.
A–2	Used for pistons & similar parts—combines high wear resistance with good strength at elevated temperature & low coefficient of thermal expansion.
A–6	Suitable for sand casting, gravity & pressure die-casting & for casting intricate patterns—has good corrosion resistance & medium strength.
A–9 (Y Alloy)	Used for cylinder blocks, cylinder heads, valve bodies, large fan bladeds, pneumatic tools *etc.* —has good castability, resistance to chemical attack & good machining & welding characteristics—can be fully heat treated.
A–18	Used for electrical & household appliances-has good castability & resistace to corrosion.
A–24	Used mainly for pressure die castings.

Table 4.7 Copper and its alloys

Type	Typical Applicatons
Brass-alloy of copper & zinc	Hydraulic cylinders, threaded fasteners in corrosive atmosphere, small bushes.
Bronze-alloy of copper & tin	Bearing material.
Phosphor-bronze, Aluminium-bronze	Bushes, worm wheels, springs, lead-screw nuts.
Sintred bronzes impregnated with oil, graphite, molybdenum disulphide	Good bearing materials.

SPECIFICATIONS OF STEELS

There are in general three grades of steels namely Structural steel, Carbon steel and Alloy steels.

Structural Steels are designated as Fe followed by a number corresponding to the tensile strength of the steel in MN/m^2 *e.g.* Fe 420 W, Fe 540 *etc.* Here 420 W means that the average ultimate strength is MN/m^2 and steel is of fusion welding quality.

Carbon Steels *e.g.* C 15, C 40, C 55 *etc.*, or C 40 means C = (40/100) percent *i.e.* 0.4%. Here 15, 40, 50 indicates the hundredth of percent of average carbon content.

Alloy Steels *e.g.* 40 Ni 2 Cr 1 Mo $\underline{28}$. In this case average Carbon content is 0.40% and Nickel 2%, Chromium 1% and Molybdenum 0.28% (a bar placed at the bottom as shown signifies hundredth of percent of Molybdenum).

AISI—SAE NUMBERING SYSTEM OF STEELS

This system was originated by the society of Automotive Engineers (SAE) and later expanded by the American Iron and Steel Institute (AISI).

In this system materials are identified by an index number of four or five digits (e.g. SAE-8740) that indicates the class of steel, principle alloy content, and approximate carbon percentage. The first figure indicates the class of steel as follows:

For Carbon	1	For Vanadium	6
For Nickel	2	For Tungsten	7
For Nickel-Chromium	3	For Steel containing	
For Molybdenum	4	Va, Mo and Ni	8, 9
For Chromium	5	except (92XX—Si-Mn Steel)	

In the case of simple alloy steels the second number generally indicates the approximate percentage of predominant alloying element, *e.g.* 31XX for Ni–1.25% and Cr–0.65% and 32XX for Ni–3.50% and Cr-1.57% and 10xx for Carbon Steels.

The last two digits indicate the approximate hundredths of percent of Carbon.

Examples :	SAE 1020	:	Plain Carbon Steel with C-0.15% to 0.25%
	SAE 3120	:	Cr-Ni Steel with Ni-1 to 1.5% and C-15% to 25%
	SAE 71360	:	Tungsten steel with Tungsten-12% to 15% and C-0.50% to 70%

BRITISH SYSTEMS OF NOMENCLATURE OF STEELS

In U.K. nomenclature systems is by EN followed by numerals like 1, 2, ...50, 60...........362 etc and sometimes with letters like A, B, C e.g. EN 30A etc. A list of some steels are given below.

EN	1	Free Cutting Steel
EN	2	Cold Forming Steel
EN	3	"20" Carbon Steel (C-0.15 to 0.25%)
EN	4	"25" Carbon Steel
EN	5	"30" Carbon Steel
EN	6	Bright Carbon Steel
EN	7	Semi-Free Cutting Carbon Steel
EN	8	"40" Carbon Steel
EN	9	"55" Carbon, 0.75% Ni
EN	10	"55" Carbon, 0.95% Ni
EN	11	"60" Carbon-Charomium Steel
EN	12	1% Ni-Steel
EN	13	Mn-Ni Mo Steel
EN	14	Carbon-Mn Steel

EN	15	Carbon-Mn Steel
EN	16	Mn-Mo Steel
EN	17	Mn-Mo Steel (Higher Mo)
EN	18	1% Chromium Steel
EN	19	1% Cr-Mo Steel
EN	20	1% Cr-Mo Steel (Higher Mo)

Effect of various alloying elements on the properties of steels

Carbon: With increase of carbon, percentage from 0.1 to 1.2 the tensile strength increases from 30 to 100 kg/mm^2, percentage elongation drops from 28 to 2, the hardness rises from 100 to 350 BHN and weldability reduces. These values are approximate.

Silicon: It gives dense casting. With its increase the elastic limit is raised but weldability and forgeability is reduced.

Manganese: It improves wear resistance.

Nickel: Tensile strength and elongation is improved. Nickel steels are highly heat resistant and they are extensively used in combination with Chromium to give Chrome-Nickel steels.

Chromium: Effect is same as that of Nickel and in addition it improves hardness.

Molybdenum: Improves strengths at high temperatures and eliminates temper brittleness.

Tungsten: It raises hardness and strength at elevated temperature.

By *suitably alloying* with one or more of these elements *special properties* can be developed.

As our market is globalised, it is useful to compare the Indian Steels with that of other countries, which is given in Table 4.8.

Table 4.8 Indian steels and their equivalents

IS Indian	BS British	AISI American	DIN Germany	JIS Japanese	GOST Russian	CSN Czech	AFNOR French
14 Mn 1S 14	En 202	C 1118	15 S 20	SUM 2	A 12	11110	15 F2
40 Mn 2 S 12	En 15 AM	C 1139	45 S 20	SUM 5	A 40 G	11140	45 M4
C 40	–	C 1040	40Mn 4	S 40 C	40	1240	–
C 45	En 8 D	C 1045	CK 45	S 45 C	M St 6	12050	XC 45F
17 Mn 1 Cr 95	En 207	5115	16 Mn Cr 5	5 Cr 21	20 ch GA	14220	16 MC 5
15 Ni 2 Cr 1 Mo 15	En 354	4317	15 Cr Ni 6	SNC M22	12 Ch N2	16220	18 NCD6
40 Ni 2 Cr 1 Mo 28	En 24	4340	36 Cr Ni Mo 4	SNC M8	40 ch	16341	–
30 Ni 4 Cr 1	En 30A	–	35 Ni Cr 18	–	–	16640	35 NC 15
T 983 Mo W 6 Cr 4 V 2	–	M2	DM 05	SKH 9	R9	19800	6–5–2
T 123 W 14 Co 5 Cr V4	–	–	EV 4 Co	–	R 14 F4	1980	–
103 Cr 1	En 31	E 52100	100 Cr 6	SUJ 1	Shch 15	14109	–

The Structural Steel sections available or produced are standardized to facilitate interchangeability and convenience in stocking. Table 4.9 and 4.10 give the details in concise form.

Table 4.9 Standard structural steel sections

a) **Rounded Bars**

Designation	:	ISRO ×, × being the diameter of round in mm eg. ISRO 5
Diameters, mm	:	5, 6, 8, 10, 12, 16, 20, 25, 28, 32, 36, 40, 45, 56, 63, 71, 80, 90, 100, 110, 125, 140, 160, 180, 200

b) **Square Bars**

Designation	:	ISSQ ×, × being the side-width in mm
Side/width, mm	:	5, 6, 8, 10, 12, 16, 20, 25, 32, 40, 50, 63, 80, 100

c) **Mild Steel Flats**

Designation	:	Width, b × thickness, t
Width, b, mm	:	10, 15, 20, 25, 30, 35, 40, 50, 60, 65, 70, 75, 80, 90, 100, 110, 120, 130, 140, 150, 200, 250, 300, 400
Thickness, t, mm	:	3, 4, 5, 6, 8, 10, 12, 16, 18, 20, 25, 32, 40

Limitation:

Width, b, mm	Available thickness, t, mm	
	Minimum	Maximum
10	3	5
15	3	8
20	3	10
25, 30	3	12
35	3	16
40, 45, 50	3	20
55, 60	3	25
65	6	32
70, 75, 80, 90, 100, 110, 120	6	40
130, 140, 150	8	40
200, 250	10	40
300	12	40
400	16	40

a) **Sheets**

Designation : length × width × thickness

Nominal thickness,
t, mm : 0.40, 0.50, 0.63, 0.80, 0.90, 1.00, 1.25, 1.40, 1.60, 1.80, 2.00, 2.24, 2.50, 2.80, 3.15, 3.55, 4.00

Nominal width,
b, mm : 600, 750, 900, 1000, 1100, 1200, 1250, 1400, 1500.

Nominal length,
I, mm : 1800, 2000, 2200, 2500

b) **Strips**

Width, b, mm : 100, 125, 160, 200, 250, 320, 400, 500, 650, 800, 950, 1050, 1150, 1250, 1300, 1450, 1550.

Limitation:

Width, b, mm	Minimum available thickness, t, mm
100 to 800	1.60
950	2.00
1150	2.24
1250	2.50
1300	2.80
1450, 1550	3.15

Table 4.10 Alloy steels (Produced in A.S.P. Durgapur)

A) Rolled Products

i) Hot Rolled Rounds:

13 to 60 mm, both incluslve in increment of 1 mm
62 to 100 mm, : 2 mm
105 to 150 mm, : 5 mm
160 to 200 mm, : 10 mm

ii) Hot Rolled Coils:

7.5 to 12 mm, : 0.5 mm
13 to 20 mm, : 1.0 mm

iii) Hot Rolled Square Edged Flats:

Thickness, mm : 10, 11, 16, 25, 32,
Width, mm : minimum – 60
maximum – 80

B) Forged Products

Rounds, Hexagons, Octagons & Squares.

	Grades	Max. size for squares mm	Max. size for round hex Oct, mm	Max. length, mm
1.	Constructional Streels	500	550	7000
2.	Stainless Steels	300	350	4000

C) Sheet & Plate Products — Stainless steels

i) Hot Rolled Plates (No. 1 Finish)

Width, mm : 750 to 1500
Thickness, mm : 5 to 25
Max, Length, mm : 4000
Max, Weight, Kgf. : 500

ii) Hot Rolled Sheets (No. 1 Finish)

Width, mm : 750 to 1000
Thickness, mm : 1.25 to 4
Max. Length, mm : 2500

iii) Cold Rolled Sheets (No. 2B & 2D Finish)

Width, mm : 750 to 1000
Thickness, mm : .8 to 2.5
Max. Length, mm : 2500

D) Discs & Rings of various grades of steels in different sizes are also manufactured.

No. 1 : Hot rolled, annealed, pickled, resquared.

No. 2B : Hot rolled, annealed, pickled, cold rolled, annealed, pickled, skip passed, resquared.

No. 2D : Hot rolled, annealed, pickled, cold rolled, annealed, pickled, resquared.

CHAPTER – 5

COMMON MANUFACTURING METHODS

The common manufacturing methods are as follows;

Casting: In this process a preshaped cavity called mould is formed wherein the metal in molten condition is poured to fill it, which after cooling, solidifies to the desired shape and size.

Forging: Here a metal blank either in cold of hot condition is subjected to a heavy force or pressure generally in a die to deform and to finally take up the desired shape. Smaller size parts may be forged manually by hammering.

Machining: The excess material of a blank in this process is removed in the form of chips by cutting on a machine tool. Various processes of cutting are turning, drilling, shaping, planning, milling, grinding etc.

Fabrication: Normally various structural shapes (plates, tubes, rolled section *etc.*) are joined by riveting and/or welding (already dealt within part A).

Casting:

General rules for castings

1. Consult the foundryman or patternmaker before making the final drawing of a to-be-cast component.
2. Equalize the thickness of all sections wherever possible. Avoid abrupt section changes.
3. Bring minimum number of sections together. Fillet all angles correctly and use gentle contours.
4. Design so, that all members increase progressively to assure solidity.
5. Use ribs, bosses, lugs and pads with discretions. Remember, these increase stress concentrations.
6. Avoid multiplicity of cores. Design inner walls of lesser thickness than main body of casting.
7. Provide vent outlet and clean out openings.
8. Replace short angles and corners with radii.

The Table 5.1 and 5.2 depict the minimum recommended thickness of casting for various metals and the allowances for contraction respectively.

Table 5.1: Minimum recommended section–thickness of casting

Metal	Thickness, mm.	Type of casting
Steel	5 to 6	Sand
Grey Cast Iron	5 to 6	Sand
Malleable Cast Iron	5 to 6	Sand
Aluminium	3	Sand
Aluminium Alloys	1.2	Die
Copper	3	Sand
Magnesium	4 to 5	Sand
Magnesium Alloys	1.2	Die
Bronze	3	Sand

Table 5.2: Allowances of contraction in casting

Metal	Allowances in mm per mm	Allowance
Cast Iron	0.01	1%
Steel	0.02	2%
Copper	0.02	2%
Magnesium	0.013	1.3 %
Aluminium	0.013	1.3 %
Brass	0.015	1.5 %

Finish allowance depends on the size and type of casting, the quality of finish desired and the casting procedure; normally, 3 mm machining allowance as a rule.

ILLUSTRATIONS OF CAST COMPONENTS

Rules for obtaining the most adequate design	Incorrect design	Correct design
A. Convenience of moulding As far as possible the casting should be arranged in one semi-bound or have one parting plane.	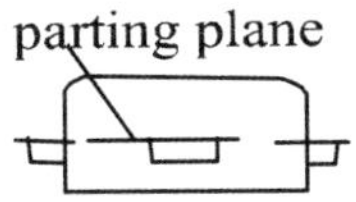	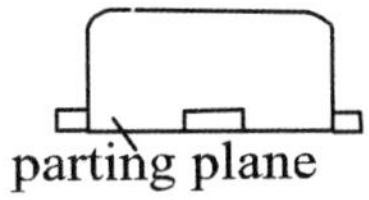
The external & internal sides of a casting should be provided with casting draft directed from the parting plane of the mould. The design of the casting should ensure moulding without detachable parts on the pattern and in the boxes and with the minimum number of cores. The observance of this principle can be checked by the following method: lines drawn from the side of the parting of the mould or the core box should not encounter on their path the external and internal contours of the components more than once.	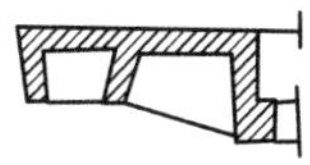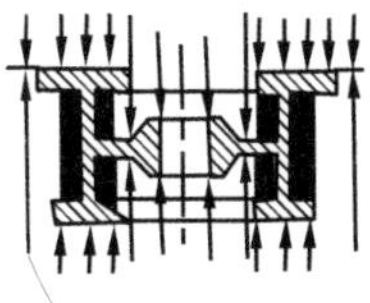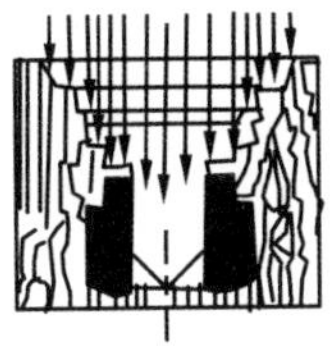	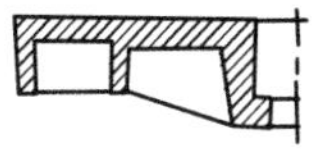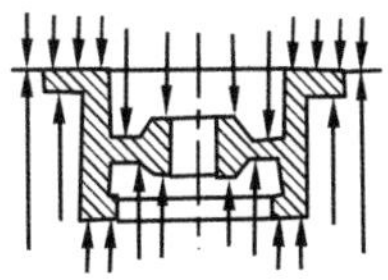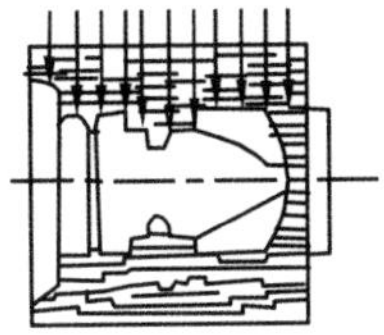
Components with complex & closed planes should be provided with an adequate number of holes for the reliable fastening of cores without Chaplets (gags).	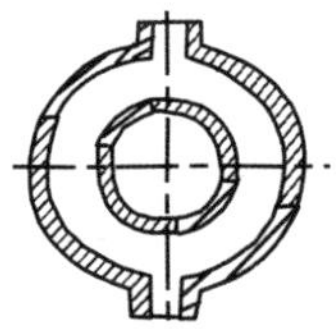	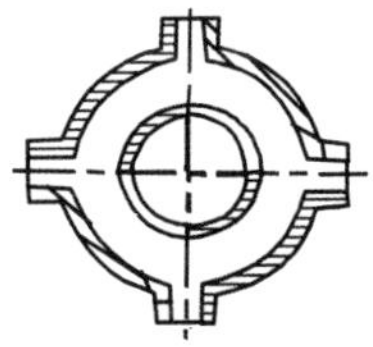
The thickness of the core in a double walled casting should not be below sum of the thickness of both walls to avoid the scouring of core necks.	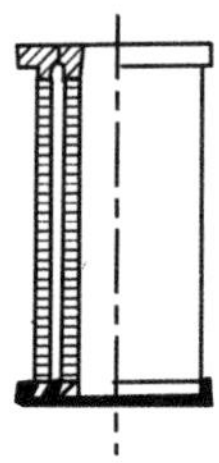	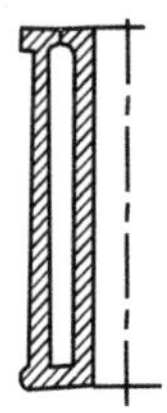

Rules for obtaining the most adequate design	Incorrect design	Correct design
B. Obtaining a good casting In casting from white cast iron, globular graphite cast iron & other grades of cast iron distinguished by the intensive formation of shrinkage cavities, there should be ensured a gradual decrease in the rate of solidification (the increase in the diameters on inscribed circles) towards the risors.	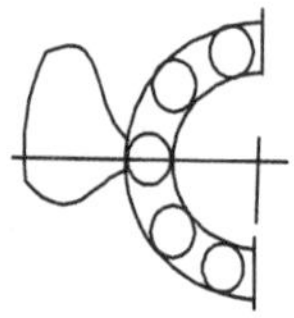	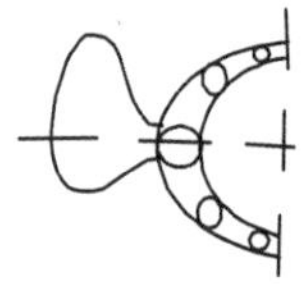
To prevent the formation of cracks curved spokes & discs should be used in flywheels & pulleys.		
Projections & bosses hindering shrinkages should be avoided especially in cast iron with intensive shrinkage to prevent the formation of cracks.	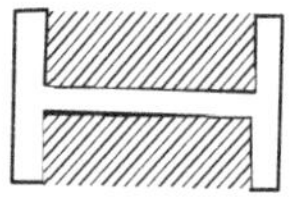	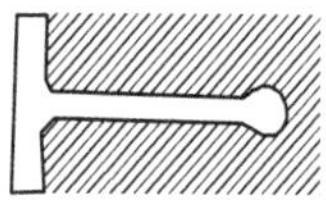
Large upper horizontal surfaces should be replaced by inclined surfaces (especially for casting subjected to enamelling) to avoid the formation of gas pits & accumulation of non-metallic inclusions.		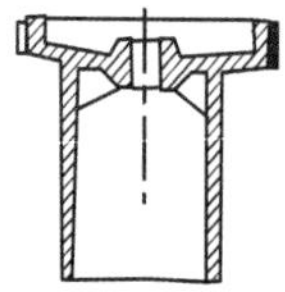
Corner joints of the external & internal surfaces of casting should be rounded off to prevent the formation of cracks, shrinkage cavities & burns. This is especially important for castings subject to enamelling.	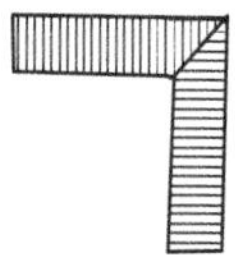	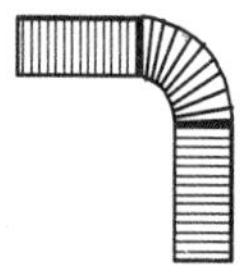
C. Convenient chipping & cleaning Casting should be provided with a sufficient number of holes of appropriate dimensions for the removal of cores & skeletons (the holes may be for production purpose only.)	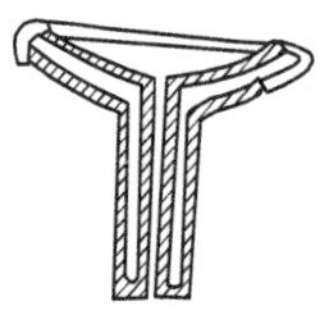	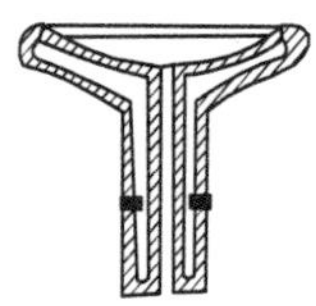
The design of castings should allow a supply of metal from the side of the external surfaces to facilitate the trimming of the sprues.	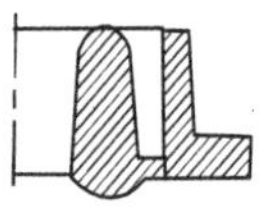	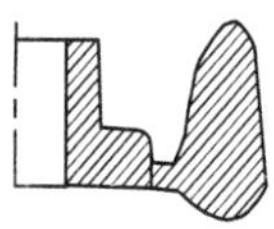

FORGING

A. Die Forging

1. Parting line, also called flash line of split line, should be so chosen that
   (i) deep impressions which promote die breakage are avoided as far as practicable.

(ii) side thrust is avoided.

(iii) sections of the forging should be balanced about the flash line; where this is not feasible, design for the simplest irregular parting line which approaches balanced condition, and

(iv) grain flow is most desirable.

2. Fillets and radii should be generous to allow easy flow of metal, *e.g.* 1.5 mm to 25 mm depending on the depth of impression in the die and on the size of the forging.
3. Changes in section must be gradual; thin isolated sections should be avoided. Sections upto 2.5 mm thick (minimum) can be forged in small parts but 3 mm preferred.
4. Deep pockets and recesses should be avoided. External bosses may have height equal to 2/3 their diameters.
5. Tolerance depends on size and is larger across the parting line than parallel to it. Minimum commercial tolerances in mm across the parting line are (+1.6, -0.2) for small parts; (+1, -0.3) for 0.5 kg forgings, (+2, -15) for 5.0 kg forgings. Close tolerances are half these values.
6. Holes of two non-parallel axes prevent removal of parts from the dies.
7. *Forging Draft:*

Axial projections or forgings are tapered (draft) so that the forging can be easily removed from die-cavities. Inside draft, θ_i, is more than the outside one, θ_o; θ_i is usually 7° to 10°and θ_o 5° to 7°

8. Machining allowance should be given as the table below.

Table 5.3: Machining Allowance

Maximum dimension, mm	Al and Mg alloy	Carbon and Low Alloy Steel	Austenitic Stainless Steel
Upto 500 (small)	1 mm	1 to 1.5 mm	0.5 to 1.5 mm
500 to 1000 (medium)	1 to 1.5 mm	1.5 to 3.5 mm	1.5 to 2 mm
Above 1000 (large)	1.5 to 3 mm	3.5 to 6 mm	2 to 5 mm

B. Tolerance to which ring forgings (done by hammersmithing) can be economically held are given in the table below:

Table 5.3: Tolerance for ring forging

Outside diameter mm	Tolerance on diameter, mm
upto 300	± 10
300 to 600	± 12
600 to 1000	± 20
above 1000	± 25

CHAPTER – 6

DRAWING OF MACHINE ELEMENTS

SHAFT COUPLING

Steel shafting is available in lengths of 6 to 10 meters. Over this length it is highly inconvenient to manufacture and transport; therefore a longer shafting is composed of smaller lengths coupled with appropriate coupling — rigid, flexible and slip type.

Rigid Coupling (Sleeve Coupling, Clamp Coupling, Flange Coupling) are used where there is very little chance to axial misalignment and flexibility of shaft axes is not desired. A rigid flange coupling and a universal coupling of Hooke's joint have been discussed in Part-A.

Flexible Coupling is suitable for the connection of two shafts which are or may become slightly out of alignment and when the driven shaft is subjected to shock and vibration. Fig.6.1 shows various views of a flexible coupling. Here leather washers have been used as flexible elements. Very often rubber bushes are also used.

Prob. 6.1 Develop the front view of the flexible coupling shown in Fig. 6.1 and provide the right hand side view in full section.

Prob. 6.2 Draw the shop details of the components of the above coupling; also prepare the drawing of one of the flanges for the use of the pattern maker.

FLEXIBLE PIPE JOINT

To join pipes going out of alignment upto 15° either ways, flexible pipe joints (Fig. 6.2) are used.

Prob. 6.3 Fig. 6.2 represents the sectional elevation of a flexible pipe joint suitable for 75 mm pipe carrying a steam pressure at 6 ata. Reproduce the given view and develop the plan.

Prob. 6.4 Prepare workshop drawings of the components, Fig. 6.2.

Fig. 6.2a Shows the details of an expansion joint suitable for steam pressure upto 1.6 MPa.

Prob. 6.5 Make an assembly drawing of the components as shown in Fig. 6.2a.

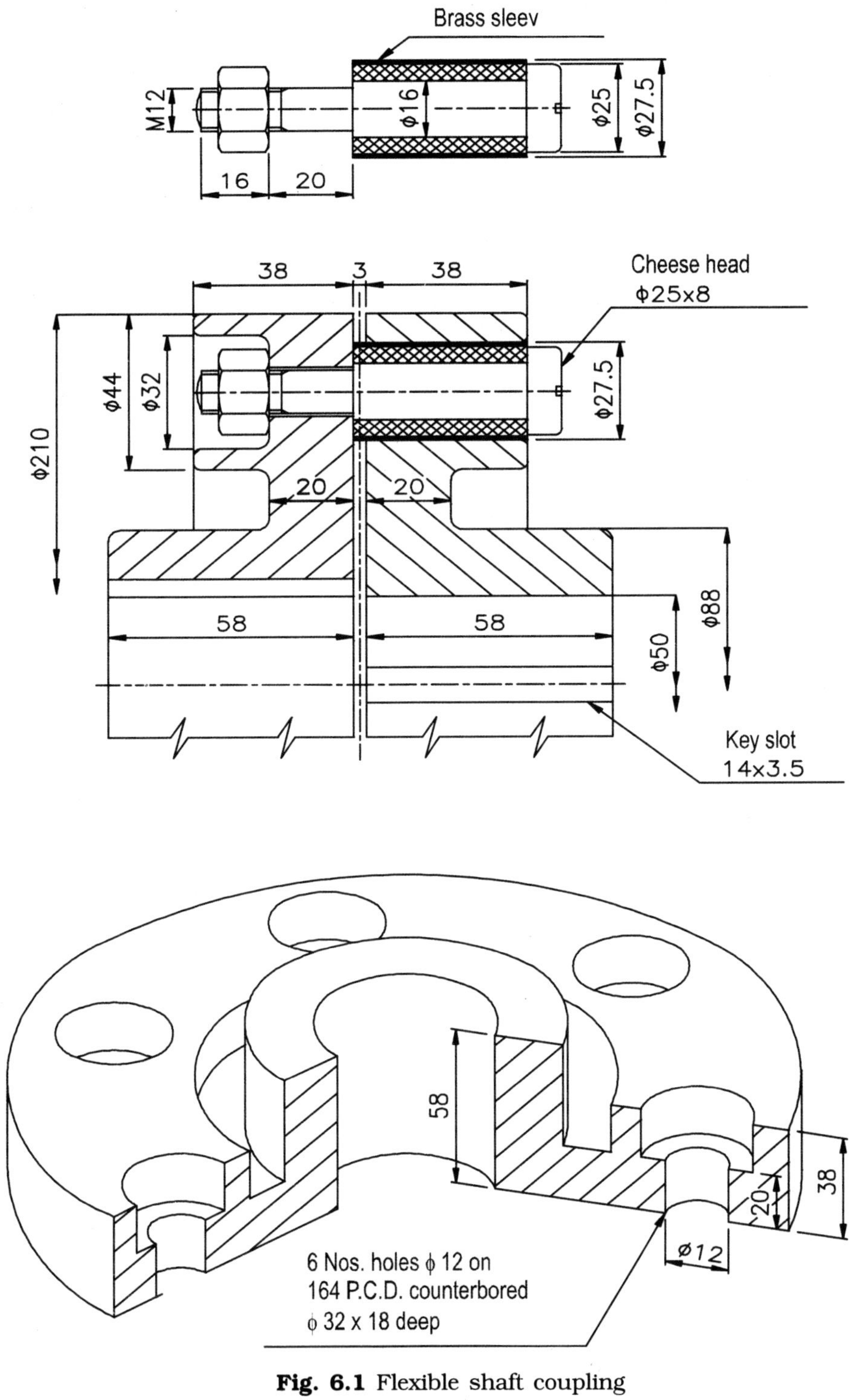

Fig. 6.1 Flexible shaft coupling

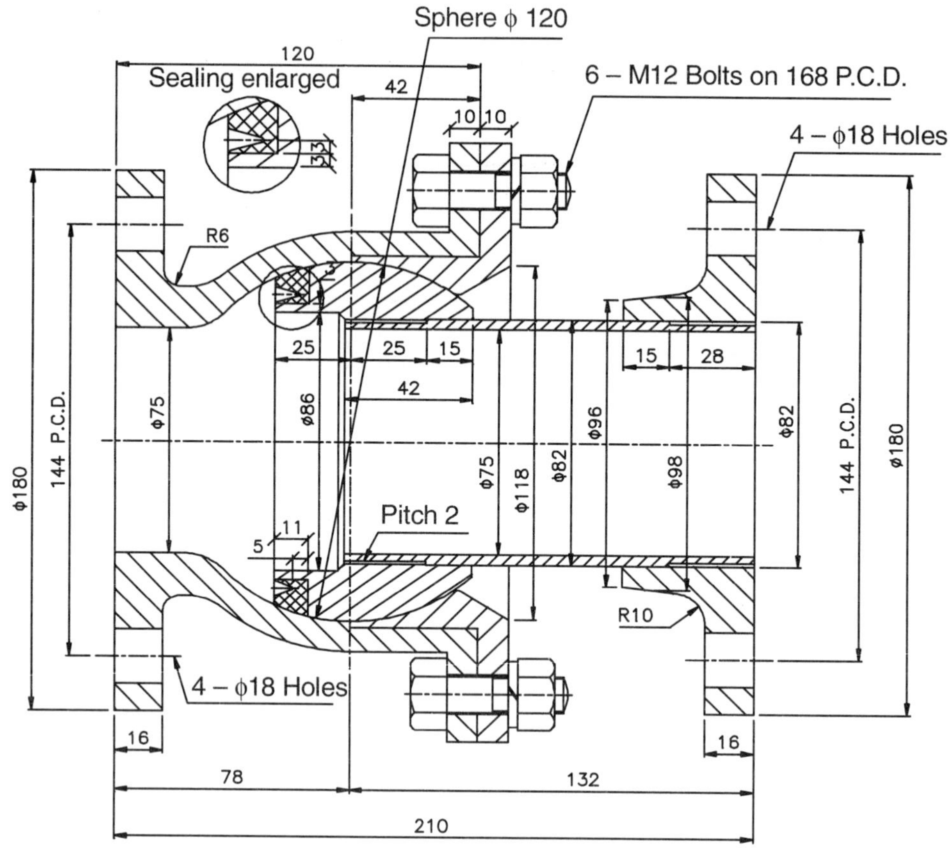

Fig. 6.2 Flexible pipe joint

PISTON

A piston is a sliding element which is driven by the fluid pressure inside a cylindrical bearing surface. The piston of a reciprocating engine or pump must be fluid (gas or liquid) tight, strong and rigid. Pistons are made an easy running fit in the cylinder and leakage past the piston is prevented commonly by leather / rubber / neoprene washer in case of liquid or piston ring in case of gas or steam. For piston construction Cast Iron, Cast Steel, Forged Steel, Aluminium, Bronze may be used depending on the speed, fluid handled and its pressure and temperature. Piston rings are made of cast iron.

Trunk Piston or Piston for Internal Combustion Engine

In single acting I.C. engines, *e.g.* oils, gas or petrol engines, the pistons are made of Cast Steel or Aluminium. These are made longer than the steam engine pistons to act as guide. In these pistons the connecting rods are normally fixed direct to the gudgeon pin in the piston, Fig. 6.3.

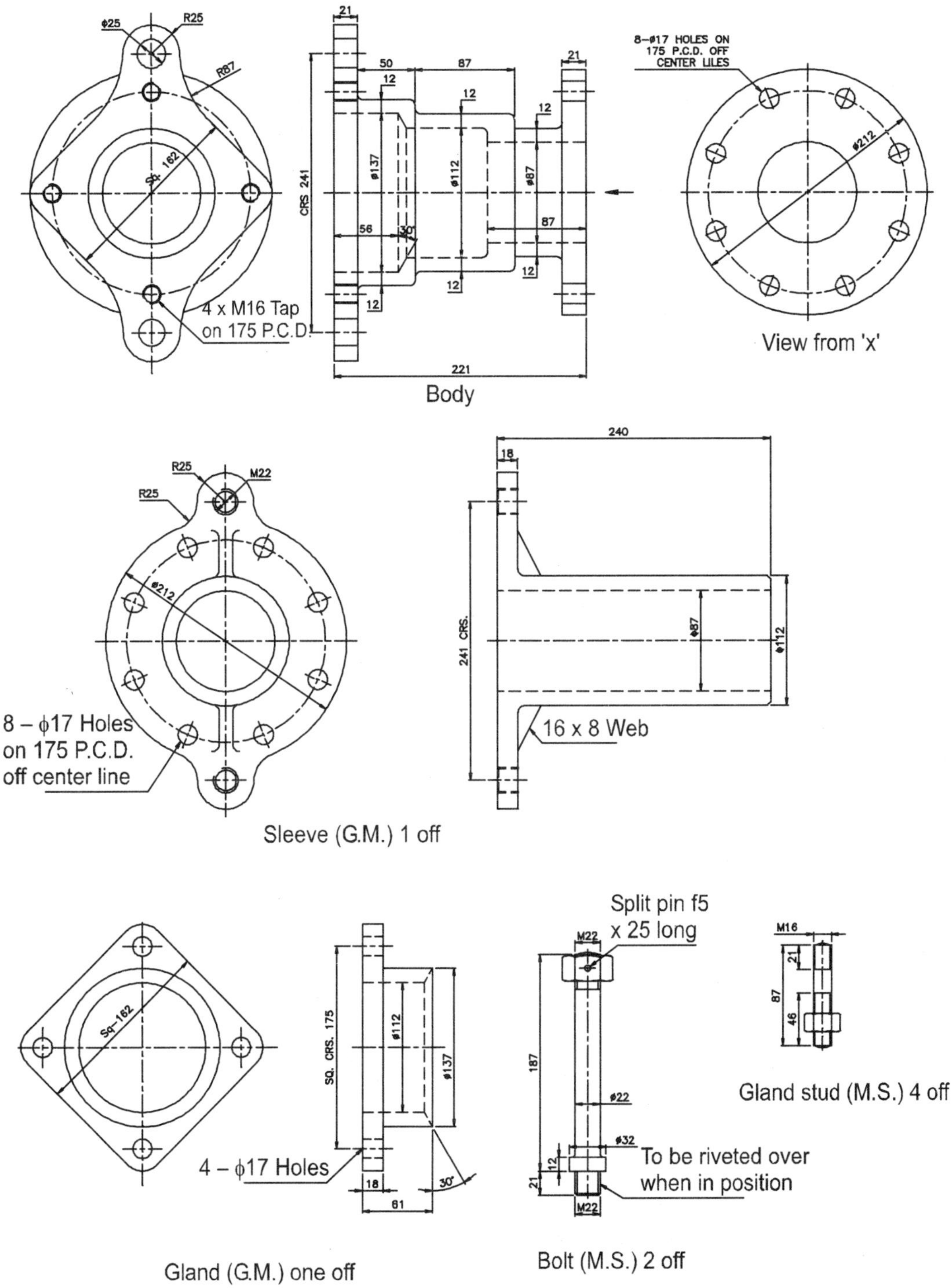

Fig. 6.2a Details of a φ 87 bore expansion joint

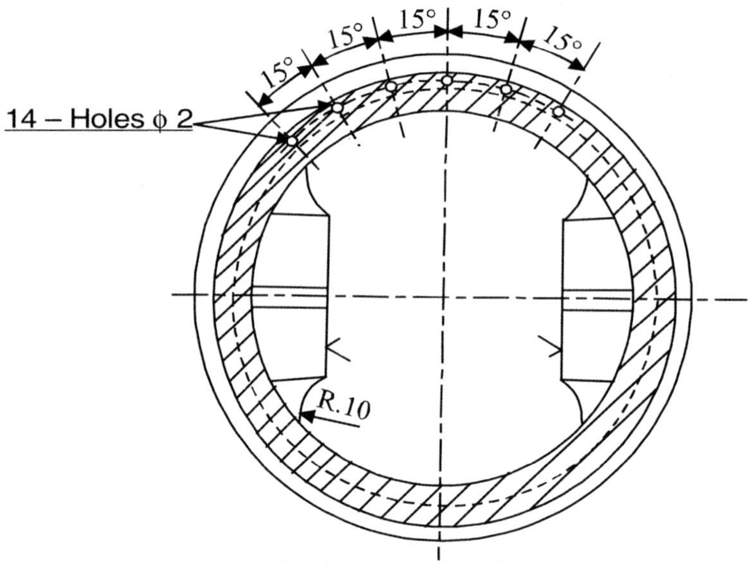

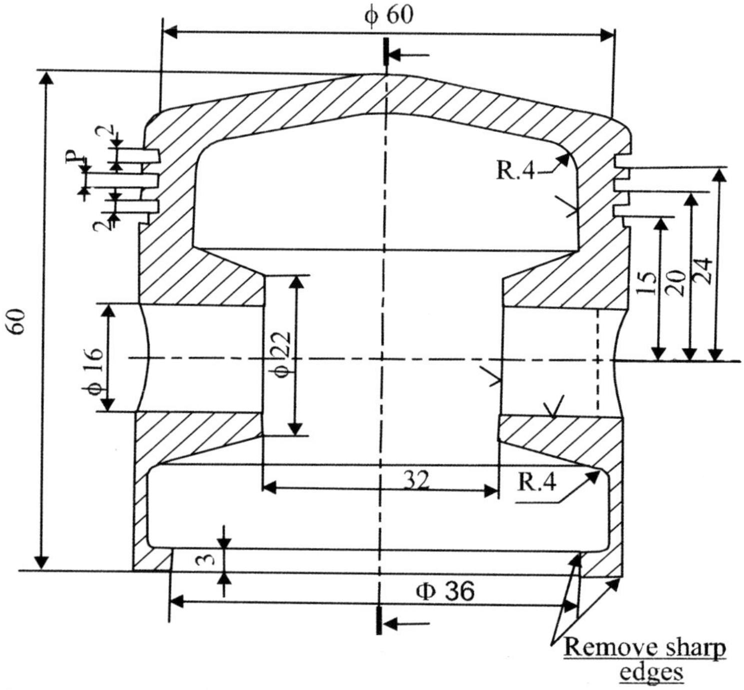

Fig. 6.3 Piston for internal combustion engine

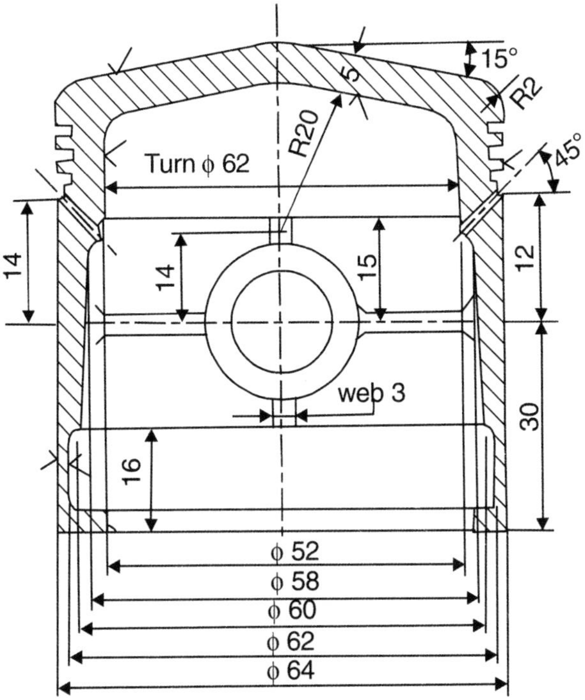

Fig. 6.3 Contd. (side view)

LOCOMOTIVE PISTON ASSEMBLY

The accompanying isometric drawing cut by two radial planes shows the locomotive piston assembly (Fig. 6.4). The cast steel body is turned to size and is provided with two piston rings. The piston is fixed on the conical portion of the piston rod by a steel nut, locked in position by a taper pin.

Prob. 6.6 Draw two views of the locomotive piston assembly, one of which should be in half section (Fig. 6.4).

Prob. 6.7 Draw details of the piston rod and piston ring indicating tolerance and surface finish.

CROSS-HEAD

Cross-head, an essential element of reciprocating machines, acts as a connecting link between the connecting rod and the piston rod. It reciprocates on guides with a straight line motion. The cross-head body or block may be made of cast steel and is provided with a spigot which fits closely with the rectangular groove milled on the flattened piston. The brasses in two halves are housed in the body and are held in position by the strip made of cast iron and cast steel. These being kept in position by means of two bolts and nuts. The slipper part of the body (210 × 115 rectangular plate portion – Fig. 6.7) slides on a guide.

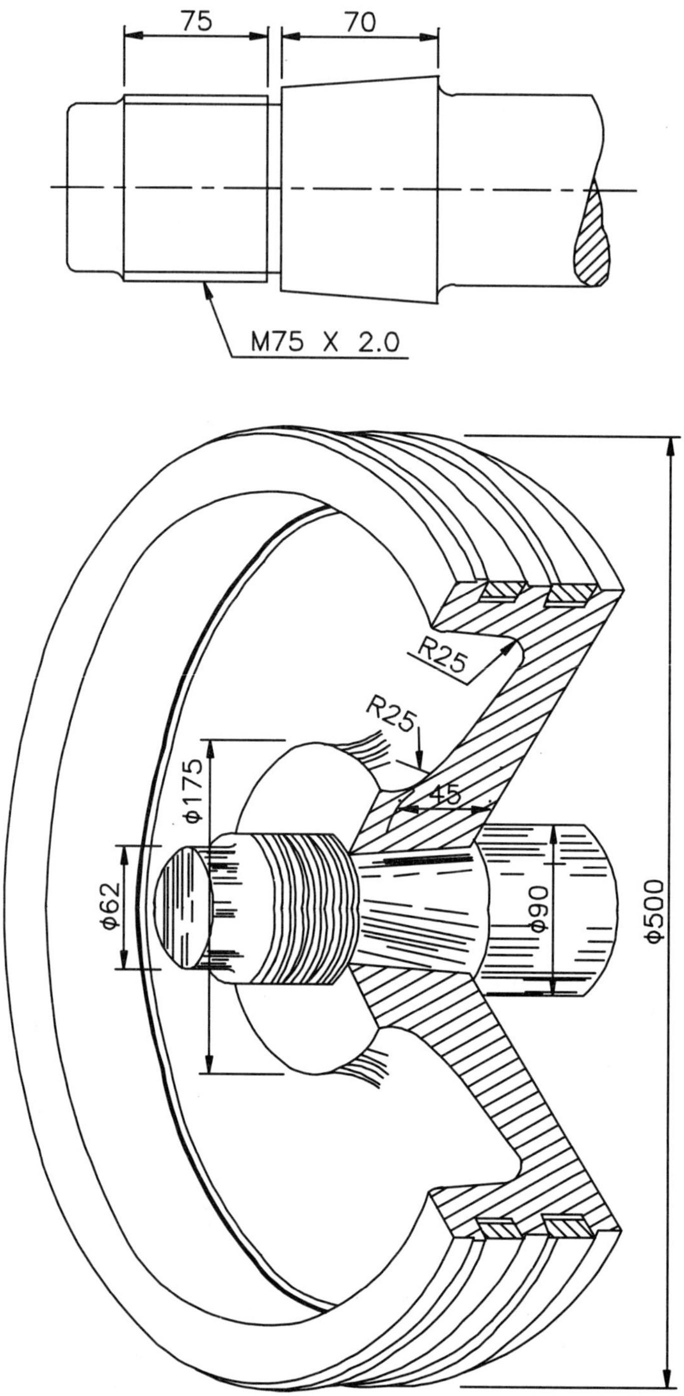

Fig. 6.4 Locomotive piston assembly

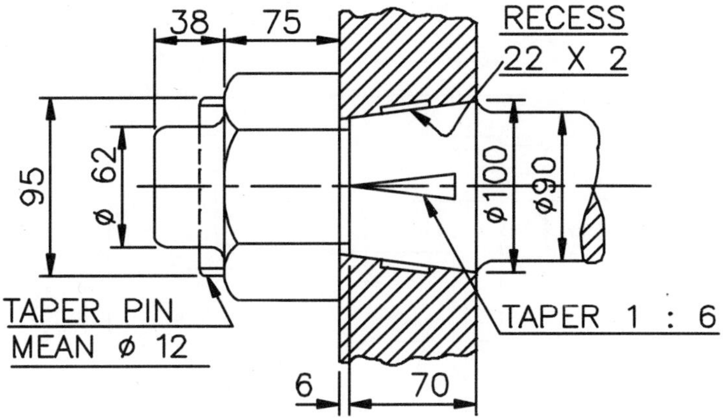

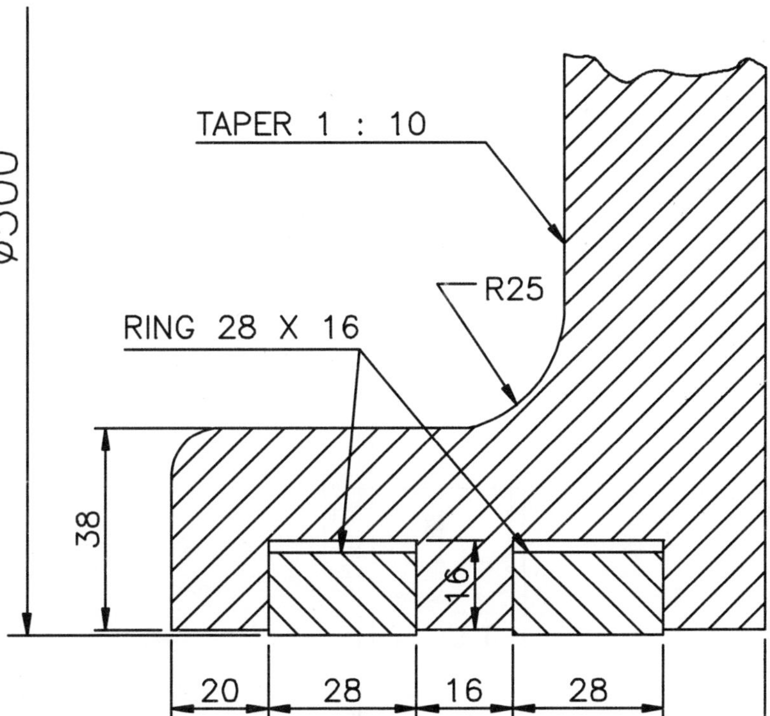

Fig. 6.5 Piston part enlarged

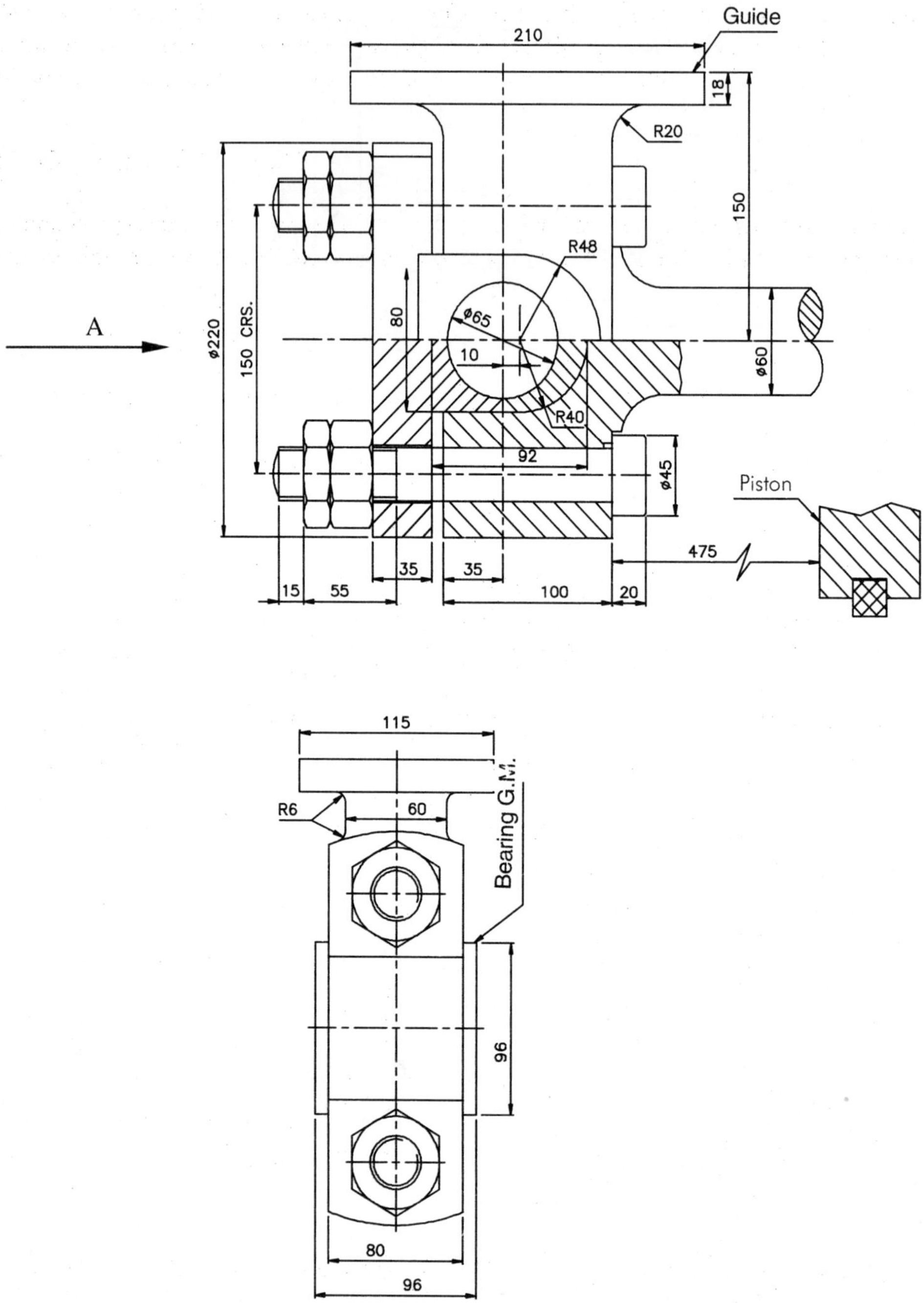

View from arrow - A

Fig. 6.6 Cross head assembly for vertical steam engine

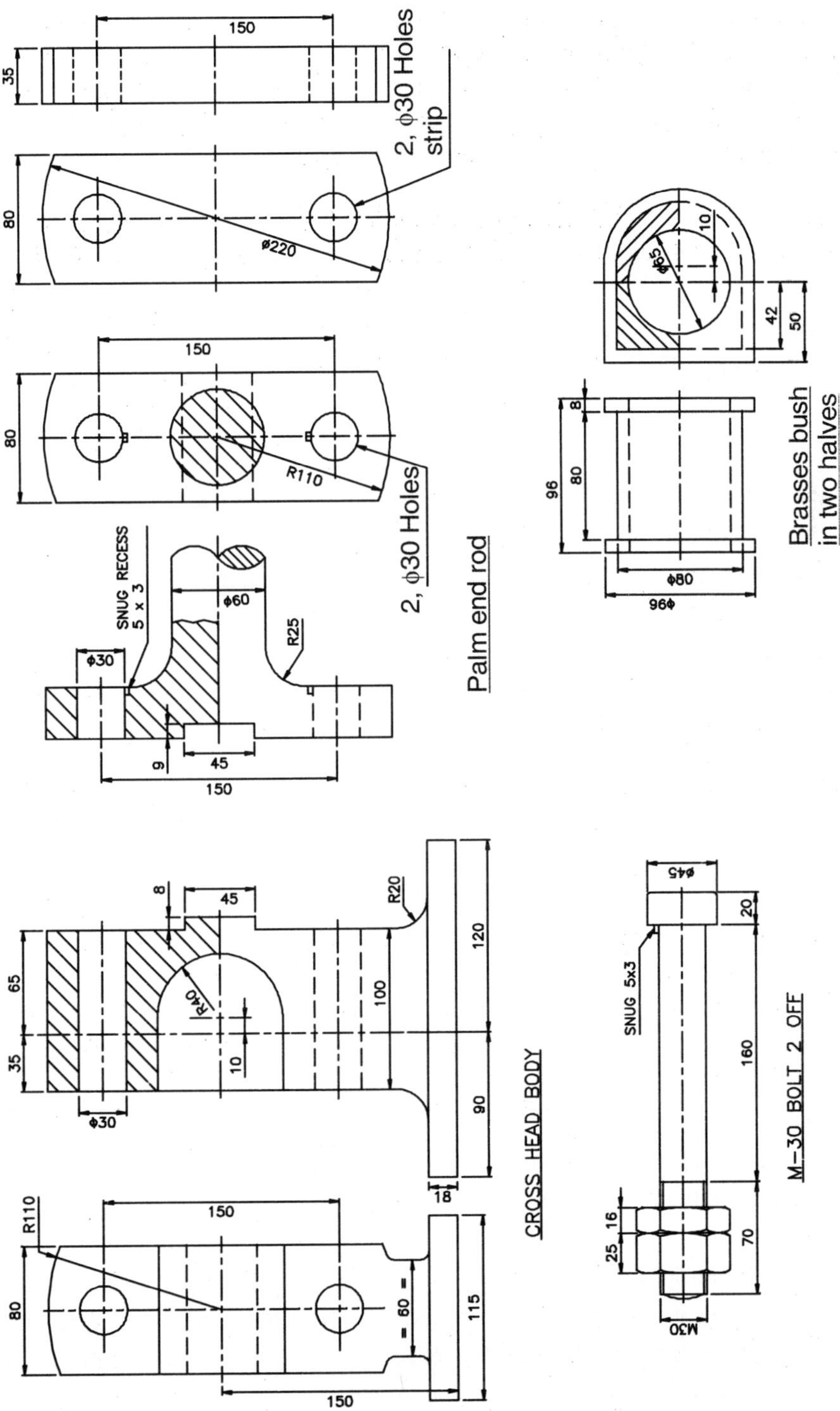

Fig. 6.7 Details of a vertical steam engine crosshead

Uses: Steam engine, reciprocating pump *etc.*

Prob. 6.8 Draw the detailed views of the assembled cross-head shown in Fig. 6.6.

Prob. 6.9 Details of a cross-head for a vertical steam engine is shown in Fig. 6.7. Draw two principal views of the assembled cross-head. The front view should be in half-screen.

Prob. 6.10 Make out a pattern drawing for the brasses giving all the necessary alowances.

Prob. 6.11 Draw tolerenced drawings of the brass part.

STUFFING BOX

Whenever the motion of a rotary or sliding part inside a pressure vessel is transmitted to other parts external to it or *vice versa*, the rod must pass through a leak-proof chamber without any loss of working fluid as well as pressure. The stuffing box (Fig. 6.8) is used to serve this end. The annular space provided round the rod is filled with packing, the materials of which depend on the fluid and its pressure & temperature. Leak-proofing is achieved by pressing the packing by drawing-in of the gland. The gland is held in position by a number of studs (normally two or three) attached to the flanges of the stuffing box body.

Metallic packing or asbestos packing in the form of ring of cord is generally used. The stuffing box shown is used for a water pump, Fig. 6.8.

Prob. 6.12 Draw the details of the stuffing box, indicating the tolerance on the mating parts, (Fig.6.8).

Prob. 6.13 Draw the pattern drawings of all the relevant components.

CONNECTING ROD

Figure 6.9 represents a Connecting Rod of a high speed vertical Steam Engine. The big end bearing is of steel casting lined with white metal. While metal is filled in the dovetailed groove and anchored at the ends the cross-head pin is shrunk fitted in the eyes of the forked end and as an additional safety two threaded dowels are used. The self lubricating arrangements for the bearing surfaces are as shown.

Note: *White Metal Lining* :

When two surfaces are in contact and in relative motion friction forces develop between the surfaces which will raise the temperature and subsequently cause seizure of the bearing. Lubricants are used to reduce friction and to prevent such seizing. In addition, bearings are occasionally lined with tin alloys like White Metal or Babbit Metal. White metal when overheated melts down and flows away; thus seizing is eliminated. Large and high speed bearings are generally lined with white metal. The liner is cast into the bearing by using a mandrel inside, which is later removed and the diameter is bored and scraped to proper size.

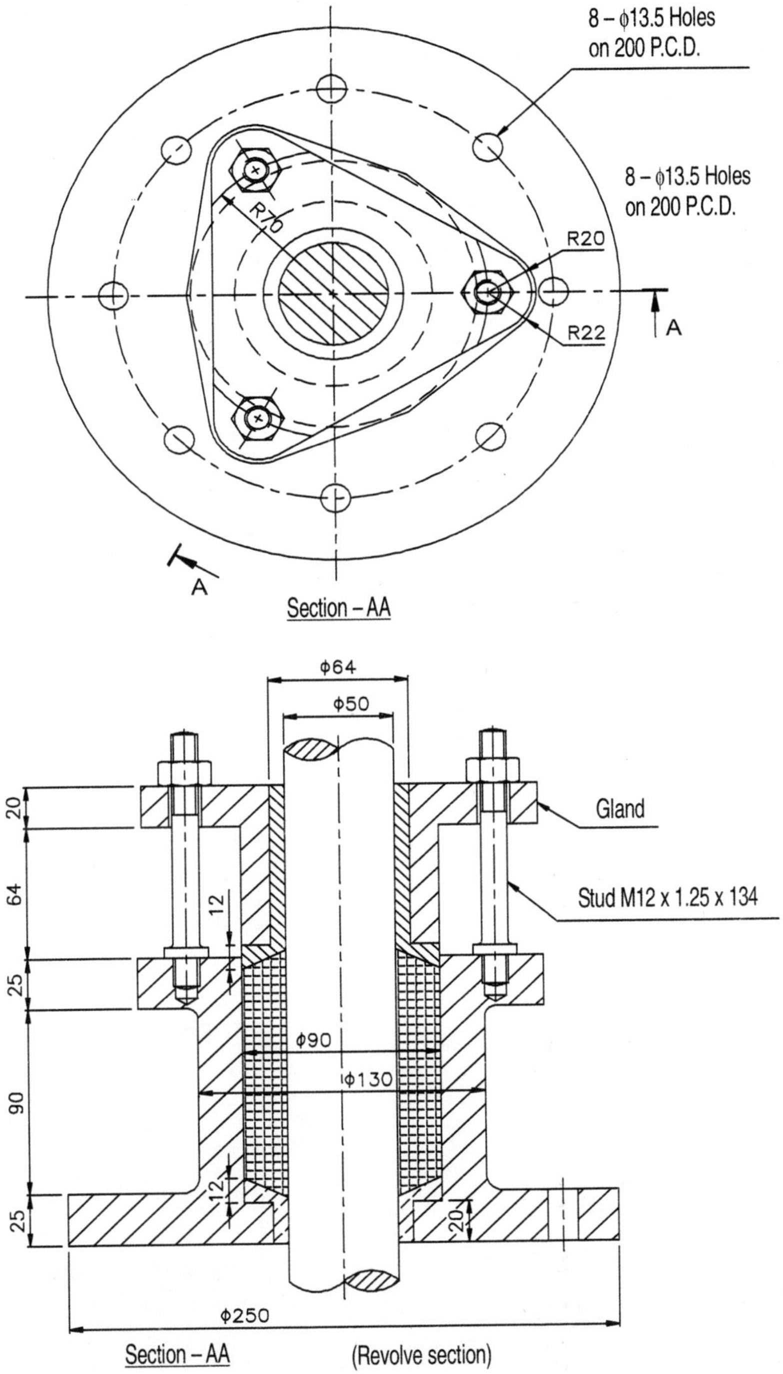

Fig. 6.8 Stuffing box

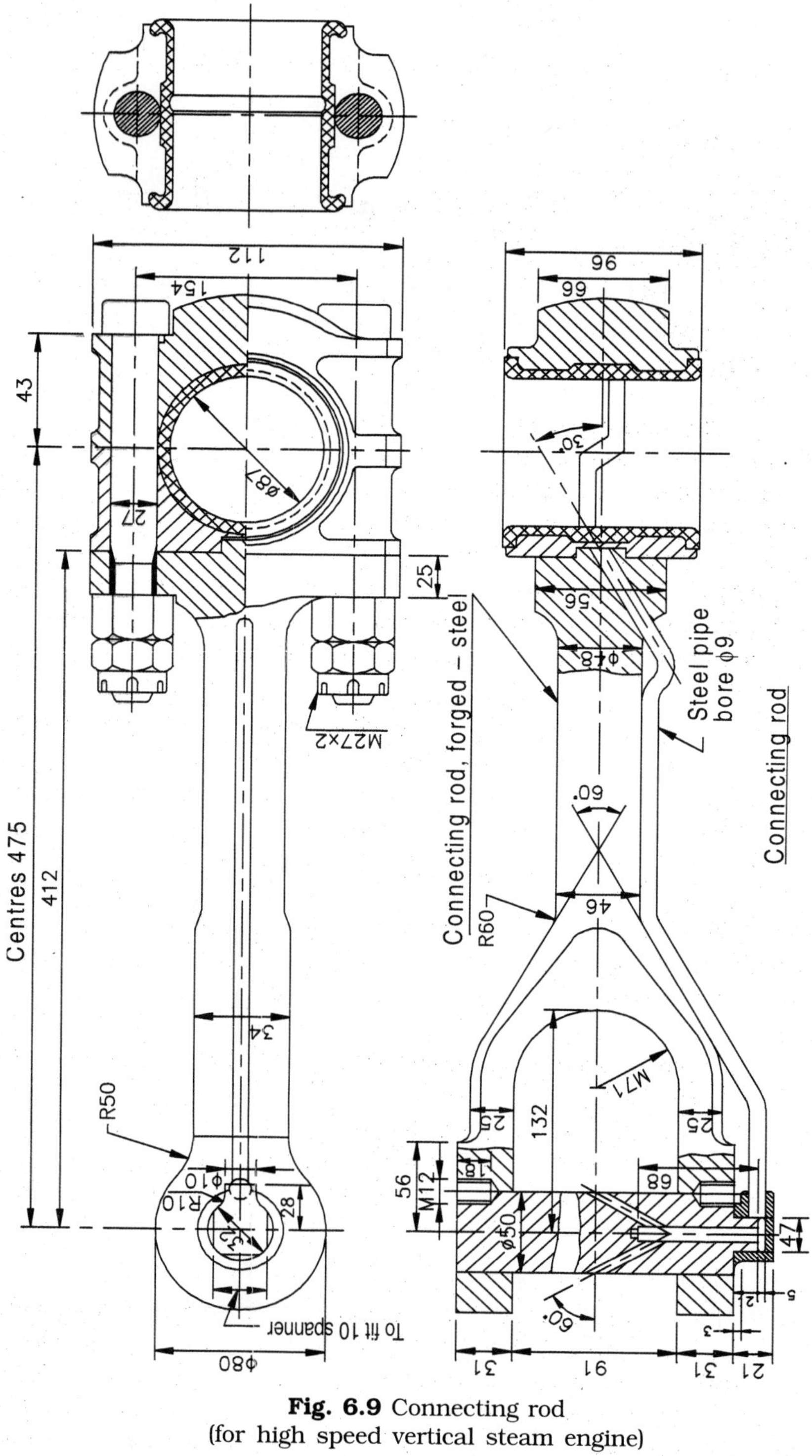

Fig. 6.9 Connecting rod
(for high speed vertical steam engine)

Prob. 6.14 Make a complete detailed drawing on the connecting rod shown in Fig 6.9.

Figure 6.10 represents a tapering I – section connecting rod for an Oil Engine. The big end bearing lined with white metal on gun metal matrix is provided with grooves to facilitate distribution or lubricating oil. The cylindrical gun metal bearing is prevented from rotation with a small pin as shown. The small end eye is provided with a G. M. bush. The big end bearing is kept in position by a steel cap and two cheese headed bolts. One side of the bolt head is machined flat to fit with the body and it prevents rotation during tightening and also keeps the width small. Castle nuts with split pins are used for locking of the bolts. Incomplete list of the parts are given below:

Item No.	Description	Material	Qty.
1.	Connecting Rod	Forged Steel	1
2.	Cap	-do-	1
3.	Bearing, big end	G.M.	1
4.	Bolt	M.S.	2
5.	Castle nut	M.S.	2
6.	Split pin	M.S.	2
7.	Pin	M.S.	2
8.	Bush, Small end	G.M.	1

Prob. 6.15 Make a complete workshop drawing of the components of the connecting rod shown in Fig. 6.10.

CRANK

Crank and Crank Shaft: Cranks of various types and shapes are used to convert oscillating or reciprocating motion into rotary motion or *vice versa*. Usually they are used in all reciprocating engines, pumps, machine tools *e.g.* shaping machine, slotting machine Engine crank shafts are commonly made of forged steel. The distance between the centers of the crank shaft and the crank pin is the crank radius, twice of which is equal to the stroke of the piston.

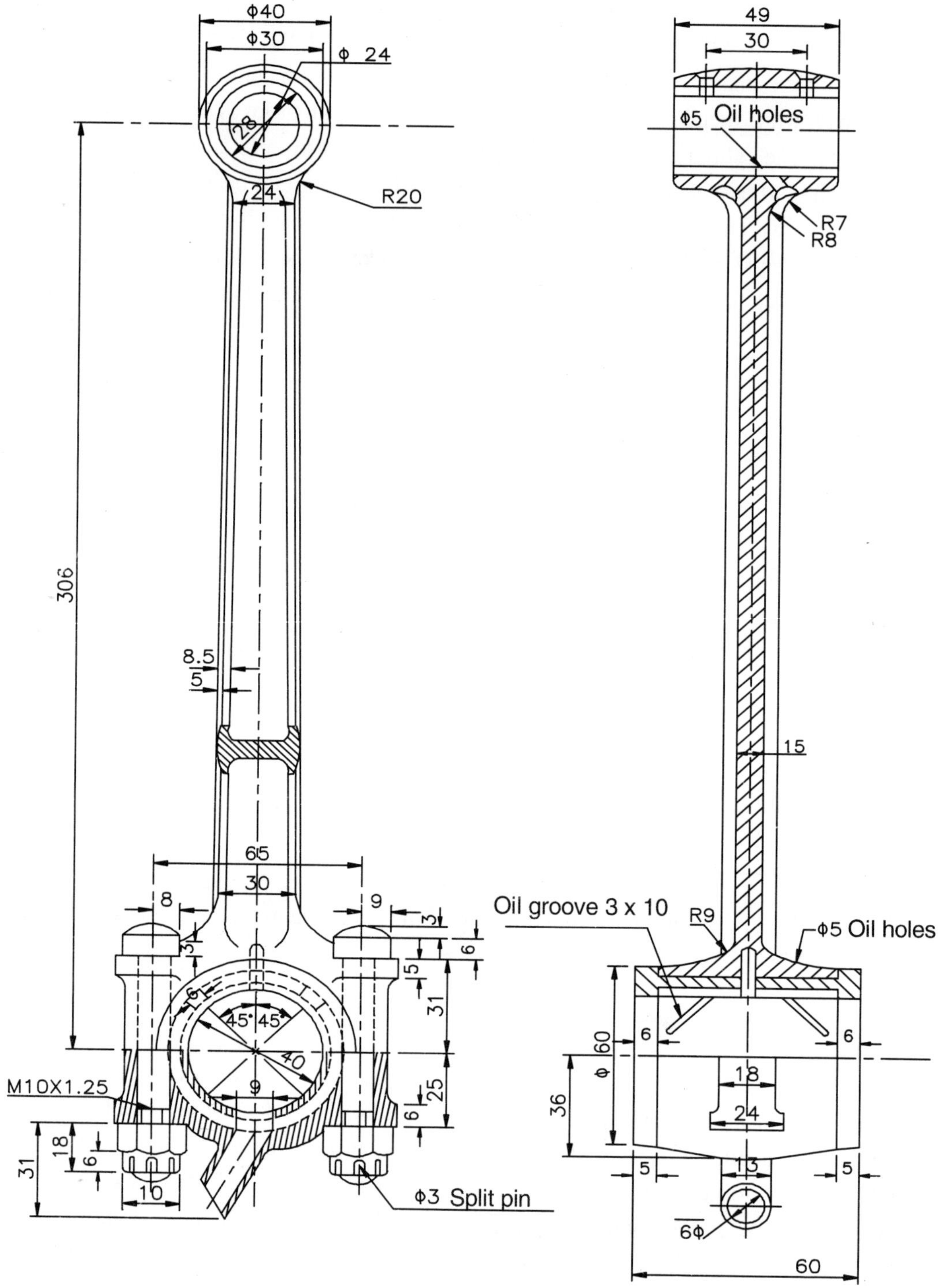

Fig. 6.10 Connecting rod for an oil engine

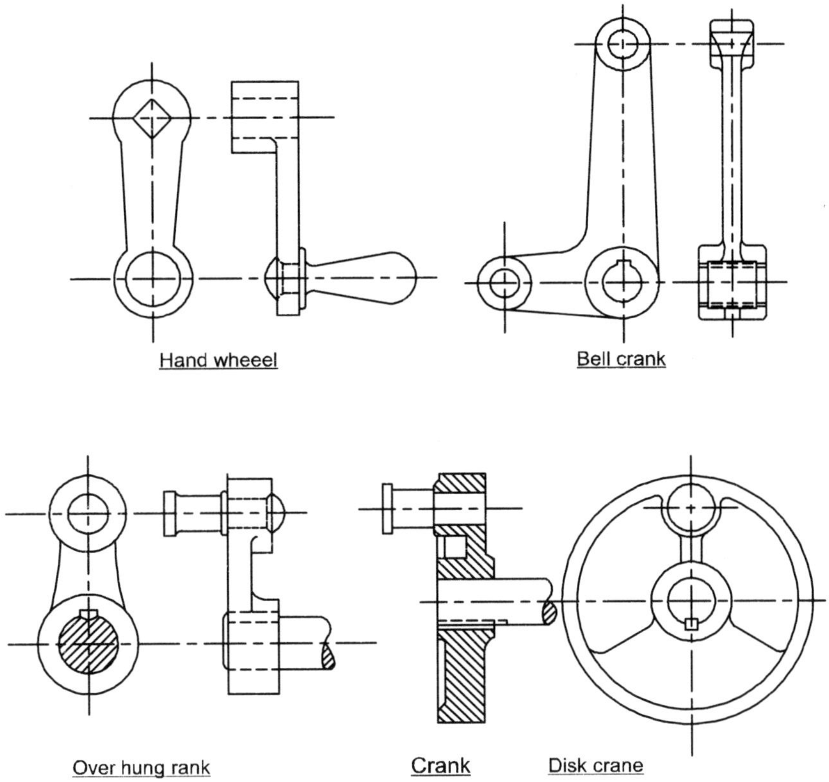

Fig. 6.11 Shows various types of cranks in use

ECCENTRIC

An eccentric is a special form of crank with small radius and the crankpin is larger in diameter than the shaft. The sheave is virtually the crankpin and the shapes are similar to the big end bearing of the connecting rod.

An eccentric is used to convert rotary motion into a reciprocating one. An eccentric sheave is, in essence, a cylindrical disc having a stepped or recessed rim and having its hole bored out of center *i.e.* eccentrically. Fig. 6.12 shows a simple one piece sheave and its two straps (made of steel castings) which are fixed by two bolts. Between the two straps, pieces or "shims" are employed. The distance from the center of the sheave to that of the crankshaft is the eccentricity. Twice the eccentricity is the throw.

It is used in pumps, steam engines *etc.* to convert the rotary motion of crankshaft into a reciprocating motion of the plunger, slide valve *etc.*

The sheave is generally made of cast iron and if mounted on the end of a shaft, it is made in one piece. For an intermediate position, which is not accessible from ends, it is made in two halves which are bolted together after being placed on the shaft. The rod end part is made of forged steel and the larger part *i.e.* the straps are made together by pin and cotter. The sheave is fixed to the shaft by a key.

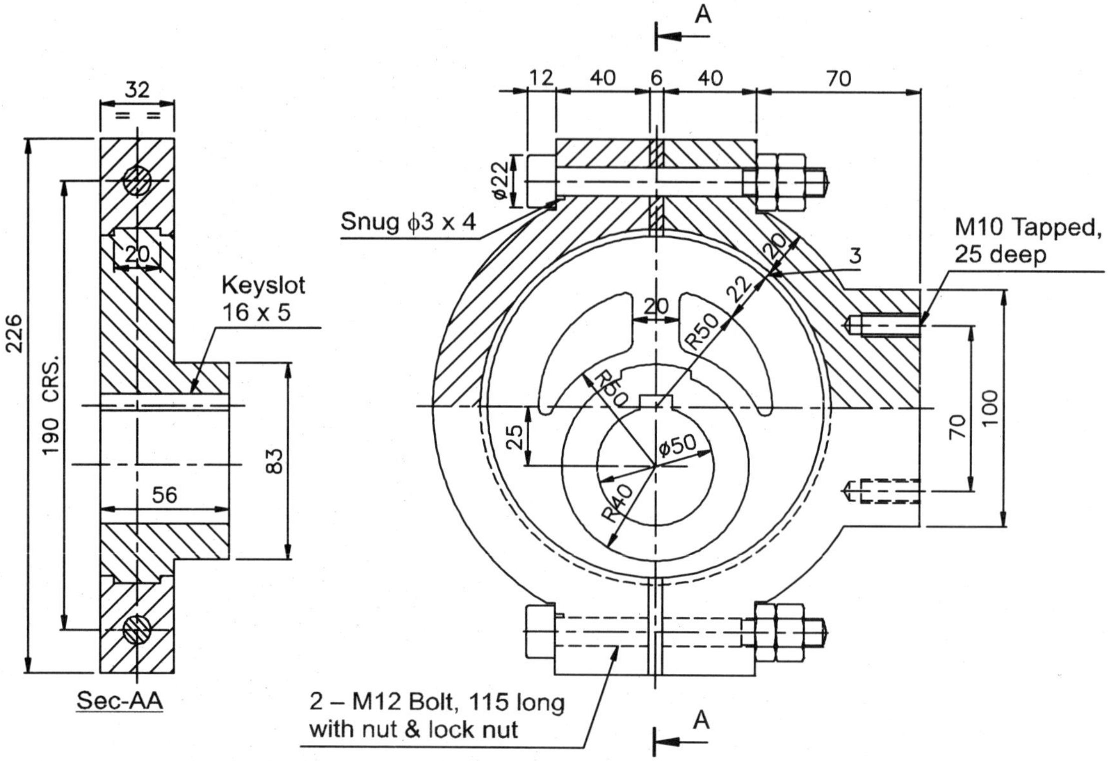

Fig. 6.12 Eccentric

Prob. 6.16 Make out a detailed drawing of the eccentric shown in fig. 6.12.

VALVE

Steam Stop Valve

Figure 6.13 illustrates the front sectional view and top view of a 75 mm steam stop valve used to control the flow of steam through the pipe line. Stop-valve should be placed at the highest point on the steam pipe line near the engine.

In general the valve body and cover are of cast iron for lower pressure of say, upto 1 MPa and of cast steel for pressure say, upto 2 MPa. The valve is made of gun metal, the spindle of bronze, the hand wheel of C.I. The studs, bolts and nuts are of M.S.

Prob. 6.17 The top view given in Fig. 6.13 is incomplete; quite a few dimensions have been purposefully left out. The students are expected to determine the dimensions while solving the problems given below:

(a) Hand out the sketch of all components, and develop the side view.

(b) Draw details of the components excluding nuts and bolts.

(c) Make the casting drawing of the cover.

(d) Specify the bolts, nuts and studs used. What is the probable material of the bridge?

Prob. 6.18 Fig. 6.14 shows the details of a 100 mm Stop Valve. Make the assembly drawing, given top view and front view in full section.

Blow-off Cock

Figure 6.15 shows the details of a 75 mm Blow-off-cock which is fitted in the lowest part of a boiler. This is necessary for periodic discharge of a portion of water from the bottom of a boiler for sediment removal. This is also used to discharge water completely from boiler for internal cleaning and inspection purpose.

The components of the blow-off cock are, in general, of gun metal, sometimes of cast iron with or without gun metal linings for the plugs to work in. However in the latter case also the plugs are made of gun metal. For steam pressure upto 1 to 2 MPa the taper of the plug varies from 1:4 to 1:8 on diameter. The cover at the top of the blow off clock provides the stuffing box.

Prob. 6.19 Make an assembly drawing of all the components and prepare the parts list of the blow-off cock shown in Fig. 6.15.

Prob. 6.20 Fig. 6.16 shows the details of a 90 mm Non-Return valve. The valve seat is screwed on to the body. The lift of the valve is controlled by the tail projected below the cover. Draw the assembly views, one of which must be in half section.

Representation of Valves

It is a usual practice to show Feed-Check Valve and Non-Return Valve in **Open position** and stop valve and safety valve in **Close position**.

Cylinder Relief Valve

Steam engine cylinders are provided with relief valves (Fig.6.17) at both ends and are loaded with compression springs. These valves release any excess pressure developed inside the cylinder at the end of piston stroke due to presence of condensed steam in the form of water. The valve is held in position by the spring, loaded as per the required pressure. The load is applied and varied with the help of the compression screw through the ϕ 30 mm G.M. washer and is locked by the locknut. The pressure from bottom opens the valve and releases extra steam through the opening provided in the casing.

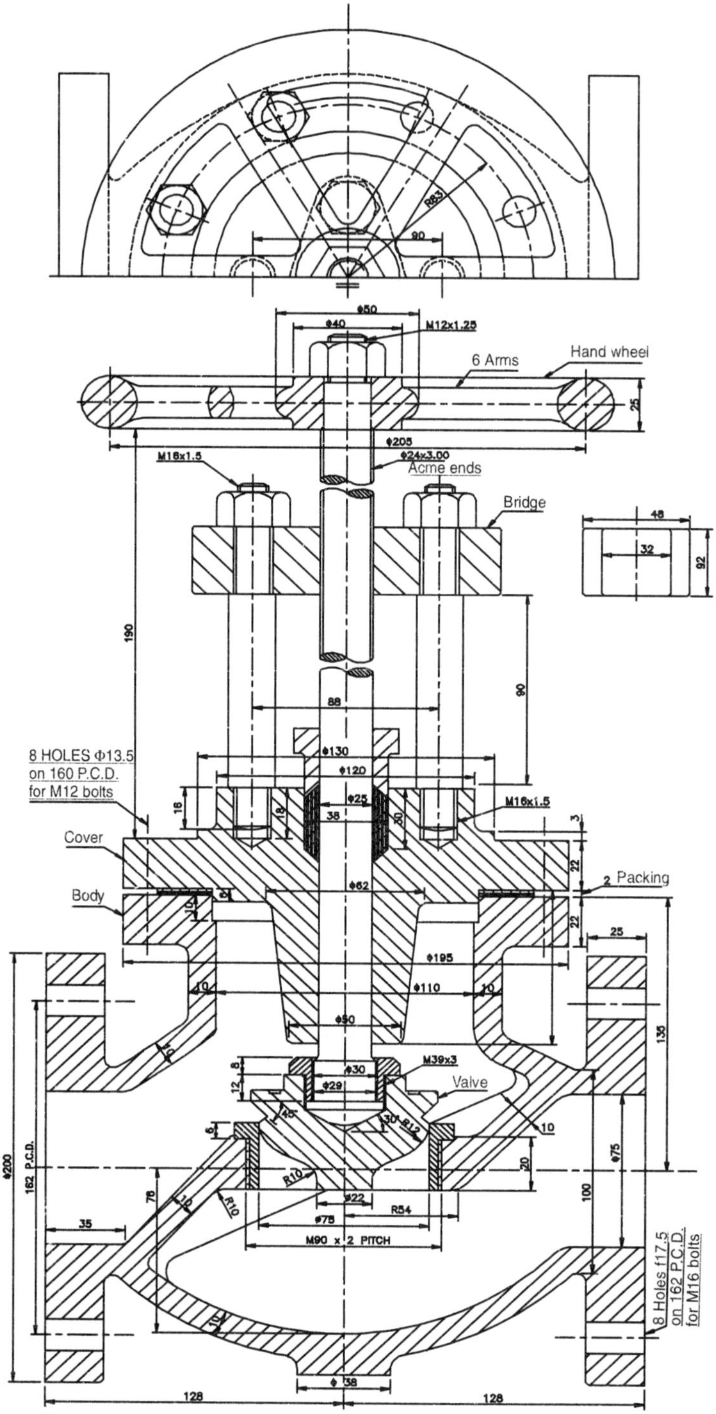

Fig. 6.13 Steam stop valve

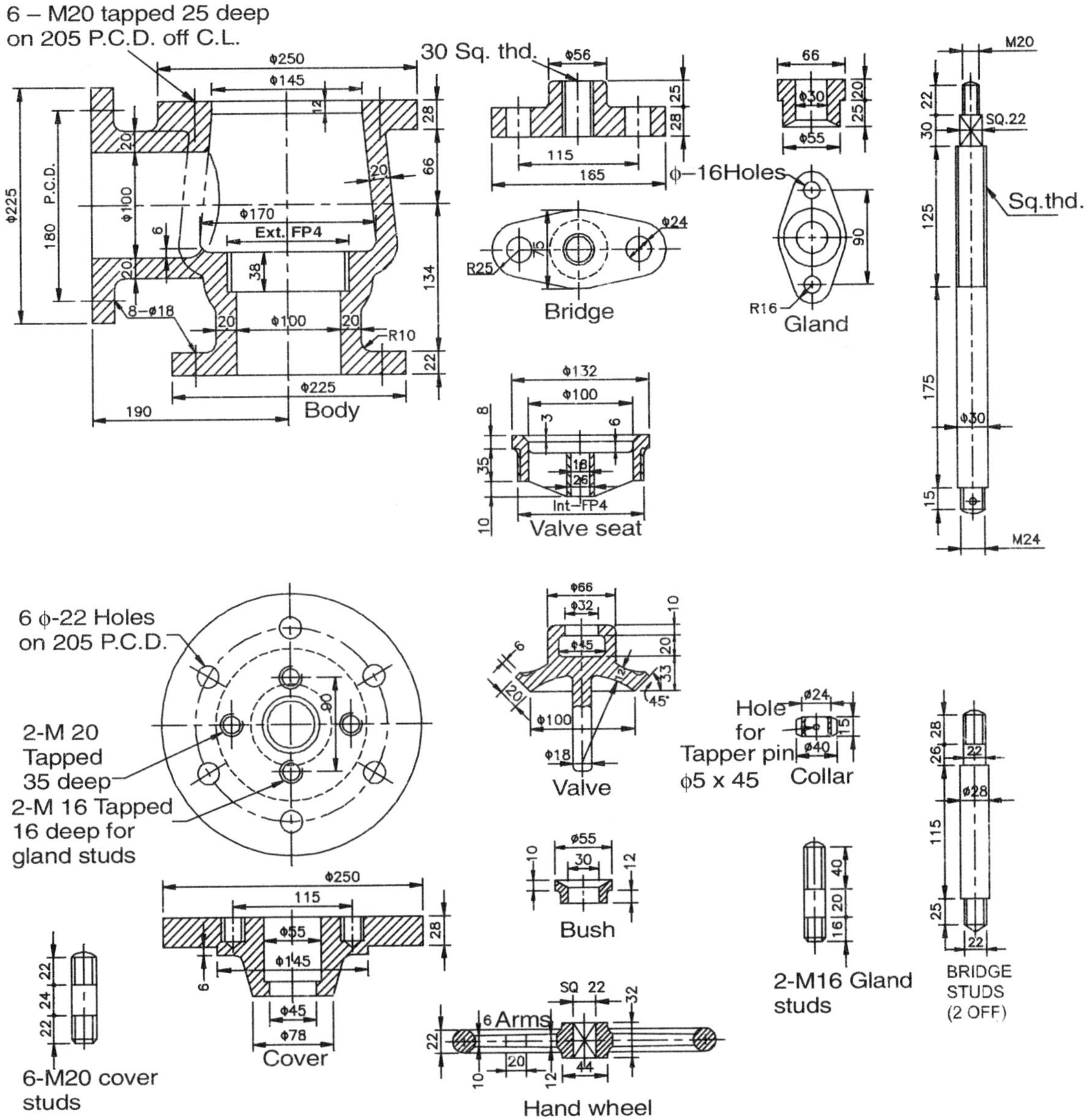

Fig. 6.14 100 mm stop valve details

Prob. 6.21 Figure 6.17 represents the details of a cylinder relief valve suitable for a 90 H.P. engine. Prepare the assembly drawing of the components providing the following views:

(i) Half-sectional elevation, (ii) Plan (iii) End view looking from left. Put only the overall dimensions and prepare the parts list.

Prob. 6.22 Fig. 6.18 shows two views of the gun metal body of a cylinder relief valve suitable for 1 MPa steam pressure. Try to develop the other components from the standpoint of functional requirement and put them into in assembly and hand out a sketch from the sense of proportion.

SAFETY VALVE

A safety valve, an essential component for all pressure vessel applications, relieves the pressure whenever the working pressure is exceeded. These valves are of disc type with narrow flat or conical seats. The working lift of a safety valve is 3 mm or less. Safety valves are of various types such as spring loaded safety valve (Fig. 6.19), lever safety valve, dead weight safety valve etc..

The high pressure steam pushes the valve upward (Fig. 6.19) when the boiler pressure exceeds the working pressure and as soon as the pressure is relieved the spring force pushes it back with impact; to reduce such impact stress the valve discs are enlarged.

Prob. 6.23 Draw the shop details of the spring loaded valve (Fig. 6.19) suitable for 1.5 MPa.

GATE VALVES

Flanged caste iron Gate Valves used by the petroleum industries have got many variations depending on the types of gate *e.g.*, wedge gate and double disk gate rising stem and non rising stem. These types of valves may handle liquid, gas or steam in pressure range of 0.7 Mpa to 3 Mpa (100 psi to 430 psi or 6.8 at m to 27 ata).

The standard sizes cover as per BS 1735:1960 the following nominal sizes:

37, 50, 62, 75, 100, 125, 150, 200, 250, 300, 400, 450, 500, 550 mm (converted to mm approximately).

MATERIALS: GATE VALVE

Body, Bonnet and Stuffing Box are made of CI, Grade FG-140, Hand wheel usually cast steel, or forged steel, MI or Sg-iron; Gaskets, compressed asbestos joining material (oil resistant); Gland packing – good quality lubricated asbestos packing; Bolting for bonnet, stuffing box, yoke and gland – carbon steel, C 30 steel.

Stem of the acme threaded screw – rolled or forged high tensile brass.

Wedges or disks – Bronze or CI.

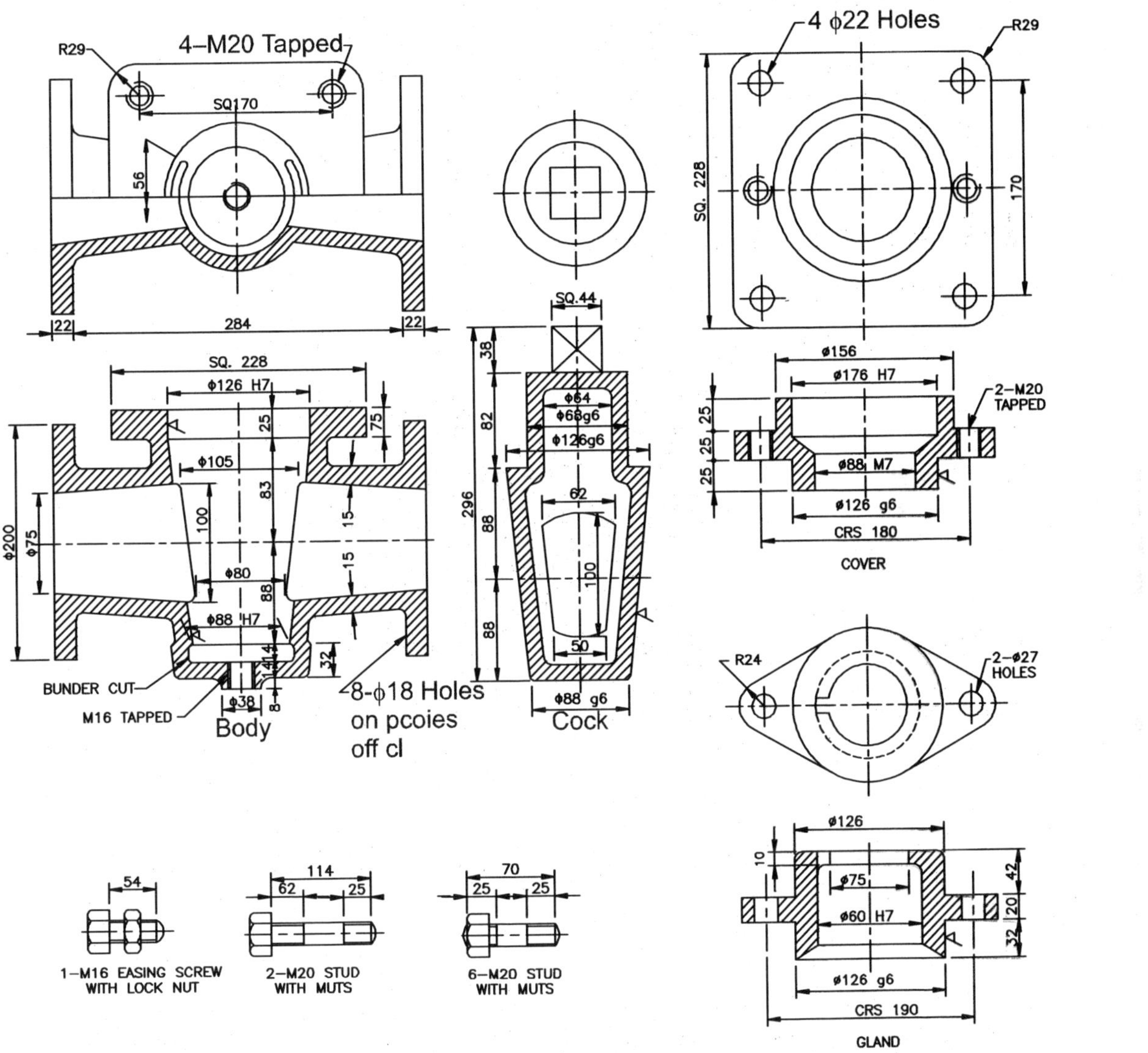

Fig. 6.15 Details of a 75 mm blow-off cock

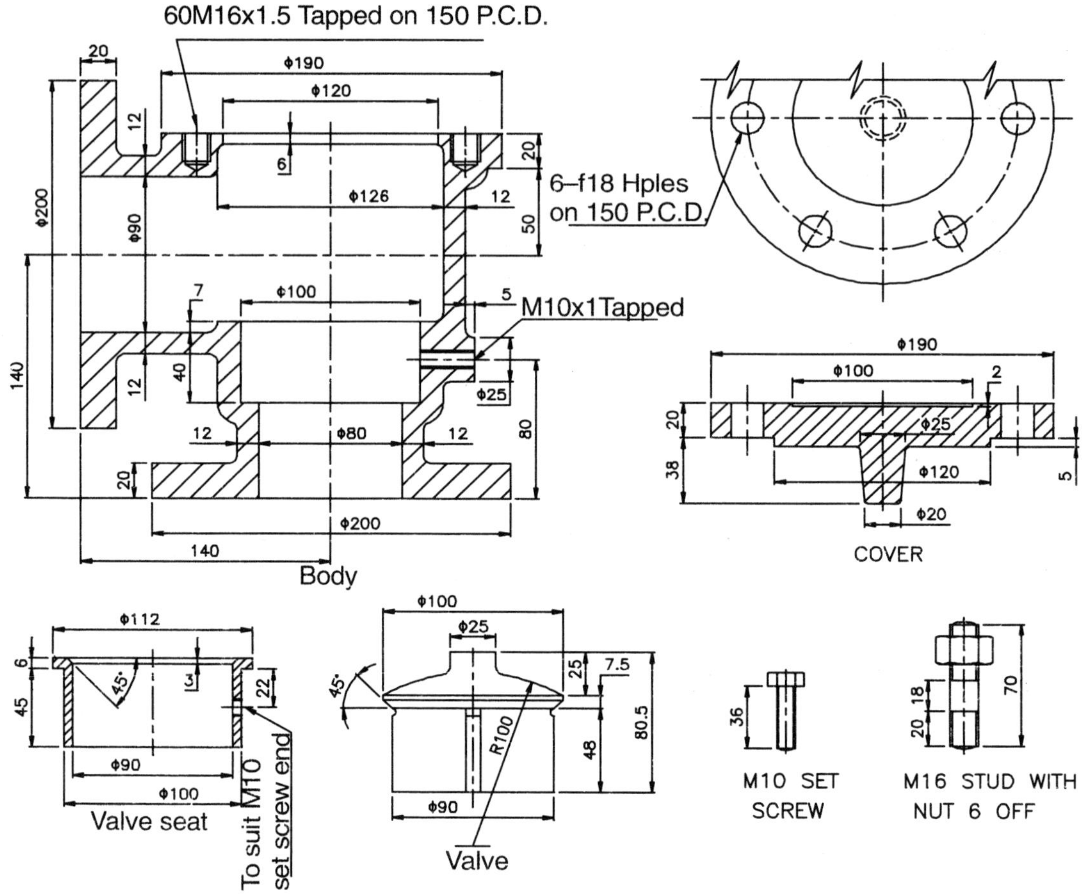

Fig. 6.16 Details of 90 mm non-return valve

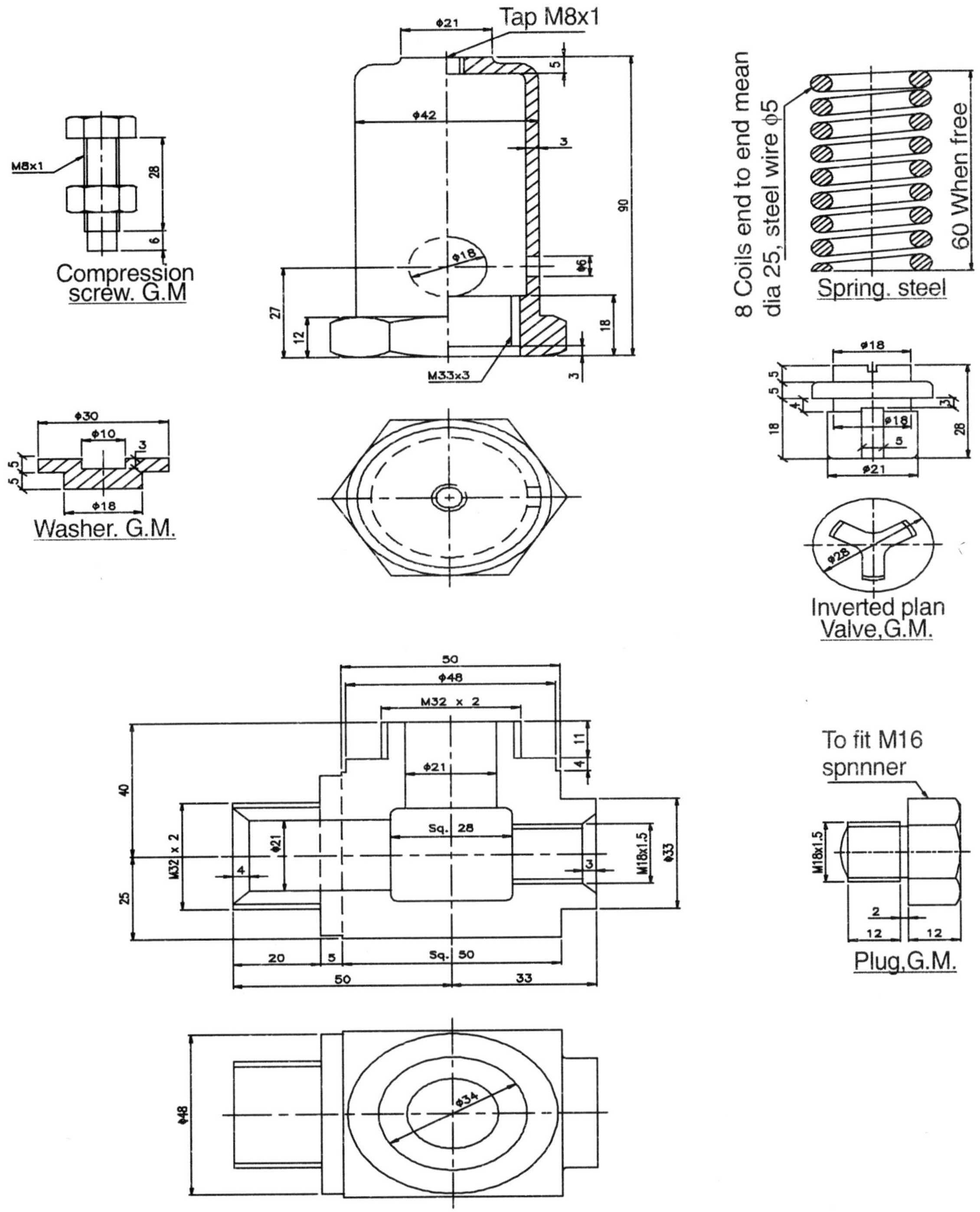

Fig. 6.17 Details of cylinder relief valve

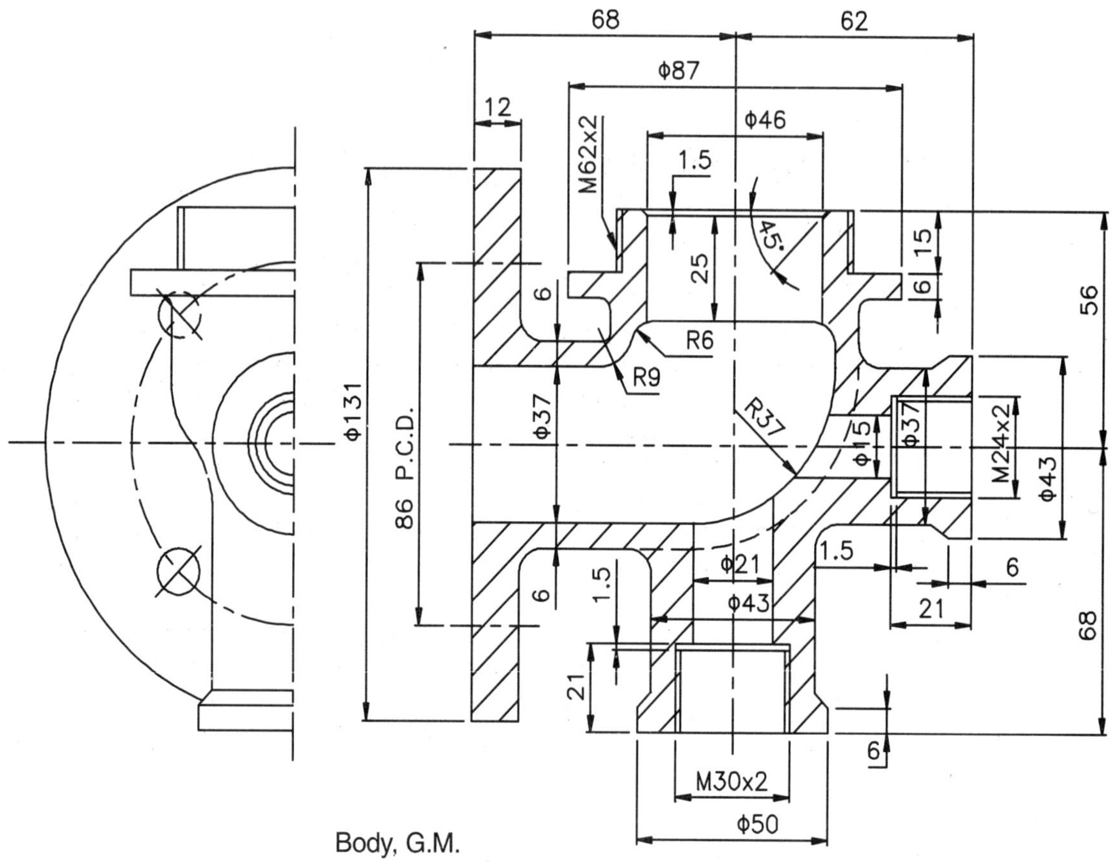

Fig. 6.18 Body of a cylinder relief value

Test Pressure upto size 300 mm 350 psi = 24 atm or 2.4 MPa, tested hydrostatically.

Some Typical Dimensions, mm

Nominal size	Inside diameter (min)	Body and bonnet thickness	Face to face	Stem diameter (min.)
50	50	8	162	16
75	75	11	200	22
150	150	14	262	28
300	300	20.5	350	37
500	500	28	450	56

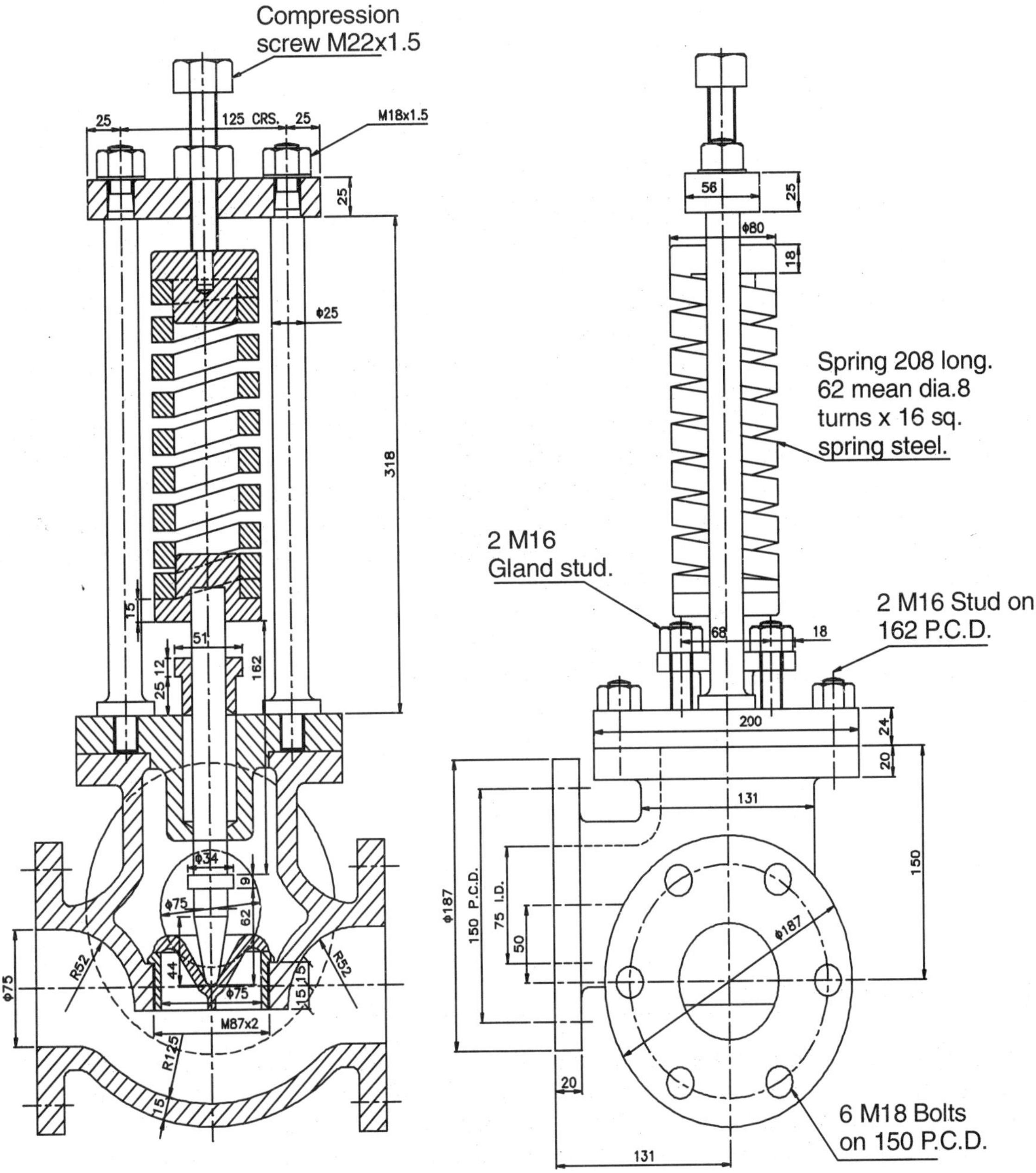

Fig. 6.19 Spring loaded safety value

Figure 6.20 shows a typical composite nonrising stem double disk type gate valve.

This means the stem (screw) rotates in place with the manual rotation of the handwheel, the upper spreader nut or busing (18,19) will move axially (vertical) depending on the rotational direction. This in turn will cause spreader to move giving away to the fluid flow.

It may by noted that the value shown is not symmetrical about its vertical axis. This is done internationally to show typical variations in individual details.

Prob. 6.23 Fig. 6.20 shows a NTS sketch of a cast iron, non-rising-stem double disk type gate valve. Make out a parts list with probable materials.

Draw the top-view and half sectional side view of the gate valve for a nominal size of 300 mm.

MACHINE TOOL PARTS

Machine tools are machines used to remove the extraneous materials in the form of chips from different blanks to transform them to desired shapes and sizes. Common machine tools are Lathe, Shaping Machine, Drilling Machine, Milling Machine *etc*.

Figure 6.21 shows the assembly view of the **Tailstock of Lathe** used to support the job or for holding drilling bit on a Lathe.

The tailstock body and the hand wheel cover are made of cast iron; tailstock spindle with square threaded screw, barrel, lathe center, screws, keys *etc.*, are of mild steel.

The left hand view (I^{st} angle) shows the profile of the body and the invisible part from left shown dotted in side view, is sectioned in plane AA and hatched for clarity. This type of sections is called **phantom section.**

Fig. 6.22 shows the details of a tailstock of a lathe.

Prob. 6.24 From the tailstock details (Fig. 6.22) of a lathe draw the two principal views of the tailstock, Prepare a parts list.

Tool Head of a Shaping Machine

Tool head details of a shaping machine are shown in Fig.6.23. The drag release plate and swivel plate are cast iron, the pin and the tool holder generally of mild steel and the clamping screw are of M.S. case hardened.

Prob. 6.25 Draw the assembled front, top and right hand side views of a tool head of shaping machine also furnish part list, Fig. 6.23.

Prob. 6.26 Draw the front views of the swivel plate and the drag release plate and their views from the left in full section and add the top views from the details given in Fig. 6.23.

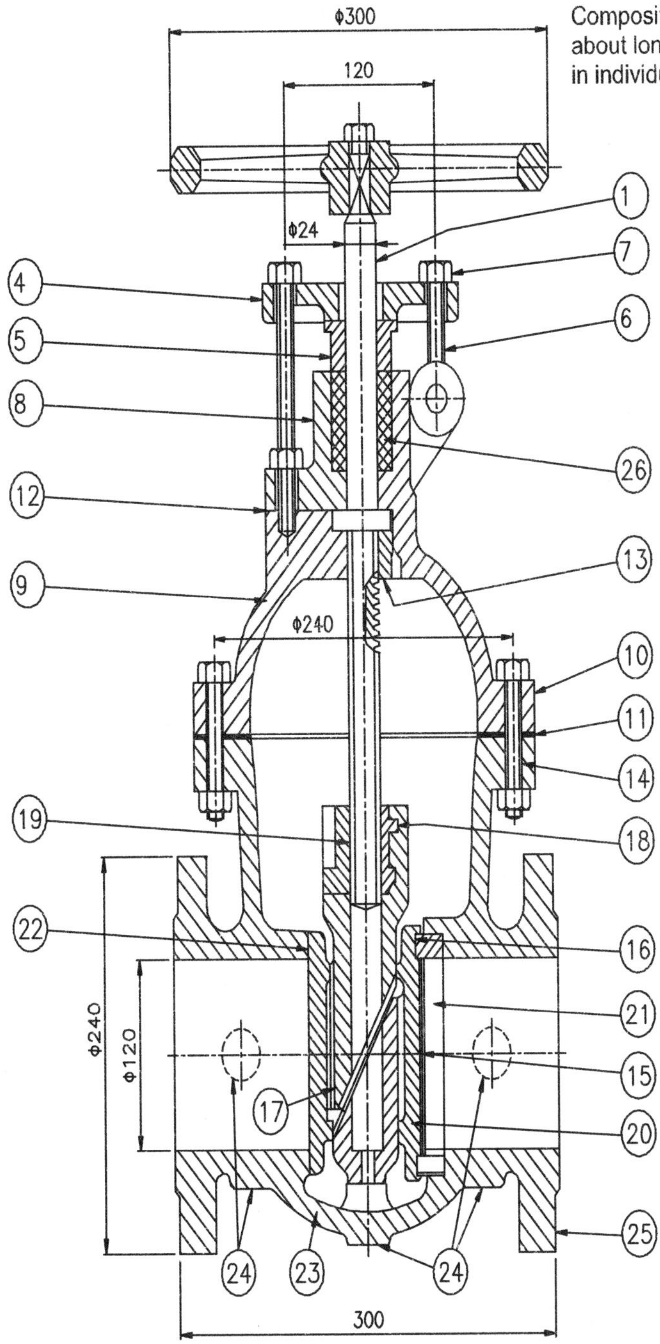

It no	Name of part
1	Stem
2	Handwheel
3	Handwheel set screw or nut
4	Gland flange
5	Gland follower.
6	Gland bolts or stud.
7	Gland bolts nuts.
8	Stuffing box.
9	Bonnet.
10	Bonnet flange.
11	Bonnet gasket.
12	Stuffing box gasket.
13	Stem bush.
14	Bonnet bolts and nuts.
15	Disks
16	Disks facing ring.
17	Upper spreader
18	Upper spreader bushing.
19	Upper spreader nut.
20	Lower spreader
21	Bottom seated ring.
22	Integral faces.
23	Body.
24	Body bosses
25	End flange.
26	Asbestos.

Fig. 6.20 Cast iron, non-rising-steam double disk type gate valve

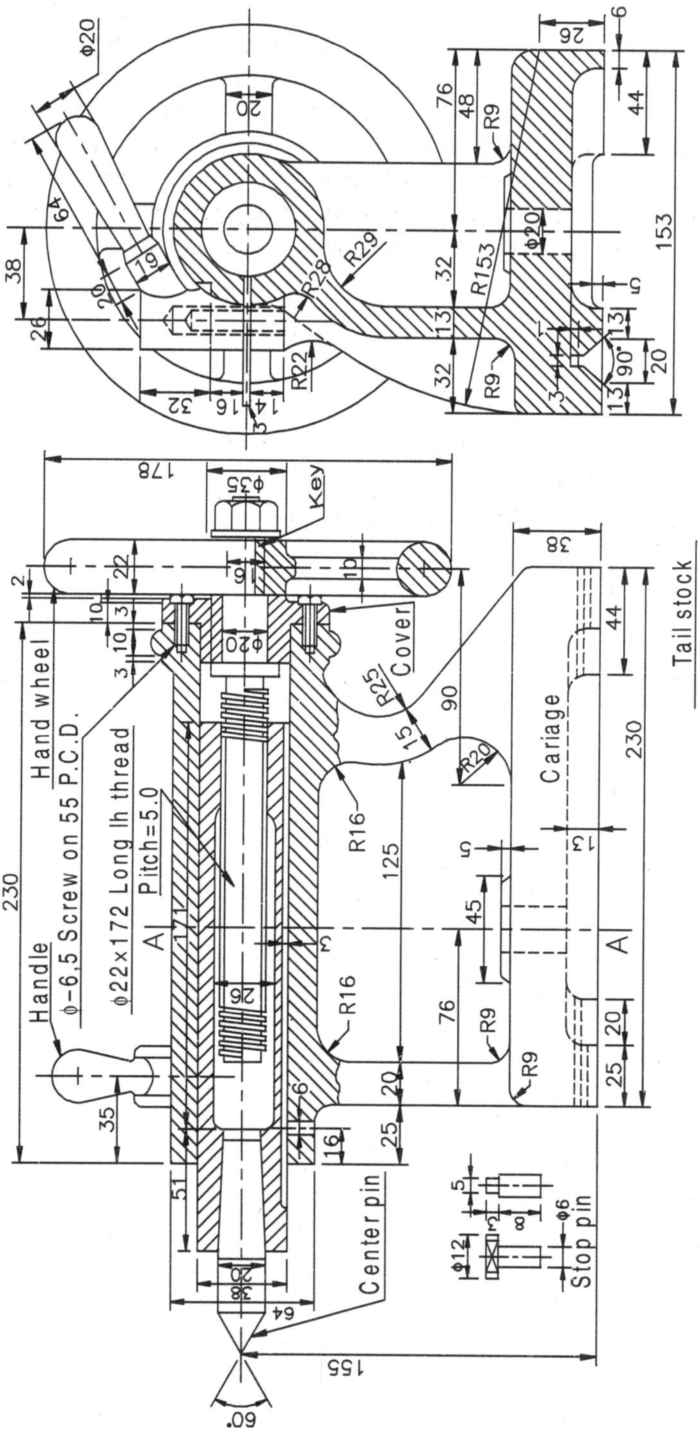

Fig. 6.21 Tail stock of a lathe

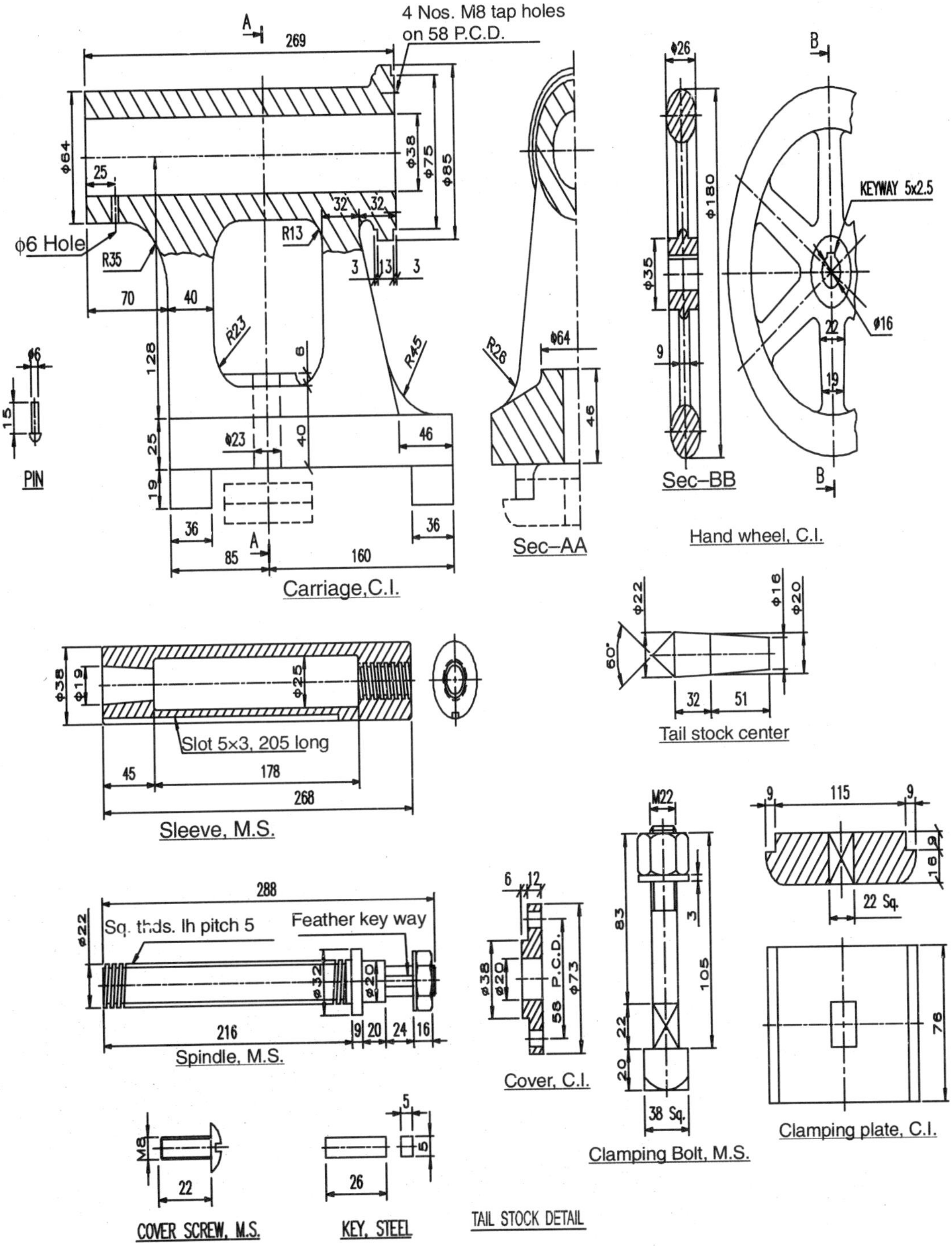

Fig. 6.22 Details of tail stock of a Lathe

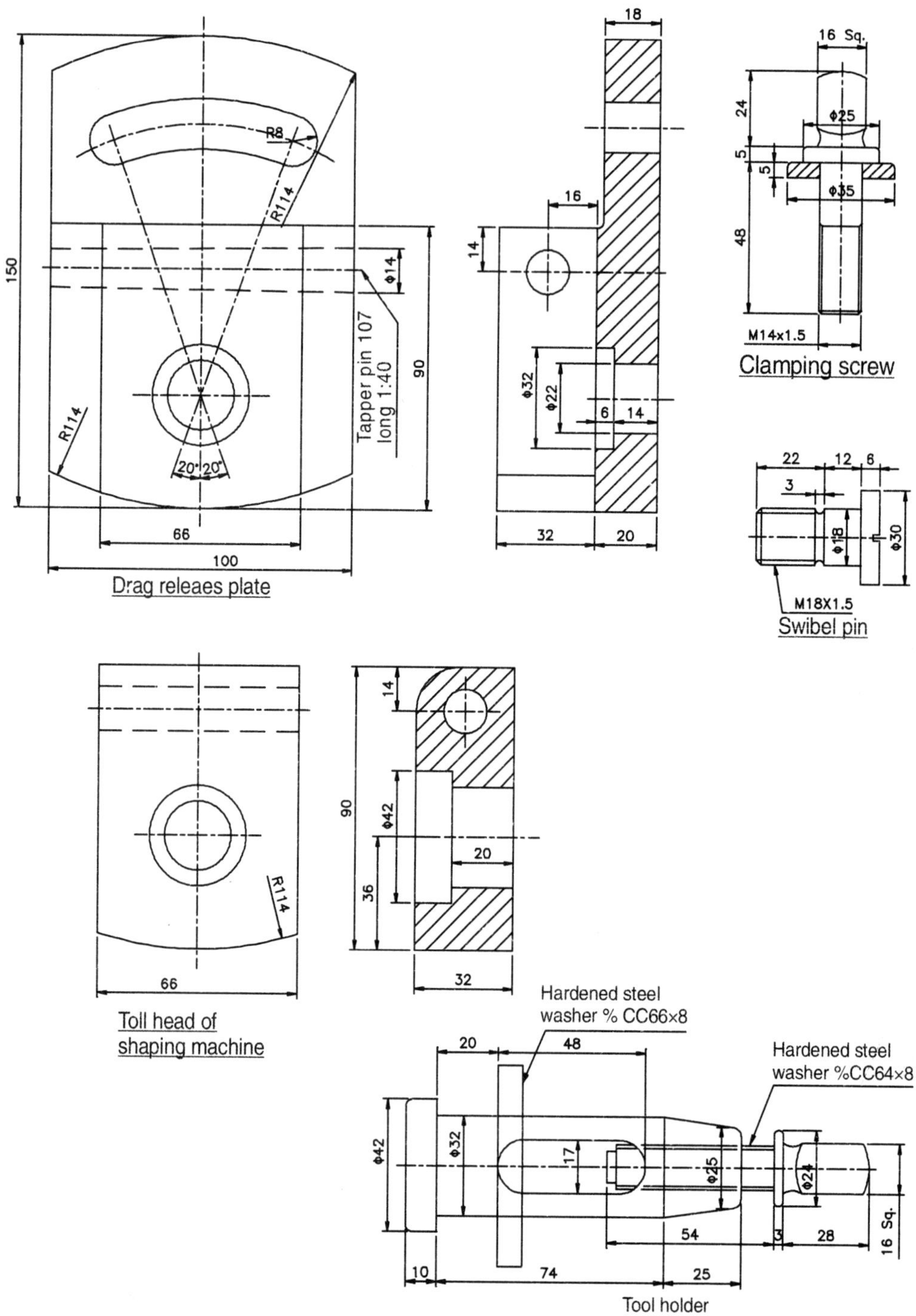

Fig. 6.23 Tool head of a shaping machine

Bearings

So far friction is concerned, bearings are of two types *e.g.*; sliding contact and rolling contact types. In the first group, journal bearings are the common types in which the supporting loads act radial to the bearing axes; journal bearings require considerable bearing widths to support the loads on the shaft. Apart from this there may be footstep or pivot type bearings in which the supporting pressure is vertically upward and parallel to shaft axis. In its simplest form a cylindrical hole bored in a cast iron block may serve as a journal or a pivot bearing. In the next stage of development a simple journal bearing is bushed with brass or gun metal. Worn bushes may be replaced.

Plummer Block

When shaft sizes are large and speeds considerable, shaft, normally of mild steel, is run in brass, gun metal or phosphor-bronze bushes supported on rigid cast iron block. The bushes are held in position by a cap bolted with the case (Fig. 6.24). The lower bush is provided with lug (in the form of projection) to prevent its rotation. Both the lower and upper bushes are flanged to restrain heir axial play. Oil is provided in the cap and in the top bush; appropriate oil grooves, not shown, are cut in the bushes to distribute the oil.

Prob. 6.27 Draw the details (mentioning materials) on a A1 sheet showing front sectional views and right hand side views of each part of the plummer block.

Prob. 6.28 Draw the principal views of the assembled plummer blocks of which two views should be in half-section, Fig. 6.24.

Prob. 6.29 Draw the pattern drawings of the cap and the bushes of the plummer block, Fig 6.24.

Rolling Contract Bearing

These bearings consist of an inner race and an outer race, and the rolling elements, *e.g.* balls, rollers, tapered or spherical, spaced certain distance apart, rolling over raceways. A separator or cage houses the rolling elements and keeps them at a fixed distance. The races and the rolling elements are of high quality hardened alloy steel.

Depending on the direction of load, rolling contact bearings are classified as radial for taking radial loads, radial-thrust for carrying both radial and axial loads and thrust bearings taking only axial roads.

The rolling bearings may be of single row, double row, or multiple row type depending on the number of rows of rolling elements. Higher rows are used to take heavier load on a smaller diameter bore.

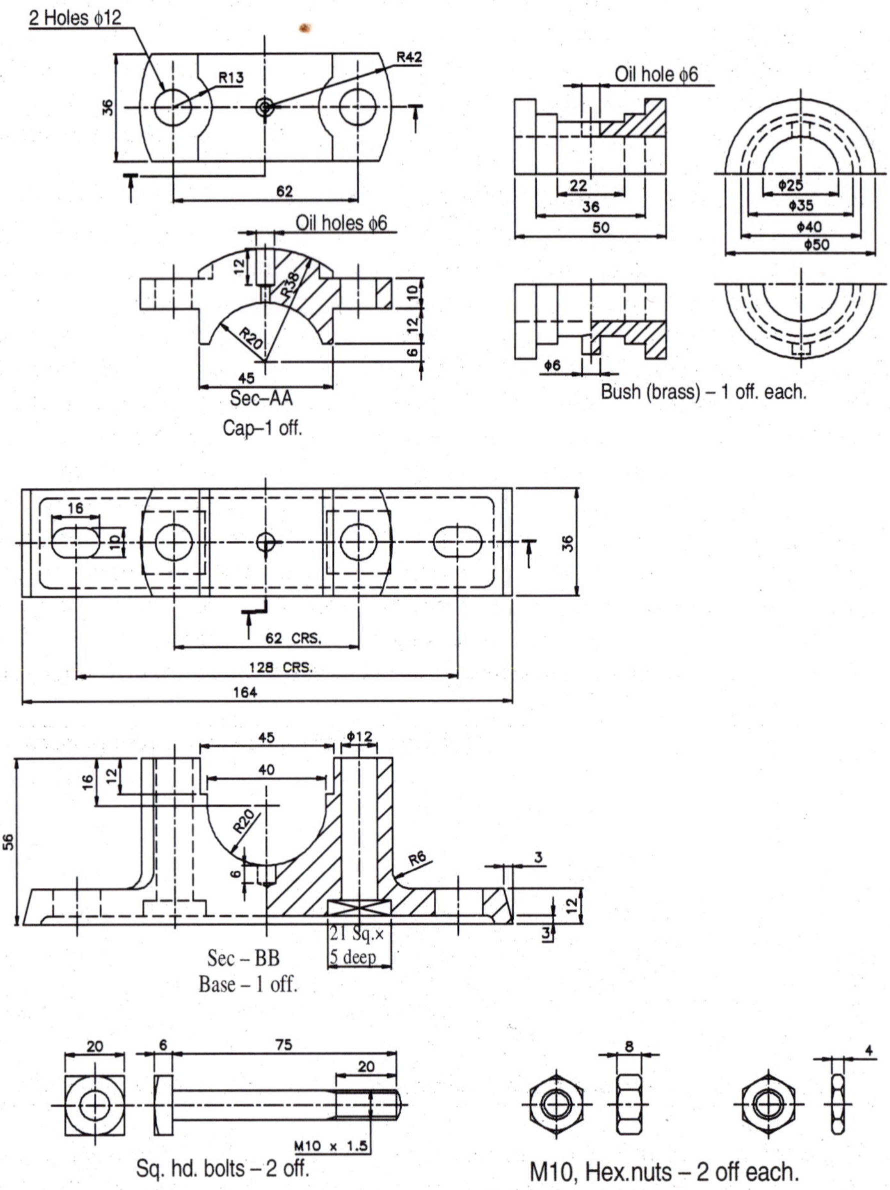

Fig.6.24 Detail of a plammer block

Figure 6.25 shows various types of bearing namely short roller bearing, tapered roller-single and double row for taking axial load in one and both directions respectively. The double row spherical ball and double row spherical barrel bearings are self-aligning, *i.e.*, they allow inclination of inner race or shaft axis with respect to outer race within 2° to 3°.

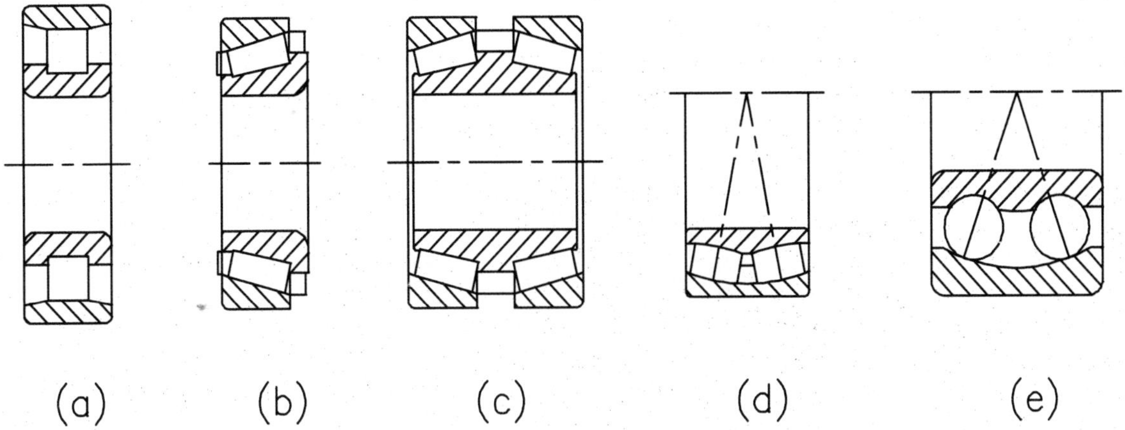

Fig. 6.25

(a) Short Roller Bearing
(b) Tapered Roller Bearing (Single Row)
(c) Tapered Roller Bearing (Double Row)
(d) Double Row Spherical Roller Bearing
(e) Double Row Spherical Ball Bearing

Mounting of Ball Bearings

The methods of mounting of bearings on shaft vary widely with the design of the shaft, type of the bearing and the type of the service of the related machine. Ball bearings are fitted on to the shaft by warming the bearing in oil to about 75° C to 80° C and driving it in position by light hammering with a soft metal hammer or mallet. The blows should only be given to the inner race.

Figure 6.26 shows the bearing arrangements used in gear boxes. Notice the input output shafts, location and fixation and prevention arrangement of the gears using collars, star washers and locknuts. Note also the use of tapered roller bearings for taking heavy axial and radial loads.

Figure 6.27 shows a sectional view of an assembled self-aligning ball bearing mounted on a shaft. The inner race is shrink-fit onto the shaft and is located against a shoulder by a locknut and a grub screw; recent practice is, however, to use lock washer in place of grub screw. The outer race is snugly fitted to the casing. The balls are retained in a cage made of sheet-steel pressed to shape. The bearing is encased by two end-covers secured to the housing by a number of screws. Felt washers are used to protect against contamination with dirt, fume etc. Grease or mineral oil may be used as lubricant depending on the application. Seals of neoprene rubber are being used currently instead of felt washers.

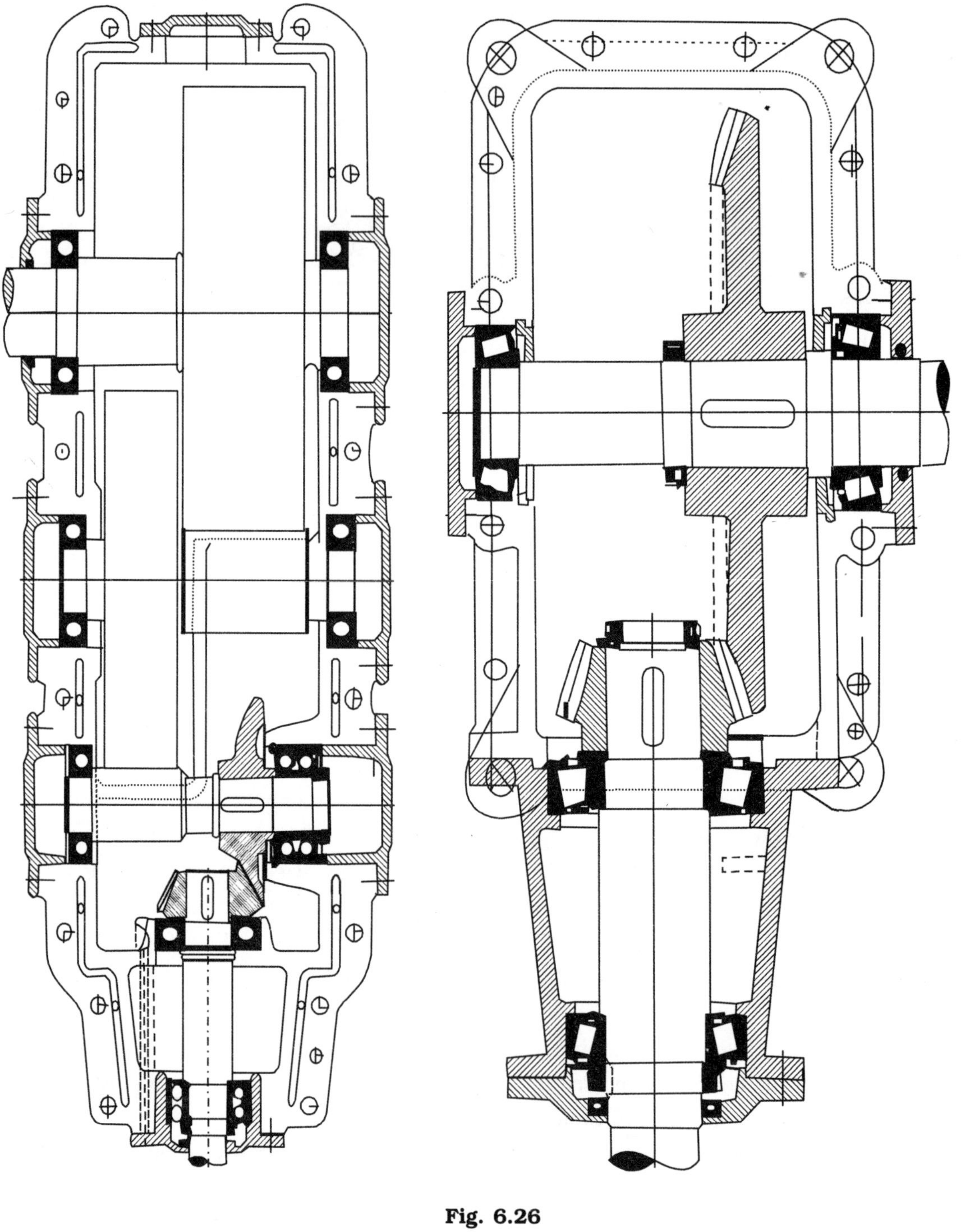

Fig. 6.26

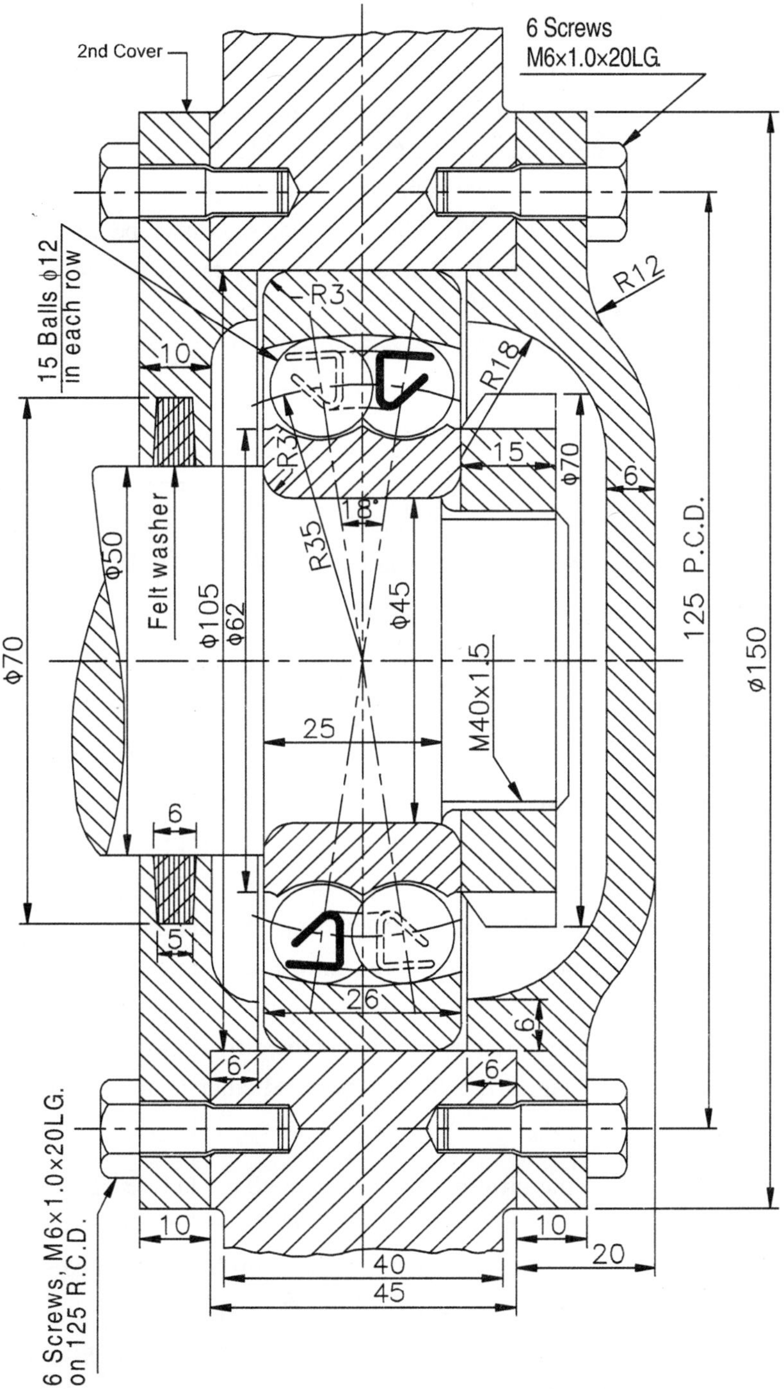

Fig. 6.27: Assembled double row spherical ball bearing mounted on a shaft with housing.

DRAWING OF GEARS

Gears are toothed wheels used to transmit power, starting from few milliwatt (in a watch) to several thousands KW in rolling mills. The gears which connect parallel shafts are called spur gears and the ones which connect intersecting shafts are called bevel gears. The gear tooth may be straight or helical. The helical teeth are quiet in operation and helps gradual load application with higher capacity over same face width. The gear materials may be cast iron for low speed gears and carbon steel for general application with medium range power and speed. Severely loaded gears are made of chrome-nickel steels.

Before one can draw a gear one has to be conversant with the gear terminology, *e.g.* module, circular pitch, addendum, dedendum, pitch circle, base circle, pressure angle, tooth thickness *etc.* Any text on machine design gives these details.
The basic definitions are given below for easy reference.

Module : Pitch circle diameter in mm, when divided by the number of teeth gives its module. These values indicate the size of the gear tooth and standard values are 0.5, 1, 1.25, 1.5, 1.75, 2, 2.5, 3.0, 3.5, 4.0, 4.5, 5.0 *etc.* in mm.

Circular pitch: It is the arc distance between two similarly situated points measured on the pitch circle of two consecutive teeth. It is obtained by dividing the perimeter of the pitch circle diameter by the number of teeth.

Pressure angle: It is the angle between the line perpendicular to the line of centers of gears passing through the pitch point and the line joining the point of contact and the pitch point. The standard value is 20°, but 22.5°, 25° are also used.

Figure 6.28 shows the two views of a CI spur wheel. The module is 14 mm and number of teeth is 33. The pitch circle diameter = πZm = 462 mm. Base circle radius = 462 cos 20° = 434 mm. Now the point d is the foot of the perpendicular drawn at the center of the circle and 'C' the pitch point. In fact the base circle radius can be obtained by drawing perpendicular to the line making the pressure angle 20° with AB which is perpendicular to the line of centers.

To Draw the Tooth Profile

The tooth working depth is 'ab' the difference of the addendum and dedendum circle radius. Lay of ac = (2/3) ab. Now with 'a' as center turn point C to C_1 on line of pressure angle. Now get $C_1 0$ = 3/4 dc = R, the radius of the arc of the circle which approximately gives the involute. Then one has to find out the circular pitch and lay it off along the pitch circle and tooth profiles will have to be drawn from both sides to give shape of the tooth. The tooth thickness at the pitch circle is given by 1.5708 m which is also equal to tooth space if there is no backlash, when m is the module.

Cast iron spur gear

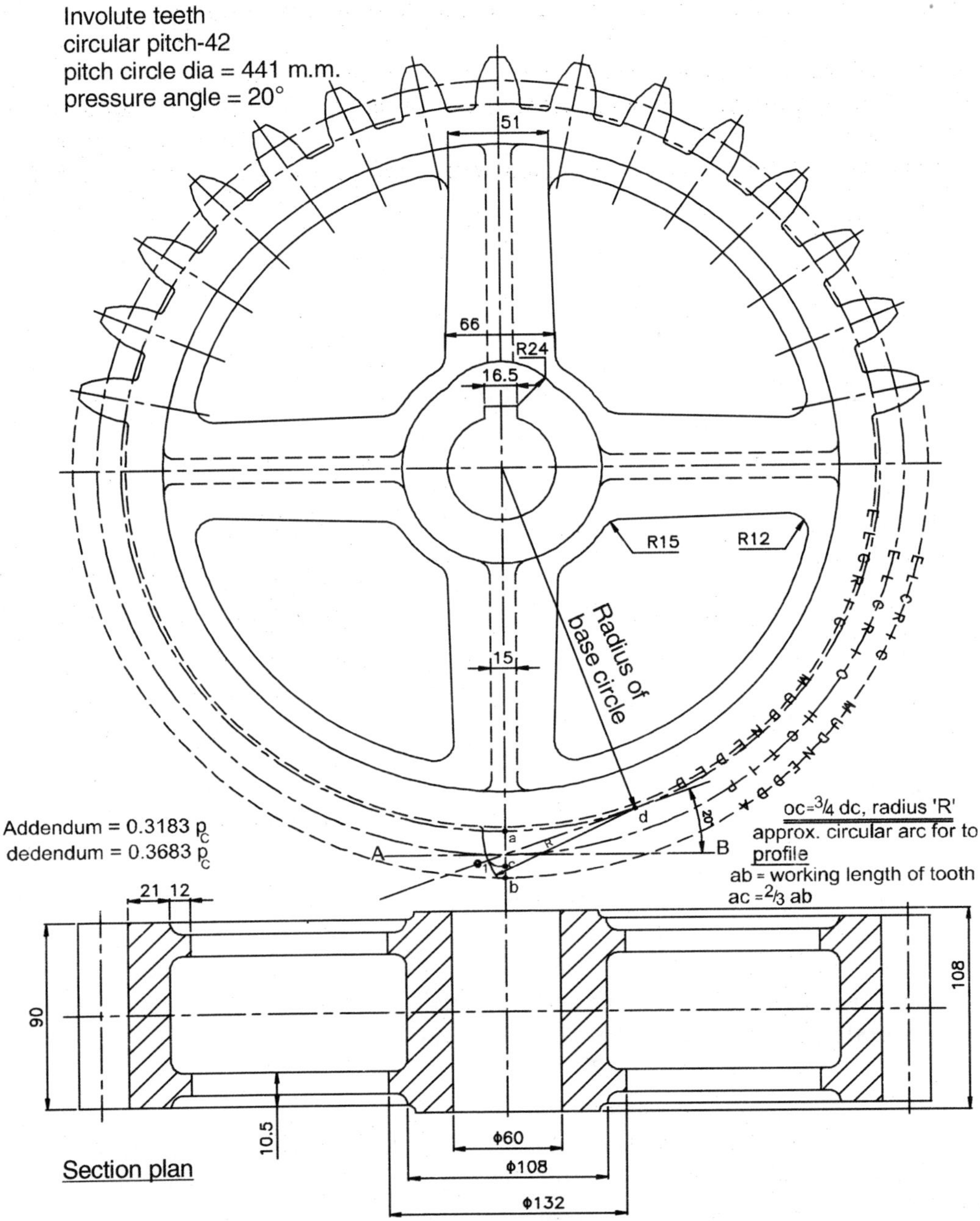

Fig. 6.28 Spur gear

Prob. 6.30 Draw the complete front view and the side view of the bearing assembly (Fig. 6.27) in half section. Scale 1:2.

Prob. 6.31 Draw the given view (Fig. 6.27) in half section and add the right and side view with the end cover removed and make the parts list.

Prob. 6.32 Draw the details of shaft, covers, locknut and a suitable lock washer (not shown). Show the bearings and shaft assembly conventionally (Fig. 6.27).

To Draw the Blanks for a Pair of Bevel Gears

A convenient way to understand the principle of design of a pair of bevel gears is to study the method of drawing the blanks from which a pair of bevel gears are to be cut. Let us assume that a module 4, twelve-tooth gear is to mesh with an eighteen tooth gear, the axes to intersect at 90°, (Fig. 6.29). Start with the point O as the point of intersection of the two axes. Draw OS, and OS_2 making the required shaft angle, q = 90°. Let 12-tooth gear be on S_1 and 18 tooth on S_2 and call them gear 1 and 2 respectively. Since module is 4, their respective pitch diameters at the base will be 48 mm and 72 mm.

From O measure along OS_1 a distance OM equal to the pitch radius of gear 2 (36 mm) and through M thus found, draw a line perpendicular to OS_1.

In like manner, make ON equal to the pitch radius of 1 and draw a line through N perpendicular to OS_2. These lines intersect at P. Make MR equal to MP and NT equal to NP. From R, P and T draw lines to O. Then the triangle ORP is the pitch cone of 1.

It will be noticed that the above construction is the equivalent of that for rolling cones. Next draw through P a line perpendicular to OP meeting OS_1 at K and OS_2 at H. Draw a line from H through T and from K through R. The cone represented by the triangle THP is called the back cone or normal cone of the gear 2 and that is represented by the triangle RKP is called the back cone of 1.

From R lay off Ra equal to the addendum that is to be used on gear 1. This being determined by the same considerations that would be used for the addendum of a spur gear. Along RO lay off Rr equal to the desired width of gear face. Through r draw a line parallel to RK. From 'a' line to O meeting this parallel at a_1. Through 'a' draw a line parallel to RP meeting KPH at a_2. From a_1 draw a line parallel to RP meeting a line draw from a_2 to O at a_3. Lay off along RK, the distance Rd equal to the dedendum and draw from d toward O meeting a_1 r at d_1. Find d_2 and d_3 in the same way that a_2 and a_3 were found. The figure add$_1$ a_1 represents the tooth. The dimensions of the hub and the position of lines FG and F_1 G_1 and of the corresponding elements on the other gear may be made anything that is desirable.

It will appear from this construction that bevel gears must be laid out in pairs.

There is a nomenclature which is peculiar to bevel gears. Some of these terms have been used in the previous discussions. Additional ones will now be defined and mostly shown in Fig.6.29.

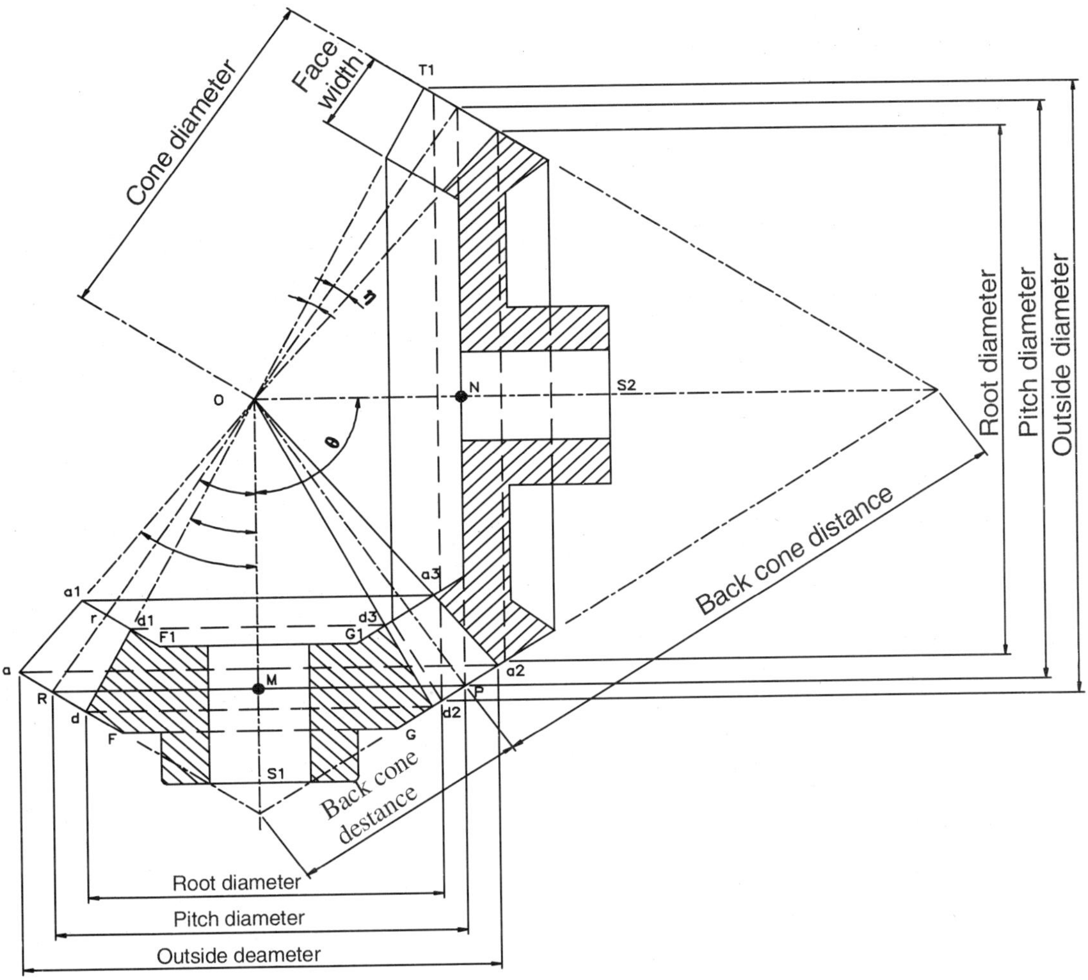

Fig. 6.29

Pitch cone is the imaginary cone upon which the teeth are made.

Cone distance is the length of the side of the pitch cone.

Pitch point, P, is the point of tangency of the pitch cones at the large ends.

Pitch, center or cone angle is the angle α, which is an element of the pitch cone makes with the axis of rotation or center line of the gear.

Face angle in the angle, 'γ' which the top surface of the tooth makes with the center line of the gear.

Addendum angle is the angle, 'η' which the top surface of the tooth makes with the cone distance; it is equal to face angle minus the pitch angle.

Root or cutting angle in the angle, 'β' which the bottom surface of the tooth makes with the cone distance and is equal to the angle minus the root angle.

Dedendum angle is the angle, σ, which the bottom surface of the tooth makes with the cone distance and is equal to the angle minus the root angle.

Back cone distance is the distance along the element of the back cone from the pitch point, ρ, to the center line of the gear. The back cone distance for gear OS_1 is KP and for gear 2 is HP.

Formative or virtual number of teeth is the number of teeth contained in a spur gear whose pitch radius is equal to the back cone distance and whose pitch is the same as that of the bevel gear. Since the supr gear is imaginary, there may be a fractional tooth in the number of formative teeth. The formative number of teeth is used in lying out and designing the teeth and in selecting the cutter of proper size.

In this section the principles of drawing the pitch surfaces of two bevel wheels are already discussed. It may be remembered that the nominal pitch of the teeth of a bevel gear is the pitch of the teeth on the base of the pitch cone, and the nominal diameter of the wheel is the diameter at the base of the cone. All straight lines on the surface of the teeth radiate from the common vertex of the pitch cones, so that the pitch of the teeth diminishes uniformly along the width of the teeth.

Prob. 6.33 Reproduce the drawing, Fig. 6.29, showing the two bevel wheels in full section for a pair of shafts intersecting at an angle of 90 degree and a speed ratio of 1:2 assume the size of base cone of the smaller wheel as 60 mm. The module is 4, face width 30 & pressure angle 20 degree. Calculate all the parameters and then draw the views in full section.

SELECTION OF MATERIAL

In the interest of economy, a company stores the following steels only.

Carbon Steels: C 10, C 45, C 65.

Alloy Steels: 40 Ni 2 Cr 1 Mo 28 (AISI 4340), 103 Cr 1, 50 Cr 1 V 23, 15 Ni 2 Cr 1 Mo 15 15 Ti6A14V, UHMWPE (Ultra High Molecular Weight Polythylene)

It may be assumed that steels are used in all the shapes and sizes that may be required. Use of other variety of steel is permitted if justified with reasons.

Prob. 6.34 For each of the following applications state which steel you would suggest and concisely explain the reasons:

(a) A coil spring for holding safety valve in position.

(b) A long cam shaft requiring low average strength but high wear resistance at the surface.

(c) A rolling contact bearing race.

(d) Stamping in large quantities.

(e) The ring gear of a cement mill.

(f) A large vertical shaft, 125 mm diameter, 2.5 meter long, which takes axial and torque loads which must have good uniform strength.

(g) The 'side gear' of a differential gear box mounted on rear axle.

(h) A helical spring in contact with food stuff in a food mixer.

(i) A low-stressed piston rod of a reciprocating pump which is to be flame-hardened to prevent excessive wear due to rubbing in at the stuffing box.

(j) The walls of large industrial cooking vessel-cum-food mixer.

(k) Supercharger gears.

(l) Collets of a heavy duty capstan lathe.

(m) A low temperature application requiring shock resistance but moderate strength in (i) corrosive atmosphere (ii) non-corrosive atmosphere.

(n) A hip joint prosthesis for human implantation which will be subjected to intermittent load 4 to 5 times the body weight and corrosive environment similar to sea water at 37° C.

(o) The liner of a knee joint prosthesis (Tibial component) to articulate against Co-Cr-Mo femoral component.

CHAPTER – 7

FORM DESIGN

Designing a machine element means putting down on paper the dimensioned shape that the part must have to properly serve the purpose for which it is meant. To lay out and draw this shape the designer must have clear understanding of the various processes of manufacturing, *e.g.* casting, machining and fabrication methods like welding, riveting *etc.*.

A machine or sub-assembly consists of several elements, which are to be assembled. It will require periodic or occasional maintenance or replacement of part/parts. Therefore the design must be so made, that these activities can be performed at ease.

The process of manufacture to be adopted is greatly influenced by the quantity required. For castings, pattern, core and mould making are involved and normally it will not be economic when a few pieces of a large part are required; welding may be economic in these cases.

Forging design also involves metal die cost for intricate shapes if hand forging is not suitable.

For stampings & blankings economic use of materials should be made by planning proper layout which will curtail the scrap losses; successive reuse possibility of the scrap should be thought of. The rigidity of sheet metal components can be increased by judicious use of beading.

Other related general considerations for forging, casting and weldment design have been discussed in some details in Chapter 4 on Manufacturing methods.

In this chapter a good number of practice problem from various live situations are given which if tackled will equip the learners with confidence in form design.

Prob. 7.1 Search out at least five errors affecting the manufacturing, or utility of the 75 vertical shaft assembly (Fig. 7.1) symmetrical about the center line. Indicate each error by an arrow and a brief note about the fault. Show in tracing the corrected drawing.

Prob. 7.2 For the lever shown (Fig. 7.2) identify the possible parting line if it is to be manufactured (i) by casting (ii) by forging.

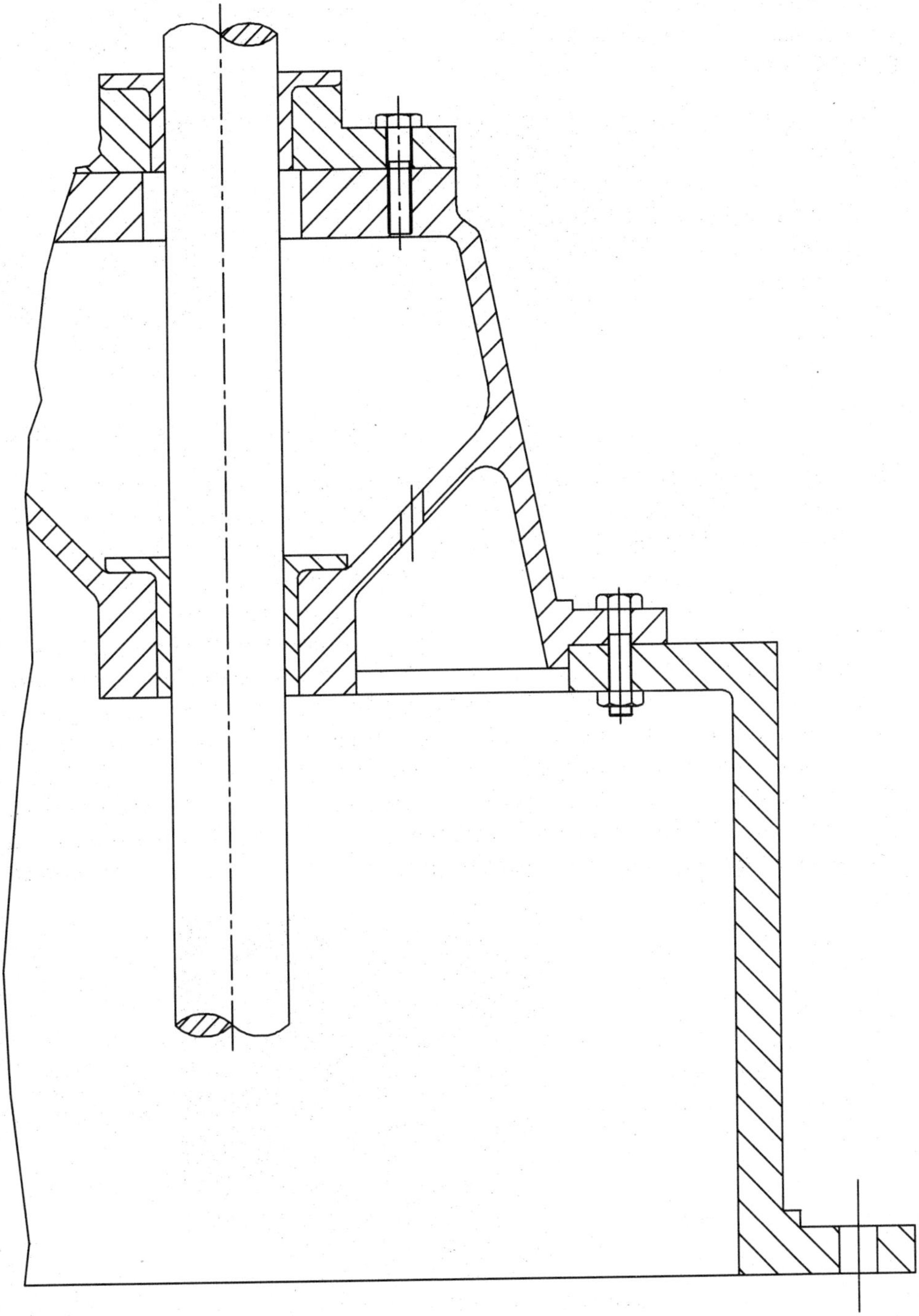

Fig. 7.1 Vertical shaft assembly

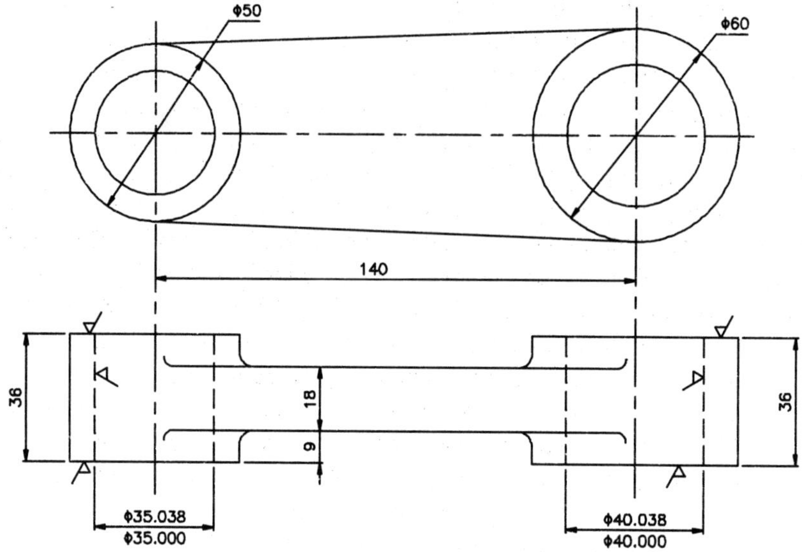

Fig. 7.2 Lever

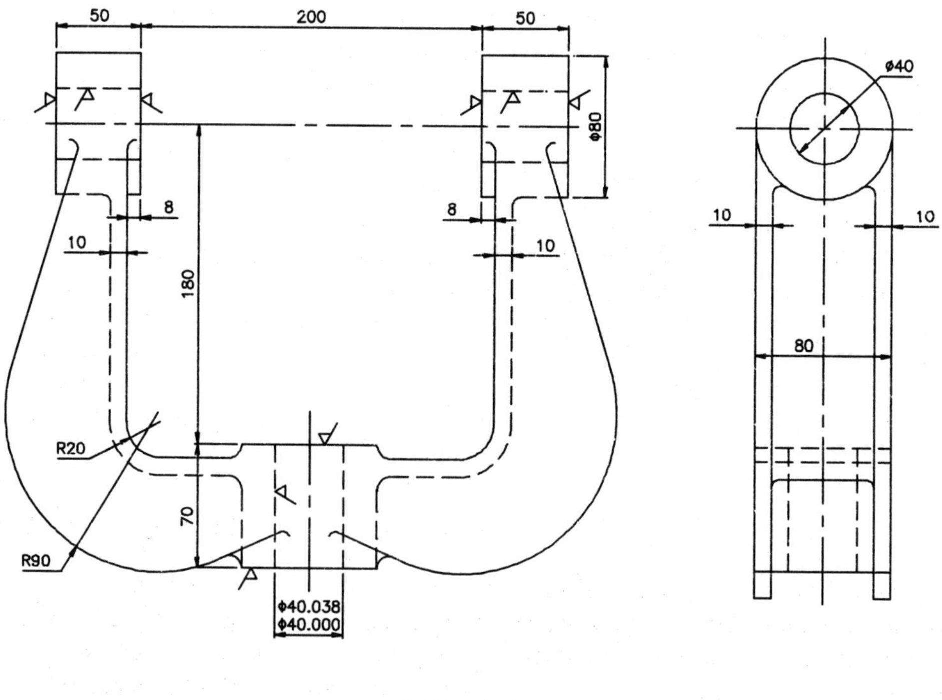

Cast Iron Body

Fig. 7.3 Forked lever cast iron

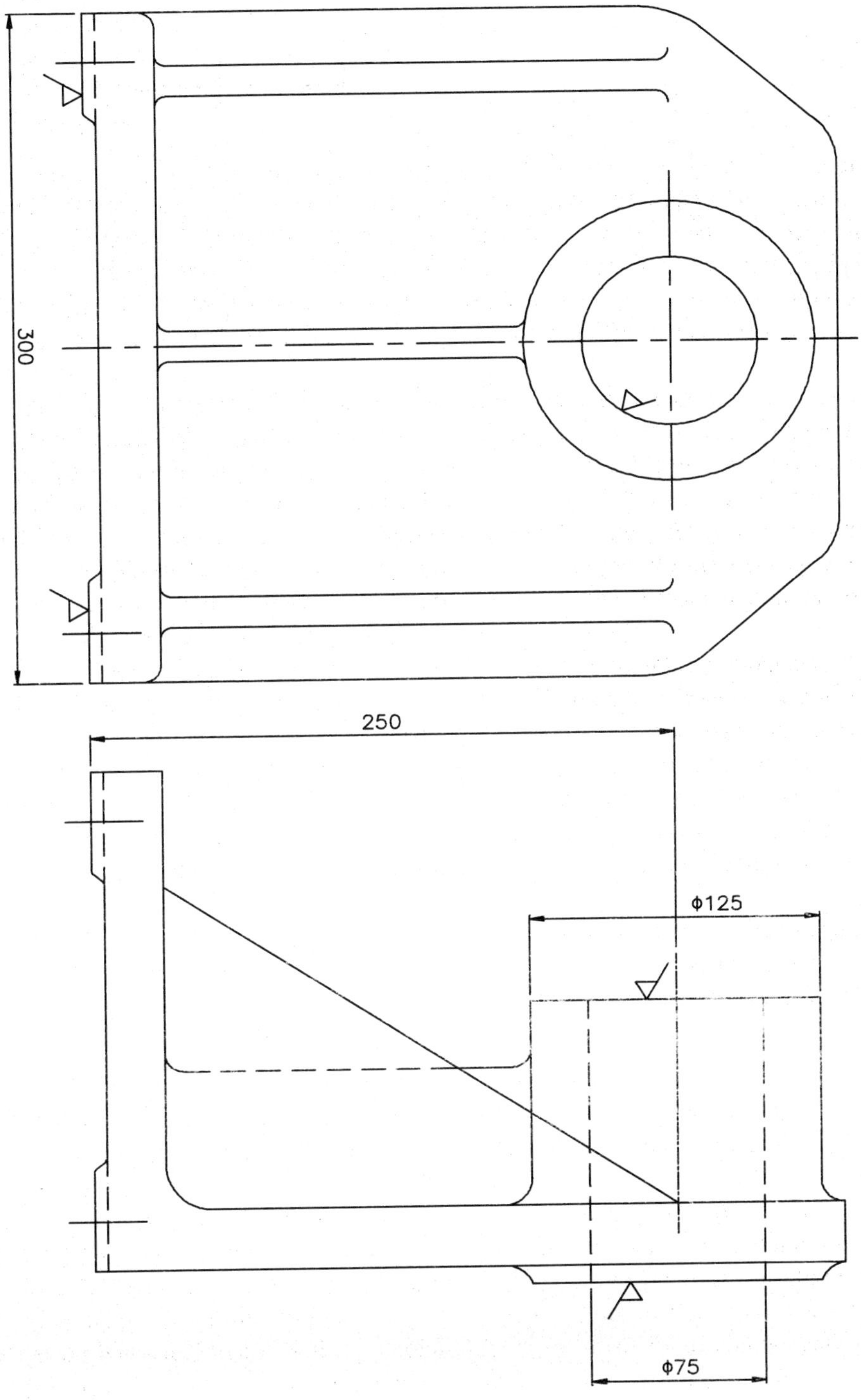

Fig. 7.4 Bracket

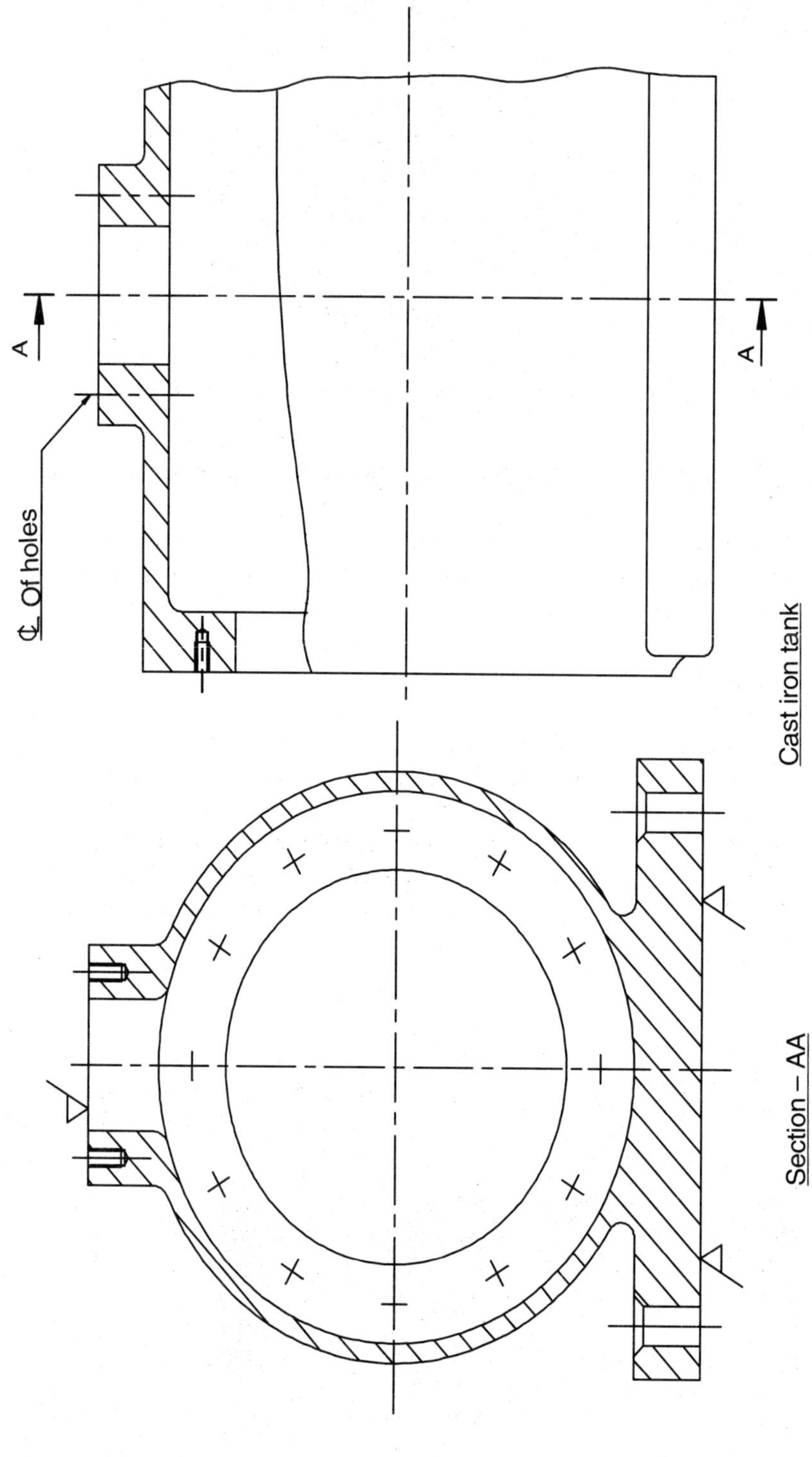

Fig. 7.5 Cast iron tank

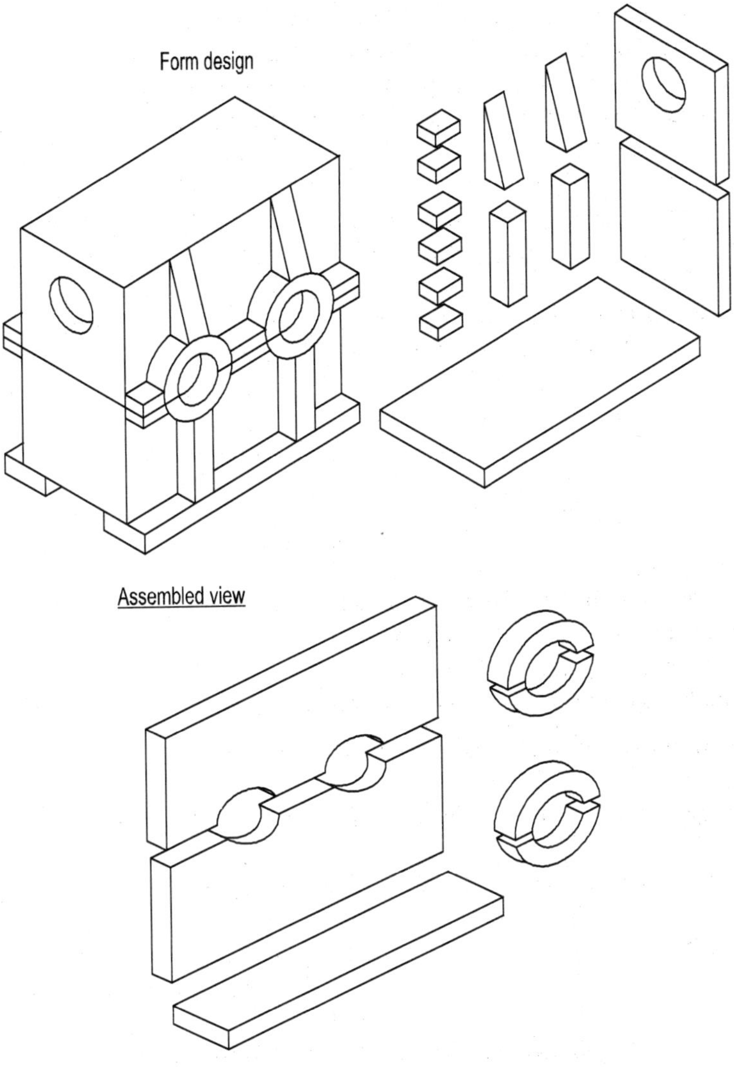

Fig. 7.6

Prob. 7.3 The cast iron body (Fig. 7.3) is to be cast. Identify the probable parting plane and also indicate the necessary sand cores. Modify the design so that the sand cores will be reduced or eliminated keeping the weight as nearly as possible equal to the existing weight of the casting.

Prob. 7.4 Fig. 7.4 shows the casting of a bracket to support a 50 mm diameter shaft. The design is proposed to be replaced by weldment. Develop the product using standard rolled plates, bars, *etc*. indicate the welds by standard symbols.

Prob.7.5 (a) Provisional design of a cast iron tank (required in large numbers) with a cylindrical opening at its top center for entry of fluid is shown(Fig. 7.5) at 1:10 scales. Holes are indicated at the ends to the attachment of end covers, at the top of the flange for the connection of a pipe and at the bottom for holding down the tank to a foundation framework. Sketch by colour pencil or soft pencil over the drawing to indicate improvement in design for better casting and assembly. Holes may be relocated.

(b) A few pieces of the same size tank is required; redesign the tank for the purpose and sketch the same.

Prob. 7.6 The assembled isometric view of a welded housing of a gear reducer (maximum dimension 1100 mm) is shown in Fig. 7.6. The design is rejected as too costly. Suggest a suitable design to reduce amount of welding and to provide a better appearances. Sketch the same on a sheet indicating the types of welds. Show also the edge preparation, castings, forgings and parts performed from rolled shapes may be used.

Prob. 7.7 Make a parts list of the internal gear reducer (Fig. 7.7). Identify faulty design features. Redesign from the point of view of manufacture, assemble and reduced machining.

Prob. 7.8 The outline of a finished steel (C 50) gear is shown by dotted lines in Fig. 7.8. Draw a forging drawing for the same showing internal and external drafts.

Prob. 7.9 The boss and rim of a fabricated steel gear of 500mm pitch diameter are shown in Fig. 7.9. The key size is 25 × 25 material being C 45 steel. Design the arms or web and complete the figure showing the joints. The windage loss should be considered while choosing the sections. The proposed gear will be used in a ball mill rotating at 50 r.p.m.

Prob. 7.10 Search and neatly correct by coloured pencil at least six errors affecting successful casting, machining and assembly of the pipe connector (Fig. 7.10). Show the fasteners in the top view.

Prob. 7.11 The impeller type water pump is shown in Fig. 7.11 purposely with many errors. Locate the errors by coloured pencil and redesign the housing improving the functional aspect and provide the necessary missing items.

Prob. 7.12 The driving end trunion of an imported 20T Ball Mill (Fig. 7.12) used to crush 12 mm mesh size rocky materials to fine powder needs replacement. Economic ground prohibits casting replacement and a fabricated design is advisable. Make out the design for the fabricated construction. Use standard rolled steel plates and sections. Calculate the saving in weight, if any. Could you estimate the difference in cost between the two designs?

Prob. 7.13 An incomplete shaft-assembly is shown in Fig. 7.13 with errors. Make out the complete corrected sketch. The sketch given is drawn to 1:2 scale.

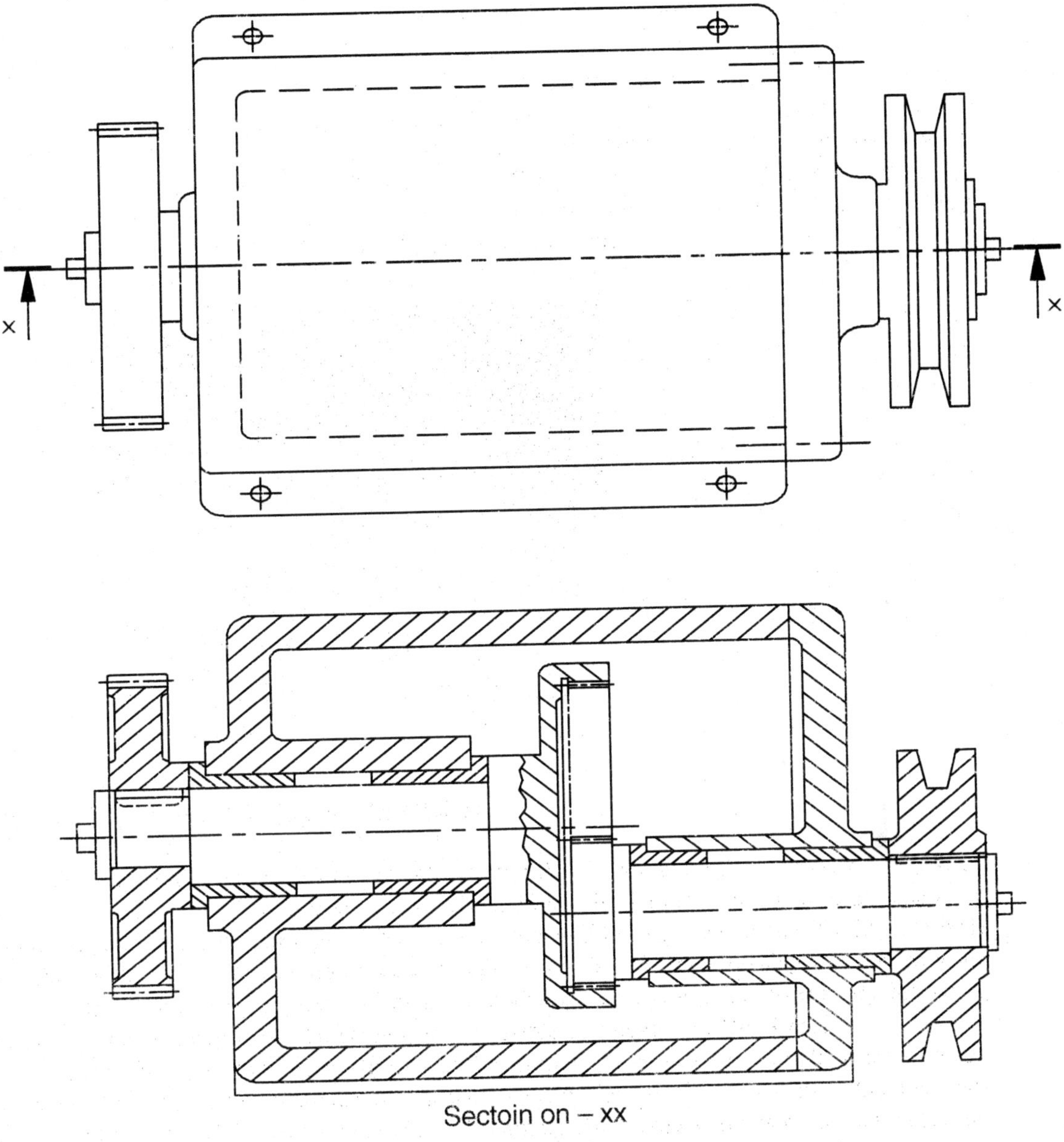

Fig. 7.7 Internal gear reducer

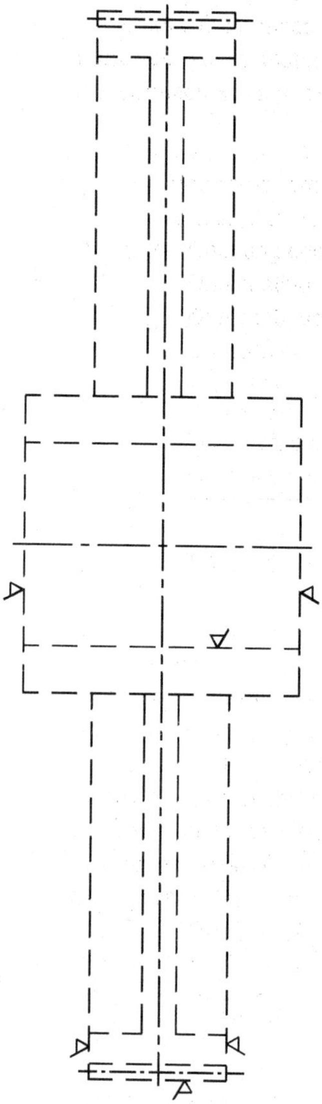

Fig. 7.8

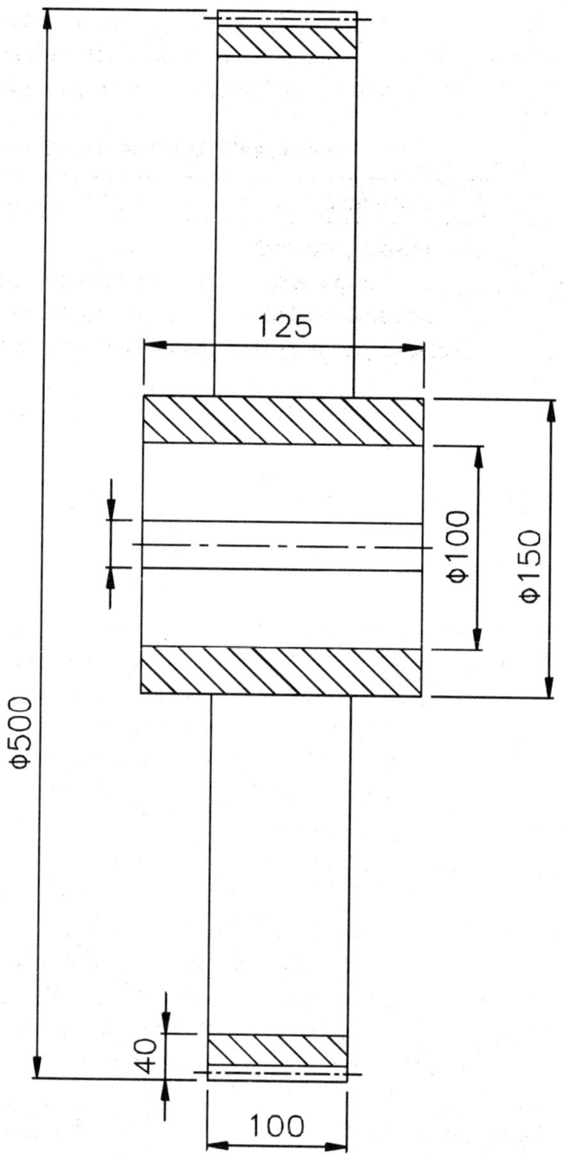

Fig. 7.9

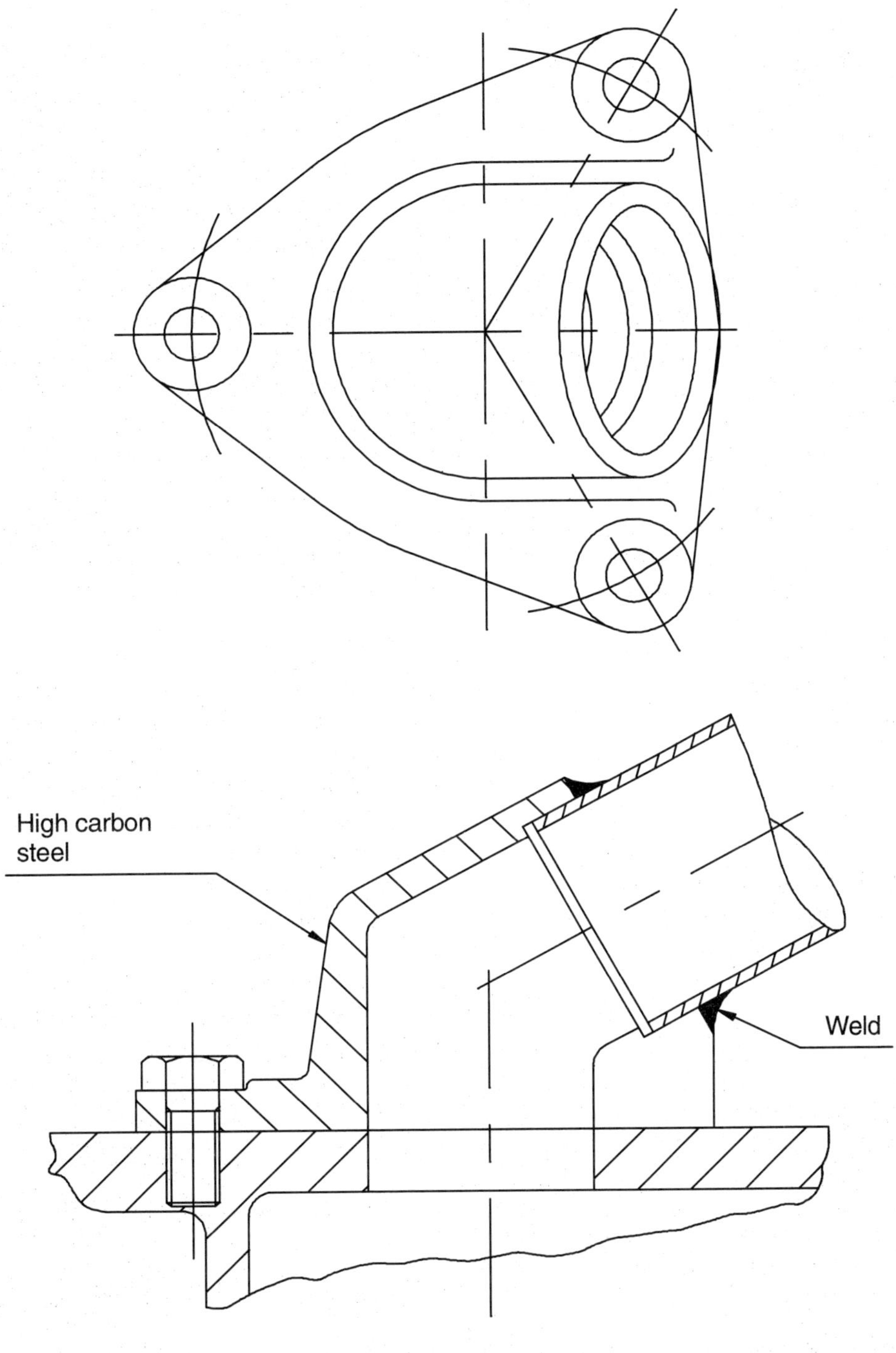

Fig. 7.10 Pipe connection

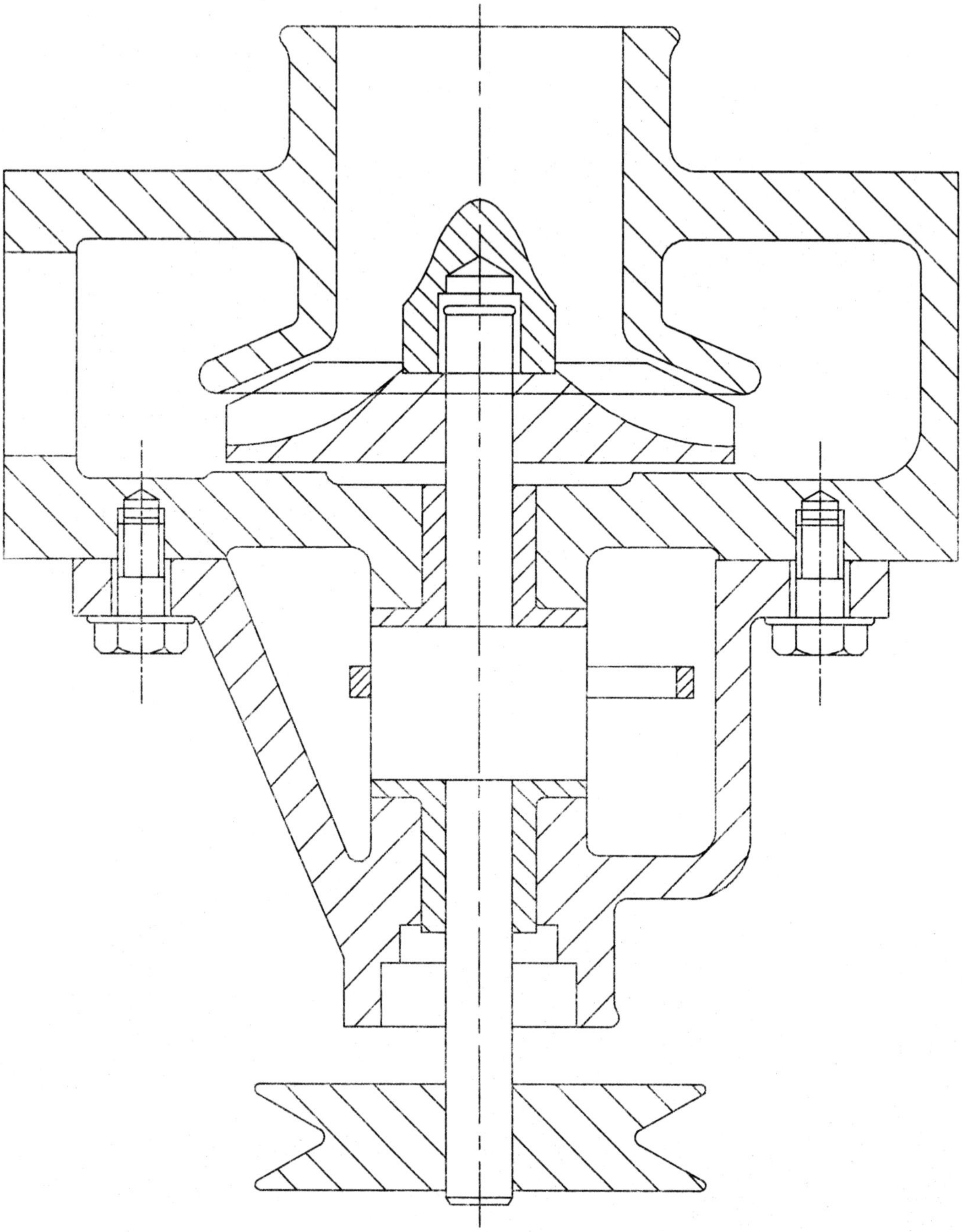

Fig. 7.11 Impeller type water pump with errors

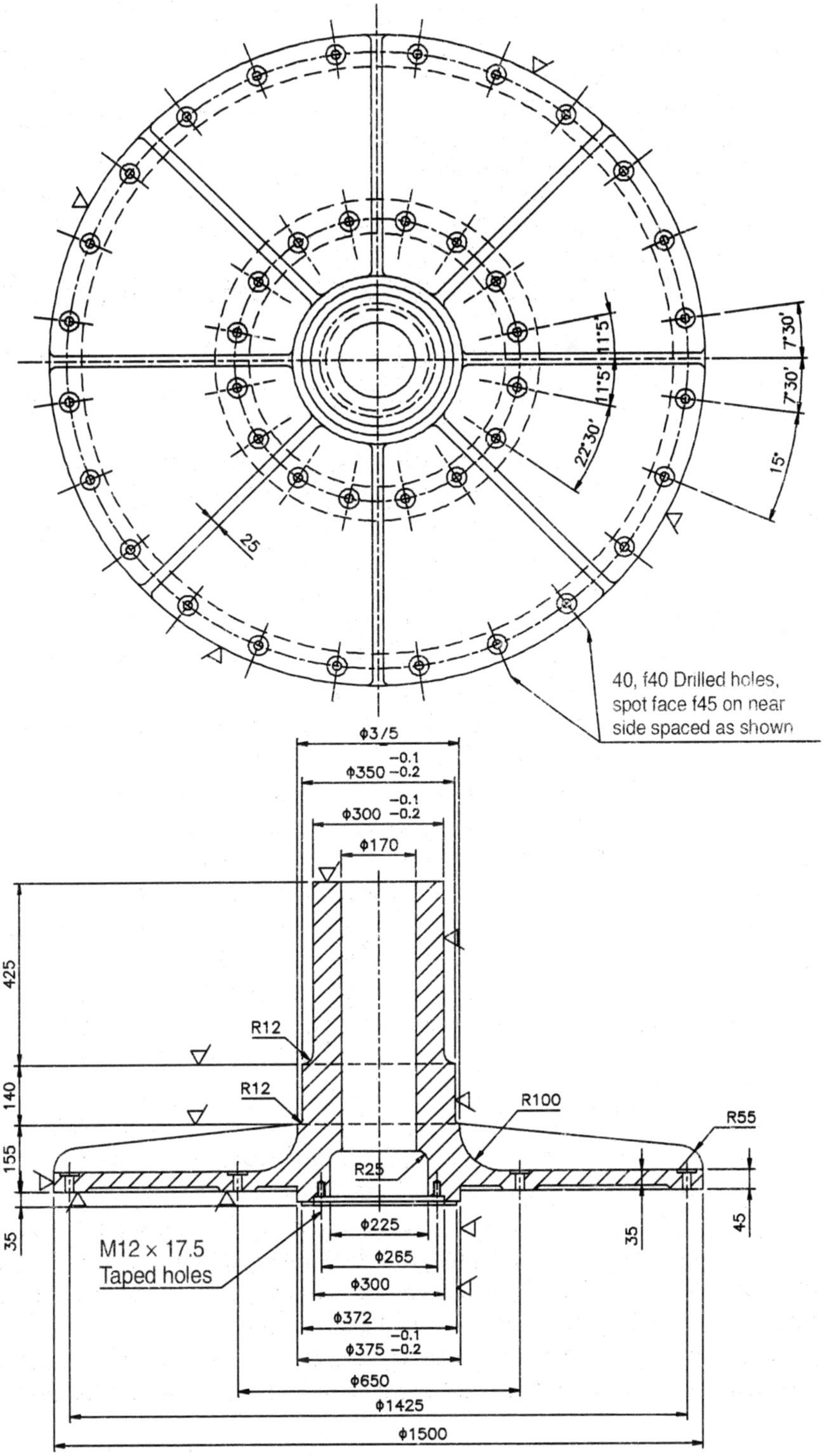

Fig. 7.12 Driving end turnion of a ball mill

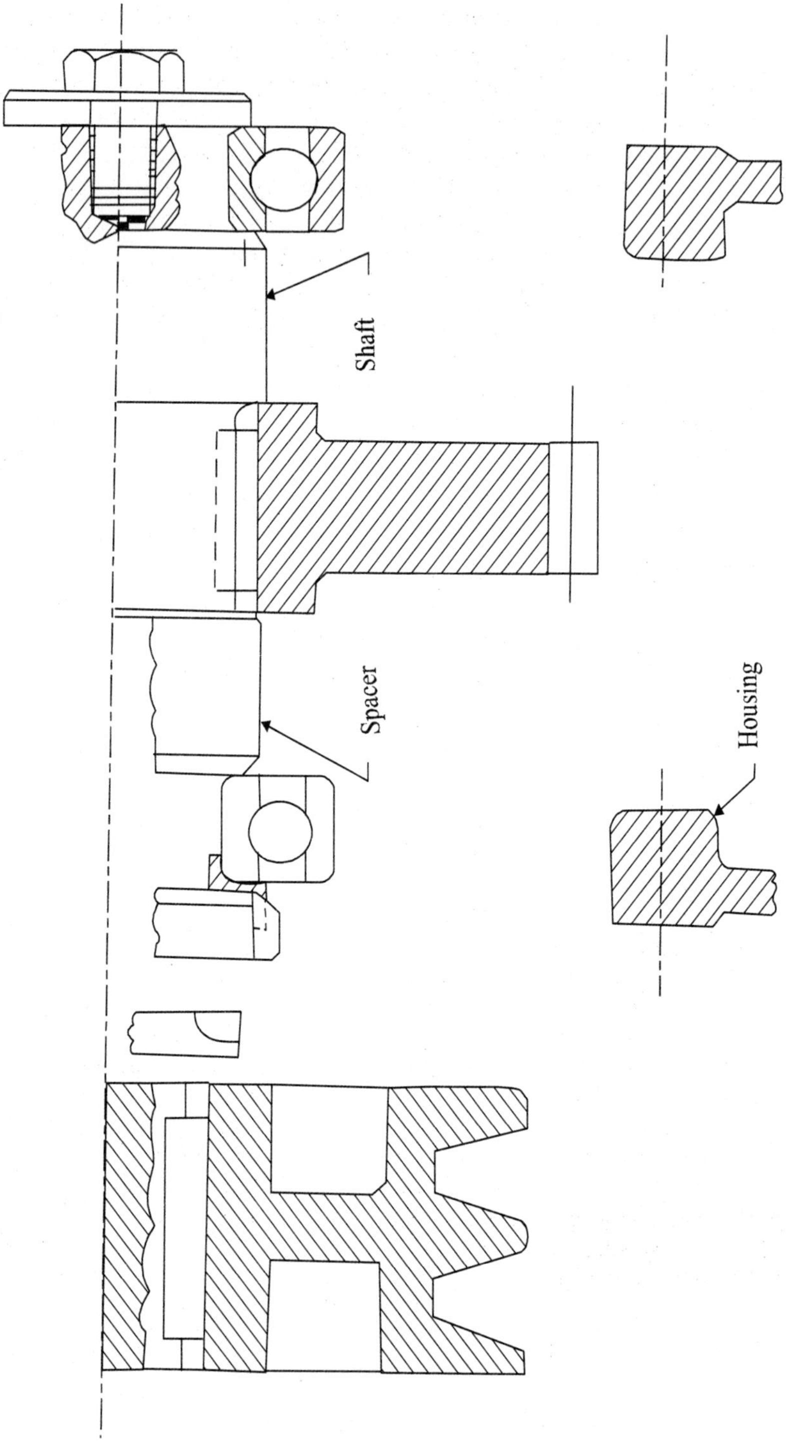

Fig. 7.13 Incomplete shaft assembly..

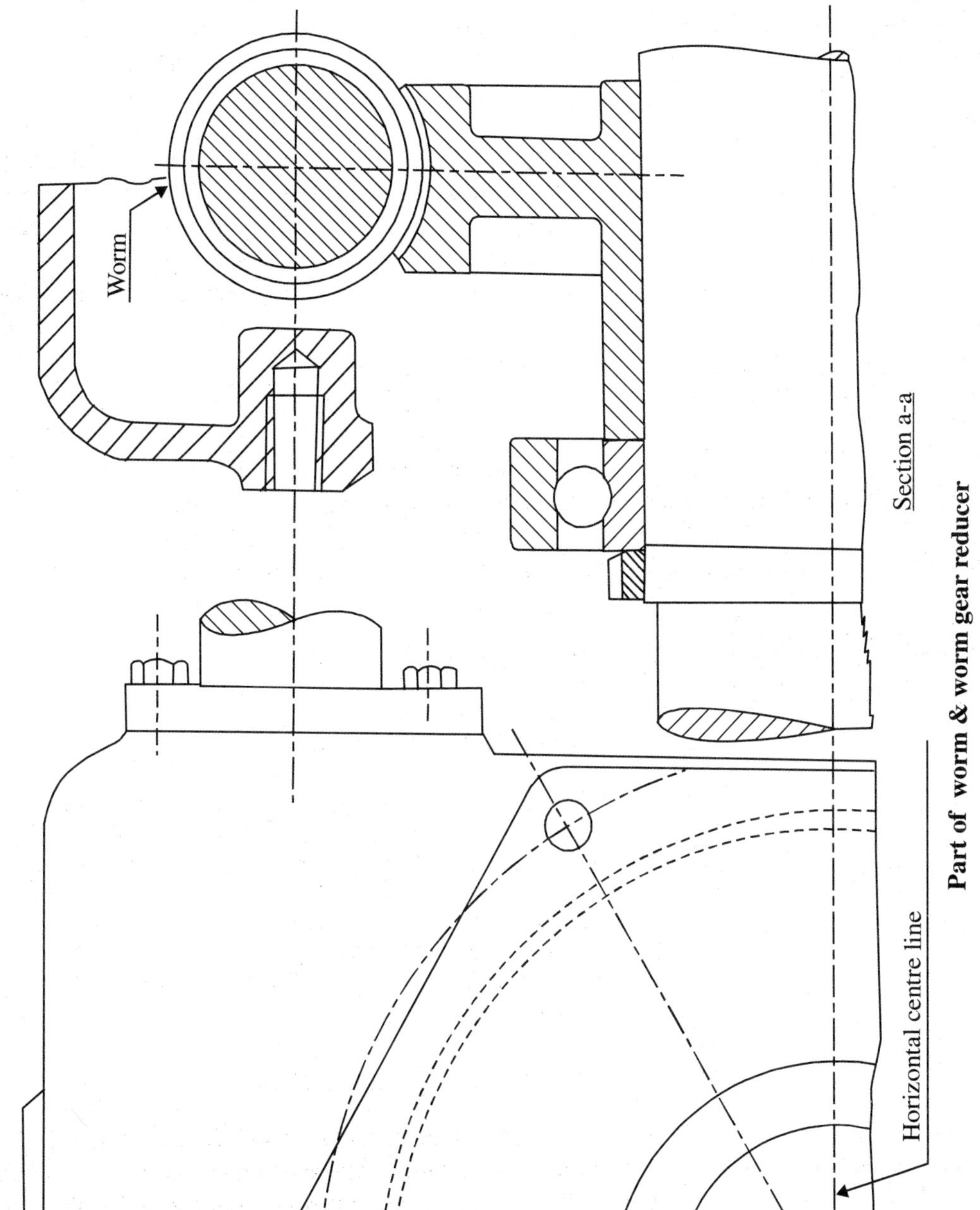

Fig. 7.14 Worm and worm gear reducer

Prob.7.14 Accompanied half sized Fig. 7.14 shows a part of a worm and a wormgear reducer and the housing for the same. Design a suitable cast end-cover to locate and fix the antifriction bearing; show also the fixing arrangement of the cover. Modify the front view as per your design.

Prob.7.15 A nearly half size view of a gear pump rotor to deliver oil to a hydrostatic thrust bearing, is shown in Fig. 7.15. Develop the end view and top view and hence sketch the housing showing details. Show also the necessary type of coupling to be used. Indicate the type of fit at least in three mating components. (*Hint*: See Fig. 2.4).

Prob.7.16 Redesign the three hole sheet metal link (Fig. 7.16) to permit successive blanking with a view to minimizing the scrap losses. Make a suitable change for improving strength and stiffness. The maximum center to center distance on the link is 100 mm and thickness is 1.5 mm.

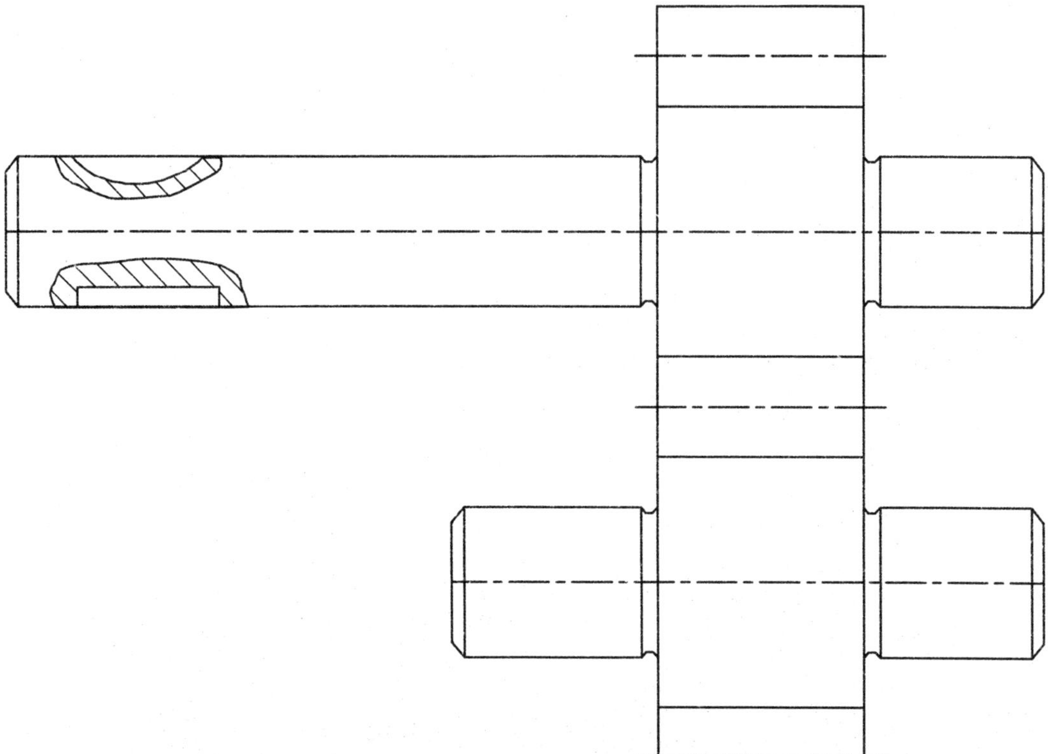

Fig. 7.15 Gear pump rotor

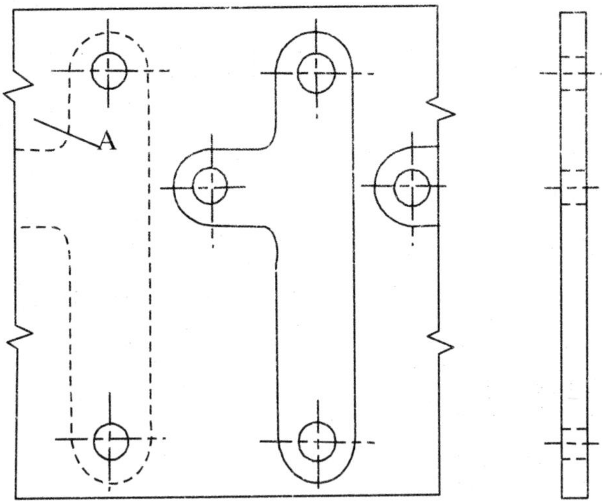

Fig. 7.16 Three hole sheet metal link

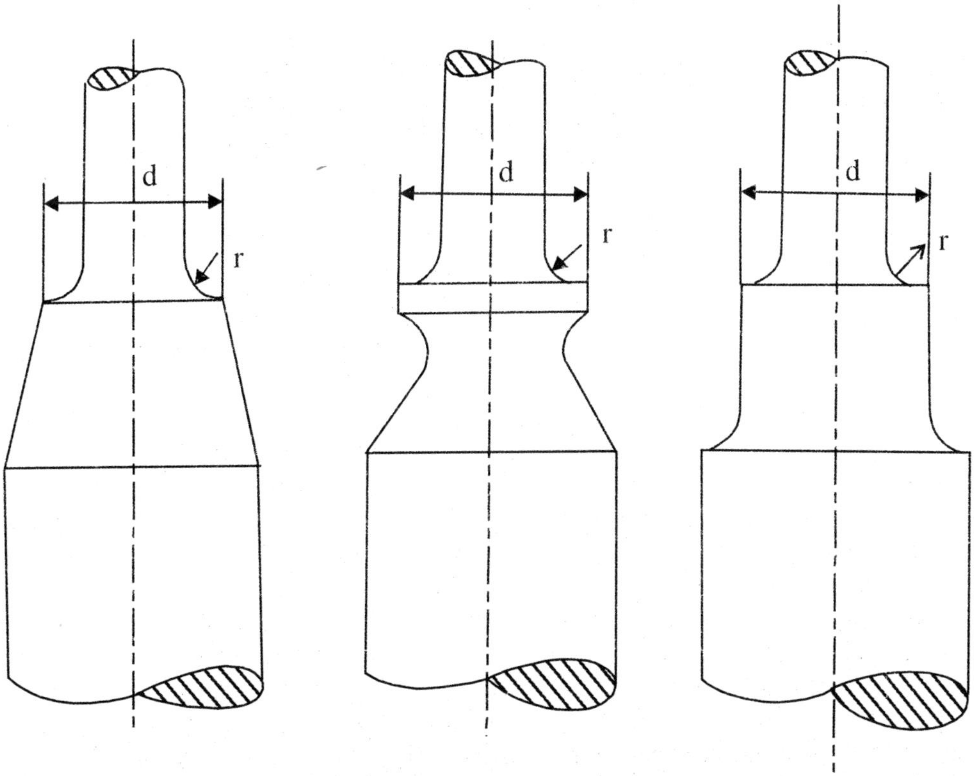

Fig. 7.17 Shaft with steps

QUALITATIVE EVALUATION OF STRESS – CONCENTRATION

Discontinuities like sudden change in section, keyways, holes, fillets *etc.* are stress raisers and at these critical sections values of the stresses are much higher than the average stress. For a better understanding of stress concentration in general, a qualitative approach known as flow line analogy, helps the designer to visualize probable stress concentration in a component of his interest.

Stress concentration in tension members may be represented by lines indicating the direction of the principal stress. If a tension piece say a flat 200 mm long, 10 mm thick, having cross section of 100 mm^2 and 50 mm^2 respectively in the left and right side from its middle then the flow lines will be parallel to its length indicating uniform stress but at the reduced end they will be closer together, indicating higher stress. Where the section changes abruptly, the force flow lines will be crowed and change the direction; the more the change in direction, the more the crowding and the more the local stress concentration.

Stepped shafts under torsion can be considered to be divided in toa number of concentric hollow shafts and a solid central shaft of such sizes that each of these elementary shaft carries an equal torque transmitted by the original solid shafts. Obviously the thickness of the shafts at the large diameter portion will be less than that at the smaller diameter portion. The thickness of the outer concentric shaft at the junction of diameter variation will be very small and high stresses will be developed.

Endurance limit in a **screw joint** is dependent to a great extent on the flow of forces in the material. The smoother the flow of force the lesser the concentration, the better the endurance strength. In an ordinary joint (Fig.7.18:3) the lower threads of a joint normally carry a much better load than the upper ones. By removing part of the threads with a conical reamer the force flow may be deflected upward, relieving the stress in the lower threads and increasing at the upper ones. Similar result may be obtained to a great extent by boring out the nut conically.

Prob. 7.17 By the flow analogy sketch over the Fig.7. 17 to show which of the three designs is best for avoiding fatigue of the shaft where the smallest diameter joins the shoulder. The radius r and shoulder diameter d must be the same in each design. Assume that the bending moment is constant along the portion shown. Indicate the best design.

Prob. 7.18 Nuts of special shape (1) and (2) are used to reduce fatigue fracture in the bolt (Fig. 7.18). Explain their effectiveness in comparison with a standard nut (3). Section lines are omitted and threads shown non-conventionally purposefully so that flow lines may be drawn.

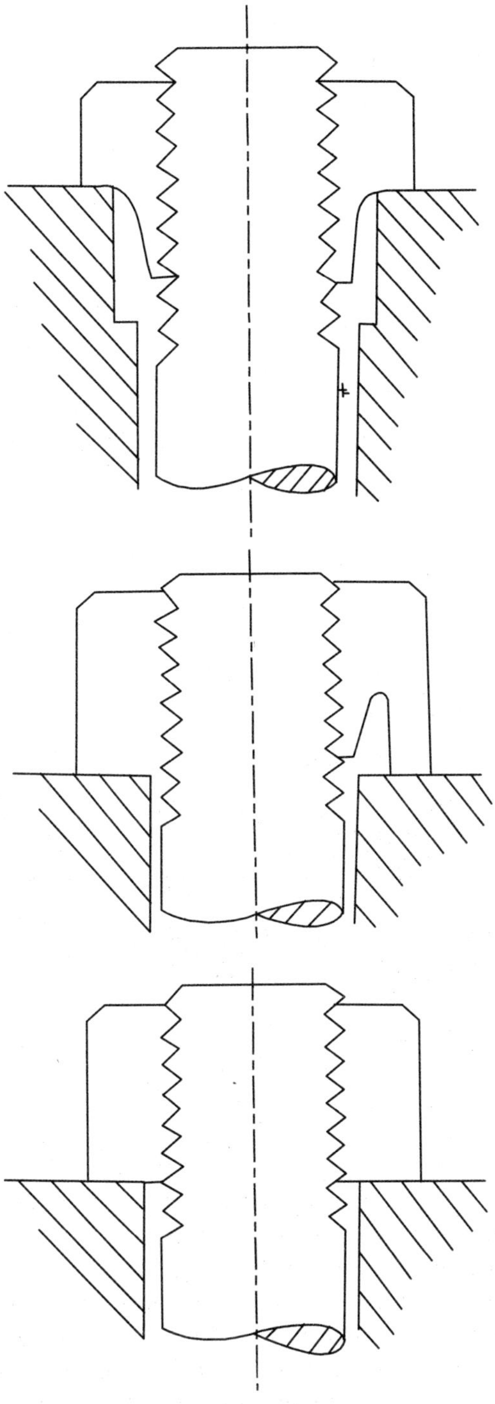

1

2

3

Fig. 7.18 Nuts of special shapes

CHAPTER – 8

DRAWING PRACTICE

Working Drawings or shop details or **detailed drawings** are prepared for the manufacture of the components; therefore all the manufacturing aspects must be considered while preparing the drawings. The parts must be so dimensioned that measurement and checking during manufacture are feasible. The mating parts should be toleranced depending on the type of fits. The itemization should be placed progressively in the drawing. The description or name of the part, its item number, material and number of pieces required per set should appear below each component drawing.

Assembly Drawings are made to show the relative arrangements of the components. Their locations and overall space requirement for accommodation of the object or machine. All the important and major dimensions of the components are given in the assembly drawing for assembly shop people. The parts are itemized and a part-list must appear in the assembly drawing preferably above the title block as per standard format.

REPRODUCTION OF DRAWING

For manufacturing operations, the shop-floor people must be given copies of drawings while the original ones should be retained in the Design/Drawing Office. For quality reproduction of drawings in multiple copies, inking of the drawing is essential.

Drawings may be linked with 'India' ink **Erographs** conveniently instead of with the common inking instruments provided with standard instrument boxes. Drawings may be drawn on drawing sheets, subsequently traced on tracing paper in ink or may be inked on the drawing sheets proper. Sometimes drawings are prepared on tracing sheets in pencils lightly and then subsequently inked.

Common Methods Used for Reproducing Copies from Tracings are:

(a) Ammonia dry printing
(b) Ferro-Prussiate printing (Blue printing)
(c) Ferro-Galic printing
(d) Photostat
(e) Xeroxing

Ammonia Printing Process

This is a widely used process for getting copies in large number, which gives prints in blue or white or bluish background on ammonia sensitized paper. In this process, the sensitized paper is placed in contact with the back of the translucent tracing with drawing either in ink or with soft lead pencil and is exposed to light preferably in a specially designed cabinet for a stipulated period. Then the exposed sensitized paper is kept for some time (say 20 to 30 minutes) in a chamber where ammonia vapour coming out of a tray full of 10% to 15% ammonia solution kept inside the cabinet, will develop the line distinctly.

Ferro-Prussiate Process

This process called blue printing, produces prints in white lines on blue background.

Similar to ammonia printing process, the tracking is placed in contract with the ferro-prussiate paper and exposed to sunlight or artificial lighting after mounting in a suitable printing frame. When the exposure is complete the paper is removed and washed gently in running water till the lines grow white. Time of exposure is a critical factor and depends on the intensity of the light, opacity of the paper and speed of the paper in case of machine printing. It is advisable to fix the timing by trial and error.

Ferro-gallic Process

It is very much similar to the ferro-prussiate method and it produces black lines on white background and ferro-gallic papers are used.

Photostat

It is a simplified form of Camera Photography used in drawing office. Exposure is given to the photosensitive paper directly from the tracing paper or cloth negative. In this process the drawings can be enlarged or reduced too.

Apart from these, Xerox method of copying is also being used for getting several copies quickly.

It may be remembered here that the **sensitized papers** mentioned in the foregoing paragraphs are all available in the market in **rolls** and some of these papers *e.g.* ferro-prussiate, ferro-gallic can be prepared very easily too.

Modern drawing offices, now-a-days are equipped with duplicating gadgets like ammonia printing machine, Photostat machine, Xerox copier etc..